# Industrial and Applied Mathematics in China

## Series in Contemporary Applied Mathematics CAM



---

1. Mathematical Finance —— Theory and Practice
   (Eds. Yong Jiongmin, Rama Cont)
2. New Advances in Computational Fluid Dynamics
   —— Theory, Methods and Applications
   (Eds. F. Dubois, Wu Huamo)
3. Actuarial Science —— Theory and Practice
   (Eds. Hanji Shang, Alain Tosseti)
4. Mathematical Problems in Environmental Science and Engineering
   (Eds. Alexandre Ern, Liu Weiping)
5. Ginzburg-Landau Vortices
   (Eds. Haïm Brezis, Ta-Tsien Li)
6. Frontiers and Prospects of Contemporary Applied Mathematics
   (Eds. Ta-Tsien Li, Pingwen Zhang)
7. Mathematical Methods for Surface and Subsurface Hydrosystems
   (Eds. Deguan Wang, Christian Duquennoi, Alexandre Ern)
8. Some Topics in Industrial and Applied Mathematics
   (Eds. Rolf Jeltsch, Ta-Tsien Li, Ian H. Sloan)
9. Differential Geometry: Theory and Applications
   (Eds. Philippe G. Ciarlet, Ta-Tsien Li)
10. Industrial and Applied Mathematics in China
    (Eds. Ta-Tsien Li, Pingwen Zhang)
11. Modeling and Dynamics of Infectious Diseases
    (Eds. Zhien Ma, Yicang Zhou, Jianhong Wu)
12. Multi-scale Phenomena in Complex Fluids: Modeling, Analysis and Numerical Simulations
    (Eds. Tomas Y. Hou, Chun Liu, Jianguo Liu)
13. Nonlinear Conservation Laws, Fluid Systems and Related Topics
    (Eds. Gui-Qiang Chen, Ta-Tsien Li, Chun Liu)

Series in Contemporary Applied Mathematics CAM 10

# Industrial and Applied Mathematics in China

*editors*

**Ta-Tsien Li**

*Fudan University, China*

**Pingwen Zhang**

*Peking University, China*

Higher Education Press

World Scientific

NEW JERSEY • LONDON • SINGAPORE • BEIJING • SHANGHAI • HONG KONG • TAIPEI • CHENNAI

**Ta-Tsien Li**
School of Mathematical Sciences
Fudan University
220, Handan Road
Shanghai, 200433
China

**Pingwen Zhang**
School of Mathematical Sciences
Peking University
5, Yiheyuan Road
Beijing, 100871
China

**Editorial Assistant:** Zhou Chun-Lian

**图书在版编目 (CIP) 数据**

中国的工业与应用数学 = Industrial and Applied Mathematics in China: 英文 / 李大潜, 张平文主编. 北京 : 高等教育出版社, 2008.12
(现代应用数学丛书)
ISBN 978–7–04–023722–1

I.中··· II.①李···②张··· III.①工程数学–英文 ②应用数学—英文 IV. TB11

中国版本图书馆 CIP 数据核字 (2008) 第 193153 号


**Higher Education Press**
4 Dewai Dajie, Beijing 100011, P. R. China, and
**World Scientific Publishing Co Pte Ltd**
5 Toh Tuch Link, Singapore 596224



**ISBN 978–7–04–023722–1**

Printed in P. R. China

# Preface

The China Society for Industrial and Applied Mathematics (CSIAM) had its 9th Annual Conference entitled "Industrial and Applied Mathematics in China" with 12 plenary talks from August 14 to 18, 2006 in Nanjing, China. Later on, in the 6th International Congress on Industrial and Applied Mathematics (ICIAM 2007) held from July 16 to 20, 2007 in Zurich, Switzerland, CSIAM organized an embedded meeting with the same title on July 18, 2007, which consists of two two-hour sessions with six lectures. Since all these talks concern the topic "Industrial and Applied Mathematics in China", we gather a large part of them in this volume for publication. We hope that the readers can get an impression on the present situation and trends of the industrial and applied mathematics in China from this volume and the researchers and graduate students in applied mathematics and in applied sciences can benefit from the mathematical models and methods with applications presented in this book.

We would like to take this opportunity to give our sincere thanks to all the speakers and, in particular, to those who gave their contribution to this volume for their kind help and support.

Ta-Tsien Li
July 2008

# Contents

# Mechanized Methods for Differential and Difference Equations*

Xiaoshan Gao, Ziming Li
*Key Laboratory of Mathematics Mechanization*
*Institute of Systems Science, AMSS*
*Academia Sinica, Beijing 100080, China*
*Email: {xgao, zmli}@mmrc.iss.ac.cn*

**Abstract**

Some recent results on the mechanized methods for differential and difference equations are surveyed. The results include: the characteristic set method for differential and difference equation systems, algorithms for computing closed-form solutions of differential and difference equations, and algorithms for solving and factoring finite-dimensional linear functional systems.

## 1 Introduction

This paper provides a survey of some recent work on differential and difference equations by researchers at the Key Laboratory of Mathematics Mechanization and their collaborators. The work under review is greatly stimulated by Wu's method for mechanical theorem-proving in differential geometries, finding closed-form solutions of differential (difference) equations, and handling analytic and discrete mathematical objects by computers.

Differential equations describe physical laws in mechanics and geometric properties of manifolds. The characteristic set method for differential equations enables us to search for physical laws and geometric properties by computers [52]. For example, Newton's gravitational law is automatically derived form Kepler's laws [51], and "Theorema Egregium" is rediscovered by computing a characteristic set of the fundamental equations of surface theory [30].

The notion of characteristic sets for differential ideals was introduced by Ritt [42]. It plays a fundamental role in differential algebra, because

*Partially supported by a National Key Basic Research Project of China (2004CB 318000).

the Hilbert Basissatz does not hold for differential ideals. The notion and algorithm of characteristic sets for polynomial and differential polynomial sets were introduced by Wu [48,50] to prove theorems in geometries and to manipulate systems of differential and algebraic equations [48,49]. Wu's work inspired a great deal of research in the communities of symbolic computation and automated reasoning. Later on, the success of Grönber bases for polynomial ideals led to methods to characterizing radical differential ideals [4]. The reader may consult [44] for more details on the recent developments of the differential characteristic set method. In this paper we briefly review Wu's scheme for differential characteristic sets and point out its recent extension to difference polynomial systems.

Integrals, special functions and combinatorial sequences are often considered as "infinite" objects. To specify them in terms of a finite amount of information on computers, one uses the differential (difference) equations annihilating these objects. For instances, automatic proofs of combinatorial identities need to find hypergeometric solutions of difference equations [37], while algorithms for symbolic integration need to compute elementary functions satisfying Risch's equation [7]. Great efforts have been made to compute closed-form solutions of linear ordinary differential (difference) equations (see, [26,39] and the references therein). There are two ways to go further: one is to look for closed-form solutions of nonlinear ordinary differential (difference) equations of some kind; the other is to develop symbolic algorithms for linear partial differential (difference) equations. We will summarize recent theoretical and algorithmic results concerning this subject.

Nonlinear differential equations arise from physics. Their analytic solutions are important for the understanding of the physical phenomena. Interesting methods to search for analytic solutions of nonlinear PDEs are given in [16, 53].

Factoring polynomials helps us to solve algebraic equations. Likewise, we want to decompose differential and difference equations into those of lower orders. There have been efficient algorithms for decomposing linear ordinary differential operators [6, 24, 25, 43]. Recent work on extending these methods to linear partial differential and difference equations [33] will be surveyed. We also mention that a decomposition algorithm for nonlinear ordinary differential equations is presented in [23].

The rest of this paper is organized as follows. In Section 2, we outline the differential characteristic method. Methods for computing rational and algebraic solutions of first-order ordinary differential and difference equations are presented in Section 3. An algebraic setting and a factorization algorithm for finite-dimensional linear functional systems are described in Sections 4 and 5, respectively.

# 2 The characteristic set method

The characteristic set method plays a central role in the theory and applications of mathematics mechanization. In this section, we will introduce its main features and applications in automated reasoning.

## 2.1 Properties of ascending chains

Let $\mathbb{K}$ be an ordinary differential field, $\mathbb{X} = \{x_1, \ldots, x_n\}$ a set of differential indeterminates, and $\mathbb{K}\{\mathbb{X}\}$ the set of *differential polynomials* in $\mathbb{X}$ with coefficients in $\mathbb{K}$. We denote $x_{i,j}$ to be the $j$-derivative of $x_i$. The *universal field* $\mathbb{E}$ over $\mathbb{K}$ is a differentially closed field containing $\mathbb{K}$ and infinitely many indeterminates. For a polynomial $D$ and a polynomial set $\mathbb{P} \subset \mathbb{K}[\mathbb{X}]$,

$$\mathrm{Zero}(\mathbb{P}) = \{\eta \in \mathbb{E}^n |\ P(\eta) = 0, \forall P \in \mathbb{P}\}$$

is called a *variety*, and $\mathrm{Zero}(\mathbb{P}/D) = \mathrm{Zero}(\mathbb{P}) \setminus \mathrm{Zero}(D)$ is called a *quasi variety*.

A set $\mathcal{A}$ of differential polynomials is called an *ascending chain* (triangular set), or simply a chain, if after renaming the indeterminates in $\mathbb{X}$ as $\mathbb{U} = \{u_1, \ldots, u_q\}$ and $\mathbb{Y} = \{y_1, \ldots, y_p\}$, we can write $\mathcal{A}$ in the following form:

$$\begin{aligned} A_1(\mathbb{U}, y_1) &= I_1 y_{1,o_1}^{d_1} + \text{ terms of lower orders and degrees in } y_1, \\ &\cdots \\ A_p(\mathbb{U}, y_1, \ldots, y_p) &= I_p y_{p,o_p}^{d_p} + \text{ terms of lower orders and degrees in } y_p. \end{aligned} \tag{1}$$

As a matter of terminologies, $o_i$ is called the order of $A_i$; $I_i$ is called the *initial* of $A_i$, $S_i = \frac{\partial A_i}{\partial y_{i,o_i}}$ is called the *separant* of $A_i$. Write $\mathbf{I}_{\mathcal{A}} = \prod_i I_i S_i$. The *dimension* of $\mathcal{A}$ is defined to be $|\mathbb{U}| = q$, which is denoted $\dim(\mathcal{A})$. The *order* of $\mathcal{A}$ is defined to be $\mathrm{ord}(\mathcal{A}) = \sum_{i=1}^{p} o_i$. The *degree* of $\mathcal{A}$ is defined to be $\deg(\mathcal{A}) = \prod_{i=1}^{p} d_i$.

We could say that the formal solutions for a chain is basically determined. Intuitively, for a set of given values of the parameters $\mathbb{U}$, the $y_i$ can be determined iteratively by solving univariate equations $A_i = 0$. In order to show the properties of chains, we first introduce several concepts. The *saturation ideal* of $\mathcal{A}$ is defined to be

$$\mathrm{sat}(\mathcal{A}) = \{P \in \mathbb{K}\{\mathbb{X}\} |\, \exists k \in \mathbb{N},\ \mathbf{I}_{\mathcal{A}}^k P \in (\mathcal{A})\}.$$

We may define a partial ordering among the chains in a nature way [42, 52]. It is known that any set of chains contains one with lowest order. A *characteristic set* of a differential polynomial set $\mathbb{P}$ is any chain of lowest ordering contained in $\mathbb{P}$.

A chain $\mathcal{A}$ is called *irreducible* if $A_1$ is an irreducible polynomial in $y_{1,o_1}$ and $A_k$ is an irreducible polynomial modulo $A_1, \ldots, A_{k-1}$.

**Theorem 2.1.** *[42, 52] Let $\mathcal{A}$ be an irreducible chain. Then* $\mathrm{sat}(\mathcal{A})$ *is a prime ideal of dimension* $\dim(\mathcal{A})$, *order* $\mathrm{ord}(\mathcal{A})$ *wrt* $\mathbb{U}$, *and degree* $\deg(\mathcal{A})$ *wrt* $\mathbb{U}$. *Conversely, a characteristic set of a prime ideal is irreducible.*

The following result shows that the dimension, order and degree of a chain are intrinsic properties.

**Theorem 2.2.** *[19, 22] Let $\mathcal{A}$ be a chain of form (1). If* $\mathrm{Zero}(\mathrm{sat}(\mathcal{A})) \neq \varnothing$, $\mathrm{Zero}(\mathrm{sat}(\mathcal{A}))$ *and* $\mathrm{Zero}(\mathcal{A}/\mathbf{I}_{\mathcal{A}})$ *are unmixed. More precisely, write* $\mathrm{Zero}(\mathrm{sat}(\mathcal{A}))$ *as an irredundant decomposition:* $\mathrm{Zero}(\mathrm{sat}(\mathcal{A})) = \cup_{i=1}^{r}\mathrm{Zero}(\mathrm{sat}(\mathcal{C}_i))$. *Then*

(1) $\mathcal{C}_i$ *is also of form (1). As a consequence,* $\dim(\mathrm{sat}(\mathcal{C}_i)) = \dim(\mathcal{A})$ *and* $\mathrm{ord}(\mathcal{C}_i) = \mathrm{ord}(\mathcal{A})$.

(2) $\deg(\mathcal{A}) \geqslant \sum_{i=1}^{r} \deg(\mathcal{C}_i)$. *Furthermore,* $\deg(\mathcal{A}) = \sum_{i=1}^{r} \deg(\mathcal{C}_i)$ *iff* $\mathcal{A}$ *is* saturated, *that is, the initials and seprants of $\mathcal{A}$ are invertible wrt $\mathcal{A}$.*

Another important property for chains is

**Theorem 2.3.** *[52] An irreducible chain admits a formal power series solution which can be computed algorithmically.*

In order to make the paper shorter, we limit to the ordinary differential case. Similar results for the partial differential case were also established, where we need to assume that the chains are either passive [49, 52] or coherent [4, 5, 27].

Similar results are also proved in the case of algebraic difference polynomials [21, 22]. However, in the difference case, we do not have algorithms to decide whether a chain is irreducible. In order to have a constructive theory, proper irreducible chains are introduced [21]. Also, Theorem 2.2 is proved only for proper irreducible chains.

## 2.2 Characteristic set method

The characteristic set method decomposes the zero set for a differential polynomial system in general form into the union of zero sets for chains. Since the zero set of a chain is considered to be known, this method gives a general tool to deal with differential equation systems.

Let $\mathbb{P}$ be a finite set of differential polynomials. Then we can perform the following operations:

$$\begin{array}{llllll} \mathbb{P} = \mathbb{P}_0 & \mathbb{P}_1 & \cdots & \mathbb{P}_i & \cdots & \mathbb{P}_m, \\ \mathcal{B}_0 & \mathcal{B}_1 & \cdots & \mathcal{B}_i & \cdots & \mathcal{B}_m = \mathcal{C}, \\ \mathbb{R}_0 & \mathbb{R}_1 & \cdots & \mathbb{R}_i & \cdots & \mathbb{R}_m = \varnothing, \end{array} \tag{2}$$

where $\mathcal{B}_i$ is a lowest chain in $\mathbb{P}_i$ with respect to a pre-selected partial ordering; $\mathbb{R}_i$ is the set of nonzero remainders of the polynomials in $\mathbb{P}_i$ wrt $\mathcal{B}_i$; and $\mathbb{P}_{i+1} = \mathbb{P}_0 \cup \mathcal{B}_i \cup \mathbb{R}_i$. In scheme (2), $\mathcal{B}_m = \mathcal{C}$ verifies

$$\mathrm{prem}(\mathbb{P}, \mathcal{C}) = \{0\} \text{ and } \mathrm{Zero}(\mathbb{P}) \subset \mathrm{Zero}(\mathcal{C}), \tag{3}$$

where prem denotes the differential pseudo-remainder. Any chain $\mathcal{C}$ verifying the property (3) is called a *Wu characteristic set* of $\mathbb{P}$.

**Theorem 2.4** (Wu's Well-ordering Principle). *[49, 52] Let $\mathcal{C}$ be a Wu characteristic set of a finite set $\mathbb{P}$ of differential polynomials. Then:*

$$\mathrm{Zero}(\mathbb{P}) = \mathrm{Zero}(\mathcal{C}/\mathbf{I}_{\mathcal{C}}) \bigcup \cup_i \mathrm{Zero}(\mathbb{P} \cup \mathcal{C} \cup \{I_i\}),$$

$$\mathrm{Zero}(\mathbb{P}) = \mathrm{Zero}(\mathrm{sat}(\mathcal{C})) \bigcup \cup_i \mathrm{Zero}(\mathbb{P} \cup \mathcal{C} \cup \{I_i\}),$$

*where $I_i$ are the initials and separants of the polynomials in $\mathcal{C}$.*

Using the well-ordering principle recursively, we obtain the following key result.

**Theorem 2.5** (Ritt-Wu's Zero Decomposition Theorem). *[42, 52] There is an algorithm which permits to determine, for a given finite set $\mathbb{P}$ of differential polynomials, a finite set of (irreducible) chains $\mathcal{A}_j$ such that*

$$\mathrm{Zero}(\mathbb{P}) = \cup_j \mathrm{Zero}(\mathcal{A}_j/\mathbf{I}_{\mathcal{A}_j}) = \cup_j \mathrm{Zero}(\mathrm{sat}(\mathcal{A}_j)).$$

Let $\mathbb{P}$ be a finite subset of $\mathbb{K}\{\mathbb{U}, \mathbb{X}\}$, and $D \in \mathbb{K}\{\mathbb{U}, \mathbb{X}\}$, where $\mathbb{U} = \{u_1, \ldots, u_m\}$ and $\mathbb{X} = \{x_1, \ldots, x_n\}$. The projection of $\mathrm{Zero}(\mathbb{P}/D)$ to $\mathbb{U}$ is defined as follows:

$$\mathrm{Proj}_{\mathbb{X}}\mathrm{Zero}(\mathbb{P}/D) = \{e \in \mathbb{E}^m \mid \exists\, a \in \mathbb{E}^n \quad \text{s.t.} \quad (e, a) \in \mathrm{Zero}(\mathbb{P}/D)\}.$$

Projection for quasi-varieties can be computed with the characteristic set method.

**Theorem 2.6** (Projection Theorem). *[19] For a finite subset set $\mathbb{P} \subset K\{\mathbb{U}, \mathbb{X}\}$ and $D \in K\{\mathbb{U}, \mathbb{X}\}$, we can compute chains $\mathcal{A}_i$ and polynomials $D_i$ in $\mathbb{K}[\mathbb{U}]$ such that*

$$\mathrm{Proj}_{\mathbb{X}}\mathrm{Zero}(\mathbb{P}/D) = \cup_{i=1}^{l} \mathrm{Zero}(\mathcal{A}_i/D_i\mathbf{I}_{\mathcal{A}_i}).$$

The concept of characteristic sets for prime ideals was introduced by Ritt [42]. The notion of characteristic sets given above, the well-ordering principle, and the current form of zero decomposition theorems were introduced by Wu [48,49,52]. An implementation of the method can be found in [46]. In order to improve the efficiency, new characteristic set methods were proposed [4, 5, 9, 10, 18, 27, 40, 45]. The characteristic set method was used to solve certain problems for analytical functions [41].

A characteristic set method for algebraic difference equation systems was proposed in [21, 22]. It is quit surprising that there are no essential progresses for the theory and algorithms of difference characteristic set methods since the early work of Ritt and his colleagues in the 1930s. In [21], an algorithm was proposed to decompose the zero set a difference polynomial system into the union of unmixed zero sets of difference polynomial systems represented by proper irreducible chains. In [22], a new resolvent theory for difference polynomial systems was proposed.

To solve a set of equations in triangular form, we need to solve univariate equations in a cascade form. The resolvent methods were introduced to reduce the solving of equation systems into the solving of one univariate equation plus a set of linear equations [13, 22].

## 2.3 Wu's method of automated geometry theorem proving and discovering

A geometry theorem is called a *theorem of equality type*, if after introducing coordinates, the theorem can be expressed in the following form

$$\forall x_i[(H_1 = 0 \wedge \cdots \wedge H_s = 0 \wedge D_1 \neq 0 \wedge \cdots \wedge D_t \neq 0) \Longrightarrow (C = 0)], \quad (4)$$

where $H_i, D_i, C$ are in $\mathbb{K}\{\mathbb{X}\}$.

For theorems of equality type, we have the following principles of mechanical theorem proving, which are consequences of Theorems 2.1 and 2.4.

**Theorem 2.7.** *[49] For a geometry statement of form (4), let $\mathcal{A}$ be a Wu-characteristic set of $\{H_1, \ldots, H_s\}$. If $\mathrm{prem}(C, \mathcal{A}) = 0$, then the statement is valid under the non-degenerate condition $\mathbf{I}_{\mathcal{A}} \neq 0$.*

Note that the non-degenerate condition $\mathbf{I}_{\mathcal{A}} \neq 0$ is generated automatically by the algorithm.

**Theorem 2.8.** *[52] Let $D = \prod_i D_i$. By Theorem 2.5, we have*

$$\mathrm{Zero}(\{H_1, \ldots, H_s\}/D) = \cup_{i=1}^{l} \mathrm{Zero}(\mathrm{sat}(\mathcal{A}_i)/D).$$

*If $\mathrm{prem}(C, \mathcal{A}_i) = 0, i = 1, \ldots, l$, then the statement is true. If $\mathcal{A}_i$ is irreducible and $\mathrm{prem}(C, \mathcal{A}_i) \neq 0$, then the statement is not valid on $\mathrm{Zero}(\mathrm{sat}(\mathcal{A}_i)/D)$.*

As an example, let us show how to prove Newton's gravitational law with Kepler's laws. The first and second Kepler laws state that each planet describes an ellipse with the sun in one focus and the radius vector drawn from the sun to a planet sweeps out equal areas in equal times. The Newton's law states that the acceleration is reversely proportional to the distance from the planet to the sun. We may use differential equations $K_1 = 0, K_2 = 0$, and $N_1 = (ar^2)' = 0$ to represent these laws:

$$\begin{aligned}
h_1 &= r^2 - x^2 - y^2 = 0,\\
h_2 &= a^2 - x''^2 - y''^2 = 0,\\
K_1 &= r - p - ex = 0 \wedge p' = 0 \wedge e' = 0,\\
K_2 &= y'x - yx' - h = 0 \wedge h' = 0,\\
d_1 &= p \neq 0 \quad \text{(The ellipse is not a line.)}.
\end{aligned}$$

Then, we need to show

$$\forall x, y, p, e, a, r[(K_1 = 0 \wedge K_2 = 0 \wedge h_1 = 0 \wedge h_2 = 0 \wedge d_1 \neq 0) \Rightarrow N_1 = 0].$$

By Theorem 2.5 ($p < e < x < y < r < a$),

$$\text{Zero}(\{K_1, p', e', h_1, h_2, n_2\}/p) = \text{Zero}(\text{sat}(\mathcal{A}_1)p),$$

where $\mathcal{A}_1$ is a chain. By computation, we have $\text{prem}(n_1, ASC_1) = 0$, which proves Newton's law.

There are two kinds of problems in differential geometry other than theorem proving. One is finding locus equations, the other is deriving geometry formulas. For a geometric configuration given by a set of polynomial equations $h_1(\mathbb{U}, x_1, \cdots, x_p) = 0, \cdots, h_r(\mathbb{U}, x_1, \cdots, x_p) = 0$, we want to find a relation between arbitrarily chosen variables $\mathbb{U}$ (parameters) and a dependent variable, say, $x_1$. Wu pointed out that the characteristic set method can be used to discover such unknown geometric formulas [51]. Actually, Newton's law can be deduced from Kepler's laws automatically in this way. More detailed accounts can be found in [10, 11, 30, 45].

The characteristic set method can be used to prove a much wider class of geometry theorems. Let $\mathbb{E}$ be a differentially closed extension of $\mathbb{K}$, say, the field of meromorphic functions [42]. A *first order formula* over $\mathbb{E}$ can be defined as follows.

1. If $P \in \mathbb{K}[\mathbb{X}]$, then $P(\mathbb{X}) = 0$ is a formula.
2. If $f, g$ are formulas, then $\neg f$, $f \wedge g$, and $f \vee g$ are formulas.
3. If $f$ is a formula, then $\exists x_i \in \mathbb{E}(f)$ and $\forall x_i \in \mathbb{E}(f)$ are formulas.

A formula can always be written as a prefix canonical form

$$\phi = Q_1 y_1 \ldots Q_m y_m \psi(u_1, \ldots, u_d, y_1, \ldots, y_m), \tag{5}$$

where $Q_k$ is a quantifier $\exists$ or $\forall$ and $\psi$ a formula free of quantifiers. For a first order formula $\phi$ of form (5), there exists a fundamental problem:

**Quantifier Elimination:** Find a formula $\theta(u_1, \ldots, u_d)$ such that $\theta$ is equivalent to $\phi$. If $d = 0$, we need to decide whether $\phi$ is valid or not.

As a consequence of Theorem 2.6, we have

**Theorem 2.9.** *There exists a decision procedure for the first order theory over a differentially closed field.*

# 3 Rational and algebraic solutions of ODEs and O$\Delta$Es

For brevity we abbreviate ordinary difference equations as O$\Delta$E.

By decomposing the zero set of a differential polynomial system into the zero sets of chains, the characteristic set method gives a complete way to describe the structure for the zero sets of equation systems. In particular, finding the solutions of differential polynomial systems can be reduced to finding those of a single differential equation or a system of equations in a single variable.

Closed-form solutions of linear ODEs and O$\Delta$Es were widely studied. On the other hand, similar results to nonlinear ODEs are very limited. In this section, we summarize some recent results on finding rational and algebraic solutions to nonlinear ODEs and O$\Delta$Es. It is interesting to see whether these results can be treated uniformity with the differential Galois theory [35].

## 3.1 Rational and algebraic solutions of algebraic ODEs

Let $P \in \mathbb{K}\{y\} \setminus \mathbb{K}$ be an irreducible differential polynomial in an indeterminate $y$ and

$$\Sigma_P = \{A \in \mathbb{K}\{y\} | SA \equiv 0 \bmod \{P\}\},$$

where $S$ is the separant of $P$ and $\{P\}$ is the radical differential ideal generated by $P$. Then $\Sigma_P$ is a prime ideal [42]. A generic zero of $\Sigma_P$ is defined to be a *general solution* of $P = 0$. In particular, an *algebraic general solution* of $P = 0$ is a general solution $\hat{y}$ which satisfies the following equation

$$G(x, y) = \sum_{i=0}^{n} a_i(x) y^i = 0, \tag{6}$$

where $a_i$ is a polynomial in $x$ with degree $\alpha_i$ and with constant coefficients, and $G(x,y)$ is an irreducible polynomial in $x,y$. When $n=1$, $\hat{y}$ is called a *rational general solution* of $P=0$.

For $\alpha_0,\alpha_1,\cdots,\alpha_n\in\mathbb{Z}_{\geqslant 0}$, we define the differential polynomial

$$\mathbb{D}_{(\alpha_0;\alpha_1,\cdots,\alpha_n)} := \det(\mathcal{A}_{(h,\alpha_1;\alpha_0)}(y)|\mathcal{A}_{(h,\alpha_2;\alpha_0)}(y^2)|\cdots|\mathcal{A}_{(h,\alpha_n;\alpha_0)}(y^n)),$$

where

$$\mathcal{A}_{(h,\alpha;k)}(y) := \begin{pmatrix} \binom{k+1}{0}y_{k+1} & \binom{k+1}{1}y_k & \cdots & \binom{k+1}{\alpha}y_{k+1-\alpha} \\ \binom{k+2}{0}y_{k+2} & \binom{k+2}{1}y_{k+1} & \cdots & \binom{k+2}{\alpha}y_{k+2-\alpha} \\ \vdots & \vdots & \cdots & \vdots \\ \binom{k+h+1}{0}y_{k+h+1} & \binom{k+h+1}{1}y_{k+h} & \cdots & \binom{k+h+1}{\alpha}y_{k+h+1-\alpha} \end{pmatrix}.$$

We have

**Lemma 3.1.** *[1] $y(x)$ satisfies an equation of the type (6) if and only if $\mathbb{D}_{(\alpha_0;\alpha_1,\cdots,\alpha_n)}(y(x))=0$. As a consequence, we give a defining differential equation for algebraic functions.*

When $n=2,\alpha_1=\alpha_2=1$ and $\alpha_0=2$,

$$\mathbb{D}_{(2;1,1)} = \begin{vmatrix} y_3 & 3y_2 & (y^2)''' & 3(y^2)'' \\ y_4 & 4y_3 & (y^2)^{(4)} & 4(y^2)''' \\ y_5 & 5y_4 & (y^2)^{(5)} & 5(y^2)^{(4)} \\ y_6 & 6y_5 & (y^2)^{(6)} & 6(y^2)^{(5)} \end{vmatrix}.$$

We have $\mathbb{D}_{(2;1,1)}(y(x))=0$ if and only if

$$(a_{2,1}x+a_{2,0})y^2(x)+(a_{1,1}x+a_{1,0})y(x)+a_{0,2}x^2+a_{0,1}x+a_{0,0}=0$$

for constants $a_{i,j}$.

The key to find a rational and algebraic function solutions is to give a degree bound for the solution. We can give these degree bounds for first order autonomous ODEs. In what follows, let $F(y,y_1)=0$ be a first order autonomous ODE. Then we have

**Theorem 3.2.** *[1] If $G(x,y)=0$ defines a nontrivial algebraic solution of $F=0$, then*

(1) $\deg(G(x,y),x)=\deg(F,y_1)$,

(2) $\deg(G(x,y),y)\leqslant\deg(F,y)+\deg(F,y_1)$.

The following example shows that the bound in (2) is optimal. Let $n>m>0$ and $(n,m)=1$. Then $G=y^n-x^m$ is irreducible. $y^n-x^m=0$ is an algebraic solution of $F=y^{n-m}y_1^m-(m/n)^m=0$. Here, $\deg(G(x,y),y)=\deg(F,y)+\deg(F,y_1)$.

For rational solutions, we could give the exact degree bound [17].

**Theorem 3.3.** *If $y = P(x)/Q(x)$ is a rational solution of $F(y, y_1) = 0$, then* $\deg(y(x)) = \deg(F, y_1)$.

These degree bounds are obtained by treating $F(y, y_1) = 0$ as an algebraic curve and the solution as a parametrization of the curve. This idea also leads to the following algorithm to find a rational solution to a first order autonomous ODE [17].

**Theorem 3.4.** *Let $y = r(x), y_1 = s(x)$ be a proper rational parametrization of $F(y, y_1) = 0$, where $r(x), s(x)$ are rational functions in $x$ with constant coefficients. Then $F = 0$ has a rational general solution iff we have the following relations*

$$ar(x)' = s(x) \quad \text{or} \quad a(x-b)^2 r(x)' = s(x),$$

*where $a, b$ are constants and $a \neq 0$. If one of the above relations is true, then replacing $x$ by $a(x+c)$ (or $b - \frac{1}{a(x+c)}$) in $y = r(x)$, we obtain a rational general solution of $F = 0$, where $c$ is an arbitrary constant.*

The above algorithm depends on the rational parametrization of algebraic curves. A more efficient algorithm is based on Hermite-Padé approximation.

Let $A(x)$ be a formal power series. If a polynomial $G(x, y)$ satisfies

$$G(x, A(x)) = O\left(x^{(n+1)(m+1)+1}\right),$$

where $m = \deg(G, x), n = \deg(G, y)$, then we call $G(x, y) = 0$ Hermite-Padé approximant to $A(x)$. We could find the algebraic solution for an first order autonomous ODEs as follows [1].

(1) Find the first $N$ terms $f(x)$ of formal power series solution of $F(y, y_1) = 0$, where

$$N = 2(\deg(F, y) + \deg(F, y_1)).$$

(2) Let $d = \deg(F, y_1)$. Construct the $(d, d, \ldots, d)$ Hermite-Padé approximant $G(x, y) = 0$ to $f(x)$.

(3) We need only to check whether $G = 0$ is a nontrivial algebraic solution of $F = 0$.

The complexity of this algorithm is polynomial in terms of the number of the multiplications in the number field.

## 3.2 Rational solutions of algebraic O$\Delta$Es

The result about rational solutions of ODEs can be extended to O$\Delta$Es. Let $\mathbb{K} = \mathbb{Q}(x)$ be the difference field with the difference operator $\mathbf{E}(x) = x + 1$, $y$ an indeterminate, and $y_n = \mathbf{E}^n y$.

Let $P \in \mathbb{K}\{y\} \setminus \mathbb{K}$ be an irreducible difference polynomial in $y$, and

$$\Sigma_P = \{A \in \mathbb{Q}(x)\{y\} | SA \equiv 0 \bmod \{P\}\},$$

where $S$ is the separant of $P$. Cohn proved that $\Sigma_P$ is a perfect difference ideal and it could be decomposed into the intersection of the principle components of $P$ [14]. Let $\Lambda$ be one of the principle components of $P$. A general solution of $P = 0$ is defined as a generic zero of one of the principle components of $\Sigma_P$. A *rational general solution* of $P(y) = 0$ is defined as a general solution of $P = 0$ with the following form:

$$\hat{y}(x) = \frac{a_n x^n + a_{n-1}x^{n-1} + \ldots + a_0}{x^m + b_{m-1}x^{m-1} + \ldots + b_0}, \tag{7}$$

where $a_i, b_j$ are constants. In particular, if $m = 0$, we call $\hat{y}(x)$ polynomial general solution. For instance, the difference equation $(y - y_1)^2 - 2(y + y_1) + 1 = 0$ has two general solutions: $y(x) = (x + c)^2$ and $y(x) = (ce^{i\pi x} + \frac{1}{2})^2$ where $c$ is an arbitrary constant.

The defining difference equations for polynomial and rational functions are given by the following lemmas [20].

**Lemma 3.5.** *Let* $\mathcal{P}_n = \sum_{i=0}^{n+1}(-1)^i\binom{n+1}{i}y_i$. *Then* $y(x) = a_n x^n + a_{n-1}x^{n-1} + \cdots + a_0$ *(*$\mathbf{E}(a_i) = a_i$*) if and only if* $\mathcal{P}_n(y(x)) = 0$.

Let

$$\mathcal{R}_{n,m} = \det(\sum_{i=0}^{n+1}(-1)^i\binom{n+1}{i}Y_i * M_i),$$

where $Y_i = \mathrm{diag}(y_i, y_{i+1}, \cdots, y_{m+i})$, $M_i = (H_{k,l}(i))_{(m+1)\times(m+1)}$,

$$H_{k,l}(i) = \frac{(i+k-n)(i+k-n-1)\cdots(i+k-n-l)(i+k-n-l-2)\cdots(i+k-n-m)}{(-1)^{m-l}(m+1-l)!(l-1)!}.$$

**Lemma 3.6.** $y(x) = \frac{a_n x^n + a_{n-1}x^{n-1} + \cdots + a_0}{b_m x^m + b_{m-1}x^{m-1} + \cdots + b_0} \Leftrightarrow \mathcal{R}_{n,m}(y(x)) = 0$ *where* $\mathbf{E}(a_i) = a_i$, $\mathbf{E}(b_j) = b_j$.

Using properties of the proper parametrization of algebraic curves, the degree bound for the rational solution can be given [20].

**Theorem 3.7.** *Let* $F(y, y_1) = 0$ *be a first order autonomous O$\Delta$E. If* $y(x) \in \bar{\mathbb{Q}}(x)\setminus\bar{\mathbb{Q}}$ *is a rational solution of* $F = 0$, *then* $\deg(y(x)) = \deg(F, y_1) = \deg(F, y)$.

Similar to the differential case, the rational solutions can be found with the help of rational parametrization of algebraic curves [20].

**Theorem 3.8.** *Let $y = r(t), y_1 = s(t)$ be a proper parametrization of $F(y, y_1) = 0$. Then $F = 0$ has a nontrivial rational solution iff $r(t)$, $s(t)$ satisfy one of the following relations:*

(1) *There exists a nonzero $a \in \mathcal{C}$ such that $r(\frac{t+1}{a}) = s(\frac{t}{a})$.*

(2) *There exist $a \neq 0, b \in \mathcal{C}$ such that $r(\frac{a}{t+1} - b) = s(\frac{a}{t} - b)$.*

It is obvious that if (1) is true, $\bar{y}(x) = r(\frac{x+c}{a})$ is a rational general solution of $F = 0$ where $c$ is an arbitrary constant. If (2) is true, $\bar{y}(x) = r(\frac{a}{x+c} - b)$ is a rational general solution of $F = 0$ where $c$ is an arbitrary constant.

# 4 Finite-dimensional partial linear functional systems

A finite-dimensional partial linear functional system consists of linear partial differential, shift, and $q$-shift operators, or any mixture thereof, and has a finite-dimensional solution space. The following is an example:

$$\begin{cases} P''(x,\, k) - \frac{2x}{1-x^2} P'(x,\, k) + \frac{k(k+1)}{1-x^2} P(x,\, k) = 0, \\ P(x,\, k+2) - \frac{(2k+3)x}{k+2} P(x,\, k+1) + \frac{k+1}{k+2} P(x,\, k) = 0\,. \end{cases} \tag{8}$$

The sequence of the Legendre polynomials $\{P(x,\, k)\}_{k=1}^{\infty}$ is a solution of (8) with the initial conditions $\{P(0,0) = 0,\ P'(0,0) = 0,\ P(0,1) = 0,\ P'(0,1) = 1\}$.

For brevity, finite-dimensional linear functional systems will be called $\partial$-finite systems in the sequel. They are also called over-determined systems in the literature. $\partial$-finite systems arise from symmetry analysis of nonlinear ordinary differential equations, theory of special functions, and combinatorics.

In this section we review a purely algebraic setting for $\partial$-finite systems, including modules of formal solutions and Picard-Vessiot extensions. The former captures the notion of $\partial$-finiteness, and makes it possible to compute the dimension of the solution space of a $\partial$-finite system; while the latter contains "all" solutions of a $\partial$-finite system, and paves a way to introduce Galois groups.

## 4.1 An algebraic setting

Let $R$ be a ring and $\Delta$ be a finite set of commuting maps from $R$ to itself. A map in $\Delta$ is assumed to be either a derivation or an automorphism. Recall that a derivation $\delta$ is an additive map satisfying the multiplicative rule $\delta(ab) = a\delta(b) + \delta(a)b$ for all $a, b \in R$. The pair $(R, \Delta)$ is called a $\Delta$-ring, and it is a *$\Delta$-field* when $R$ is a field.

For a derivation $\delta \in \Delta$, an element $c$ of $R$ is called a *constant* with respect to $\delta$ if $\delta(c) = 0$. For an automorphism $\sigma \in \Delta$, $c$ is called a *constant* with respect to $\sigma$ if $\sigma(c) = c$. An element $c$ of $R$ is called a *constant* if it is a constant with respect to all maps in $\Delta$. The set of constants of $R$, denoted by $C_R$, is a subring. The ring $C_R$ is a subfield if $R$ is a field.

Let $(F, \Delta)$ be a $\Delta$-field. By reordering the indices, we can always assume that

$$\Delta = \{\delta_1, \ldots, \delta_\ell, \sigma_{\ell+1}, \ldots, \sigma_m\}$$

for some $\ell \geqslant 0$, where the $\delta_i$'s are derivation operators on $F$ and the $\sigma_j$'s are automorphisms of $F$. The *Ore algebra* ( [12]) over $F$ is the polynomial ring $\mathcal{S} := F[\partial_1, \ldots, \partial_m]$ in $\partial_i$ with the usual addition and a multiplication as follows:

$$\partial_i \partial_j = \partial_j \partial_i, \quad \partial_s a = a\partial_s + \delta_s(a), \quad \partial_t a = \sigma_t(a)\partial_t,$$

for any $1 \leqslant i, j \leqslant m$, $1 \leqslant s \leqslant \ell$, $\ell < t \leqslant m$ and $a \in F$.

Remark that $\partial_i(a)$, where $a$ is an element of a $\Delta$-ring, is meant to be $\delta_i(a)$ if $\partial_i$ is associated to a derivation operator $\delta_i$, and to be $\sigma_i(a)$ if $\partial_i$ is associated to an automorphism $\sigma_i$; while $\partial_i a$, where $a$ is an element of the Ore algebra $\mathcal{S}$, means the product of $\partial_i$ and $a$.

**Definition 4.1.** *Let $(F, \Delta)$ be a $\Delta$-field.* A linear functional system *over $F$ is a system of the form $A(\mathbf{z}) = 0$ where $A$ is a $p \times q$ matrix with entries in the Ore algebra $\mathcal{S}$ and $\mathbf{z}$ is a column vector of $q$ unknowns.*

**Example 4.2.** *The system (8), satisfied by the Legendre polynomials, can be rewritten as $A(z) = 0$ where*

$$A = \left(\partial_x^2 - \frac{2x}{1 - x^2}\partial_x + \frac{k(k+1)}{1 - x^2},\ \partial_k^2 - \frac{(2k+3)x}{k+2}\partial_k + \frac{k+1}{k+2}\right)^\tau,$$

*with $\partial_x$ the differentiation with respect to $x$ and $\partial_k$ the shift operator with respect to $k$.*

Let $F$ be a $\Delta$-field. A commutative ring $R$ containing $F$ is called a *$\Delta$-extension* of $F$ if all the maps in $\Delta$ can be extended to $R$ in such a way that all derivations (resp. automorphisms) of $F$ become derivations (resp. automorphisms) of $R$ and the extended maps commute pairwise.

By a solution of a linear functional system $A(\mathbf{z}) = 0$ over $F$, we mean a vector $(s_1, \ldots, s_q)^\tau$ over some $\Delta$-extension of $F$ such that $A(s_1, \ldots, s_q)^\tau = 0$, *i.e.*, the application of the matrix $A$ to the vector is zero.

## 4.2 Modules of formal solutions

Let $F$ be a $\Delta$-field and $\mathcal{S} = F[\partial_1, \ldots, \partial_m]$ be the corresponding Ore algebra. In the differential case, an $\mathcal{S}$-module is classically associated to a linear functional system [34, 39]. In the difference case, however, $\mathcal{S}$-modules may not have appropriate dimensions, as illustrated by the following counterexample.

**Example 4.3.** *Let $\sigma \neq \mathbf{1}$ be an automorphism of $F$ and $\mathcal{S} = F[\partial]$ be the corresponding Ore algebra. The equation $\partial(y) = 0$ cannot have a fundamental matrix $(u)$ in any difference ring extension of $F$, for otherwise, $0 = \partial(u) = \sigma(u)$, thus $u = 0$. Therefore $\partial(y) = 0$ has only trivial solution. However, the $\mathcal{S}$-module $\mathcal{S}/\mathcal{S}\partial$ has dimension one as an $F$-vector space.*

In [38, page 56], modules over Laurent algebras are used instead to avoid the above problem. It is therefore natural to introduce the following extension of $\mathcal{S}$: let $\theta_{\ell+1}, \ldots, \theta_m$ be indeterminates independent of the $\partial_i$. Since the $\sigma_j^{-1}$ are automorphisms of $F$, $\overline{\mathcal{S}} = F[\partial_1, \ldots, \partial_m, \theta_{\ell+1}, \ldots, \theta_m]$ is also an Ore algebra in which the $\theta_j$ are associated to the $\sigma_j^{-1}$. Note that $\partial_j \theta_j$ is in the center of $\overline{\mathcal{S}}$, since

$$(\partial_j \theta_j) a = \partial_j \sigma_j^{-1}(a) \theta_j = \sigma_j(\sigma_j^{-1}(a)) \partial_j \theta_j = a \partial_j \theta_j, \text{ for all } a \in F \text{ and } j > \ell.$$

Therefore the left ideal $T = \sum_{j=\ell+1}^{m} \overline{\mathcal{S}}(\partial_j \theta_j - 1)$ is a two-sided ideal of $\overline{\mathcal{S}}$, and we call the factor ring $\mathcal{L} = \overline{\mathcal{S}}/T$ the *Laurent-Ore algebra* over $F$. Writing $\partial_j^{-1}$ for the image of $\theta_j$ in $\mathcal{L}$, we can write $\mathcal{L}$ (by convention) as $\mathcal{L} = F[\partial_1, \ldots, \partial_m, \partial_{\ell+1}^{-1}, \ldots, \partial_m^{-1}]$ and view it as an extension of $\mathcal{S}$. For linear ordinary difference equations, $\mathcal{L} = F[\sigma, \sigma^{-1}]$ is the algebra used in [38]. For linear partial difference equations with constant coefficients, $\mathcal{L}$ is the Laurent polynomial ring used in [36, 54].

When revisiting Example 4.3 with Laurent-Ore algebras, we get that the left ideal generated by $\partial$ in $\mathcal{L} = F[\partial, \partial^{-1}]$ is $\mathcal{L}$, therefore the dimension of $\mathcal{L}/(\mathcal{L}\partial)$ over $F$, which is zero, equals that of the solution space of $\partial(y) = 0$ in any difference ring extension.

Let $F$ be a $\Delta$-field, and $\mathcal{S}$ and $\mathcal{L}$ be the corresponding Ore and Laurent-Ore algebras. We have the following theorem.

**Theorem 4.4.** *Let $A \in \mathcal{S}^{p \times q}$ and $M = \mathrm{coker}_{\mathcal{L}}(A)$. Then $\mathrm{sol}_N(A(\mathbf{z}){=}0)$ and $\mathrm{Hom}_{\mathcal{L}}(M, N)$ are isomorphic as $C_F$-vector spaces for any $\mathcal{L}$-module $N$.*

The proof of Theorem 4.4 reveals that the vector $\mathbf{e} := (\mathbf{e}_1, \ldots, \mathbf{e}_q)^\tau \in M^q$ specified above is a "generic" solution of the system $A(\mathbf{z}) = 0$ in the sense that any solution $(s_1, \ldots, s_q)^\tau$ of that system in $N$ is the image of $\mathbf{e}$ under the map in $\mathrm{Hom}_R(M, N)$ sending $\mathbf{e}_i$ to $s_i$. Thus $\mathrm{coker}_{\mathcal{L}}(A)$ describes the properties of all the solutions of $A(\mathbf{z}) = 0$ "anywhere". This motivates us to define

**Definition 4.5.** *Let* $A \in \mathcal{S}^{p\times q}$*. The* $\mathcal{L}$*-module* $M = \mathcal{L}^{1\times q}/(\mathcal{L}^{1\times p}A)$ *is called the* module of formal solutions *of the system* $A(\mathbf{z}) = 0$*. The dimension of* $M$ *as an* $F$*-vector space is called the* linear dimension *of the system. The system is said to be* of finite linear dimension, *or simply,* $\partial$-finite, *if* $0 < \dim_F M < +\infty$.

Note that we choose to exclude systems with $\dim_F M = 0$ in the above definition since such system has only trivial solution in any $\mathcal{L}$-module, particularly, in any $\Delta$-extension of $F$.

One can compute the dimension of a module of formal solution by Gröbner bases in Laurent-Ore algebra (see [47, 55]).

## 4.3 Picard-Vessiot extensions

A $\partial$-finite system can be reduced to a normal form defined below:

**Definition 4.6.** *A system of the form*

$$\delta_i(\mathbf{z}) = A_i\mathbf{z}, \quad 1 \leqslant i \leqslant \ell, \qquad \sigma_i(\mathbf{z}) = A_i\mathbf{z}, \quad \ell+1 \leqslant i \leqslant m, \tag{9}$$

*where* $A_i \in F^{n\times n}$ *and* $\mathbf{z}$ *is a column vector of* $n$ *unknowns, is called an* integrable system *of size* $n$ *over* $F$ *if the following compatibility conditions are satisfied:*

$$\begin{aligned} \delta_i(A_j) &= \delta_j(A_i), & 1 \leqslant i < j \leqslant \ell, \\ \sigma_i(A_j)A_i &= \sigma_j(A_i)A_j, & \ell < i < j \leqslant m, \\ \sigma_j(A_i)A_j &= A_iA_j + \delta_i(A_j), & 1 \leqslant i \leqslant \ell < j \leqslant m. \end{aligned} \tag{10}$$

*The integrable system (9) is said to be* fully integrable *if the matrices* $A_{\ell+1}, \ldots, A_m$ *are invertible.*

Using Ore algebra notation, we write $\{\partial_i(\mathbf{z}) = A_i\mathbf{z}\}_{1\leqslant i\leqslant m}$ for the system (9) where the action of $\partial_i$ is again meant to be $\delta_i$ for $i \leqslant \ell$ and to be $\sigma_i$ for $i > \ell$. Observe that the conditions (10) are derived from the condition $\partial_i(\partial_j(\mathbf{z})) = \partial_j(\partial_i(\mathbf{z}))$ and are exactly the matrix-analogues of the compatibility conditions for first-order scalar equations in [28]. For a linear ordinary difference equation, we often assume that its trailing coefficient is nonzero, while, for a first-order matrix difference equation, we assume that its matrix is invertible. These assumptions lead to the condition on invertibility of $A_{\ell+1}, \ldots, A_m$ in Definition 3.1.

**Example 4.7.** *Let* $F = \mathbb{C}(x, k)$*,* $\delta_x$ *be the differentiation with respect to* $x$ *and* $\sigma_k$ *the shift operator with respect to* $k$*. Then* $\mathcal{A} : \{\delta_x(\mathbf{z}) = A_x\mathbf{z}, \sigma_k(\mathbf{z}) = A_k\mathbf{z}\}$ *is a fully integrable system where*

$$A_x = \begin{pmatrix} \frac{x^2-kx-k}{x(x-k)(x-1)} & \frac{x^2-kx+3k-2x}{kx(x-k)(x-1)} \\ \frac{k(kx+x-x^2-2k)}{(x-k)(x-1)} & \frac{x^3+x^2-kx^2-2x+2k}{x(x-k)(x-1)} \end{pmatrix}$$

*and*

$$A_k = \begin{pmatrix} \frac{k+1+kx^2-xk^2-x}{(x-k)(x-1)} & -\frac{k+1+kx-k^2-x}{k(x-k)(x-1)} \\ \frac{x(k+1)(k+1+kx-k^2-x)}{(x-k)(x-1)} & \frac{(k+1)(x^2-2kx-x+k^2)}{k(x-k)(x-1)} \end{pmatrix}.$$

We will first define the notion of Picard-Vessiot extensions of fully integrable systems, and then generalize it to $\partial$-finite systems. Recall that a square matrix with entries in a commutative ring is said to be *invertible* if its determinant is a unit in that ring.

Let $F$ be a $\Delta$-field and $\{\partial_i(\mathbf{z}) = A_i\mathbf{z}\}_{1\leqslant i\leqslant m}$ be a fully integrable system of size $n$ over $F$. We define

**Definition 4.8.** *An $n \times n$ matrix $U$ with entries in a $\Delta$-extension of $F$ is a* fundamental matrix *for the system $\{\partial_i(\mathbf{z}) = A_i\mathbf{z}\}_{1\leqslant i\leqslant m}$ if $U$ is invertible and $\partial_i(U) = A_iU$ for each $i$, i.e., each column of $U$ is a solution of the system.*

A two-sided ideal $I$ of a commutative $\Delta$-ring $R$ is said to be *invariant* if $\delta_i(I) \subset I$ for $i \leqslant \ell$ and $\sigma_j(I) \subset I$ for $j > \ell$. The ring $R$ is said to be *simple* if its only invariant ideals are $(0)$ and $R$.

**Definition 4.9.** *A* Picard-Vessiot ring *for a fully integrable system is a ring $E$ such that:*

(i) *$E$ is a simple $\Delta$-extension of $F$.*

(ii) *There exists some fundamental matrix $U$ with entries in $E$ for the system such that $E$ is generated by the entries of $U$ and* $\det(U)^{-1}$ *over $F$.*

Definitions 4.8 and 4.9 are natural generalizations of their analogues in the purely differential case [39, (pages 12, 415)] and the ordinary difference case [38, (Errata)].

The existence of fundamental matrices and Picard-Vessiot extensions for fully integrable systems is stated in the following [8].

**Theorem 4.10.** *Every fully integrable system over $F$ has a Picard-Vessiot ring $E$. If $F$ has characteristic* 0 *and $C_F$ is algebraically closed, then $C_E = C_F$. Furthermore, that extension is minimal, meaning that no proper subring of $E$ satisfies both conditions in Definition 4.9.*

Consequently, if $F$ has characteristic zero and an algebraically closed field of constants, then all the solutions of a fully integrable system in its Picard-Vessiot ring form a $C_F$-vector space whose dimension equals the size of the system.

**Example 4.11.** *Consider the system $\mathcal{A}$ in Example 4.7. Note that the change of variable*[1] $\mathbf{z} = M\mathbf{y}$ *where*

$$M = \begin{pmatrix} \frac{x-k}{x} & x^2 \\ (x-k)k & x^2 k \end{pmatrix},$$

*transforms $\mathcal{A}$ into another* F*system* $\mathcal{B} : \{\delta_x(\mathbf{y}) = B_x\mathbf{y}, \sigma_k(\mathbf{y}) = B_k\mathbf{y}\}$ *with* $B_x = \begin{pmatrix} 1 & 0 \\ 0 & 0 \end{pmatrix}$ *and* $B_k = \begin{pmatrix} 1 & 0 \\ 0 & k \end{pmatrix}$. *It suffices to find a Picard-Vessiot ring of $\mathcal{B}$. We get that* $V = \begin{pmatrix} e^x & 0 \\ 0 & \Gamma(k) \end{pmatrix}$ *is a fundamental matrix for $\mathcal{B}$, and thus $MV$ is for $\mathcal{A}$. Moreover, $F[e^x, \Gamma(k), e^{-x}, \Gamma(k)^{-1}]$ is a Picard-Vessiot extension for $\mathcal{A}$.*

Let $A(\mathbf{z}) = 0$ with $A \in \mathcal{S}^{p\times q}$ be a system of linear dimension $n$ and $M$ be its module of formal solutions with an $F$-basis $\mathbf{b}_1, \ldots, \mathbf{b}_n$. Suppose that $\partial_i(\mathbf{b}_1, \ldots, \mathbf{b}_n)^\tau = B_i(\mathbf{b}_1, \ldots, \mathbf{b}_n)^\tau$ where $B_i \in F^{n\times n}$ for $1 \leqslant i \leqslant m$. By a straightforward verification, $\{\partial_i(\mathbf{x}) = B_i\mathbf{x}\}_{1\leqslant i\leqslant m}$ is a fully integrable system, which is called the *integrable connection* of $A(\mathbf{z}) = 0$ with respect to the basis $\mathbf{b}_1, \ldots, \mathbf{b}_n$ of $M$. $\partial$-finite and fully integrable systems are connected by the next proposition whose proof is given in [8, Proposition 2] and [47, Proposition 2.4.12].

**Proposition 4.12.** *Let $A, \mathbf{b}_1, \ldots, \mathbf{b}_n, B_1, \ldots, B_m$ be as above, and $B$ be the stacking of the blocks $(\partial_i \cdot \mathbf{1}_n - B_i)$. Then*

(i) $\operatorname{coker}_{\mathcal{L}}(A) \cong_{\mathcal{L}} \operatorname{coker}_{\mathcal{L}}(B)$.

(ii) *Let $\{\mathbf{e}_1, \ldots, \mathbf{e}_q\}$ be the set of $\mathcal{L}$-generators of $M$ satisfying $A(\mathbf{e}_1, \ldots, \mathbf{e}_q)^\tau = 0$ and $P \in F^{q\times n}$ be given by $(\mathbf{e}_1, \ldots, \mathbf{e}_q)^\tau = P(\mathbf{b}_1, \ldots, \mathbf{b}_n)^\tau$. Then, for any $\Delta$-extension $E$ of $F$, the correspondence $\xi \mapsto P\xi$ is an isomorphism of $C_E$-modules between* $\operatorname{sol}_E(\{\partial_i(\mathbf{x}) = B_i\mathbf{x}\}_{1\leqslant i\leqslant m})$ *and* $\operatorname{sol}_E(A(\mathbf{z}) = 0)$.

Remark that the inverse of the correspondence in Proposition 4.3 (ii) is given by $\eta \mapsto Q\eta$, where $Q$ is a matrix in $\mathcal{L}^{n\times q}$ such that $(\mathbf{b}_1, \ldots, \mathbf{b}_n)^\tau = Q(\mathbf{e}_1, \ldots, \mathbf{e}_q)^\tau$. From Proposition 4.12 (ii), all the solutions of the system $A(\mathbf{z}) = 0$ can be obtained from those of its integrable connection $\{\partial_i(\mathbf{x}) = B_i\mathbf{x}\}_{1\leqslant i\leqslant m}$, and vice versa. Figure 1 illustrates such a relationship, and it also suggests reducing the problem of solving $\partial$-finite systems to that of solving fully integrable systems.

Proposition 4.12 allows us to generalize the notion of Picard-Vessiot extensions from fully integrable systems to $\partial$-finite ones.

[1] Which can be found, for example, by computing the hyperexponential solutions of the system ( [28, 47]).

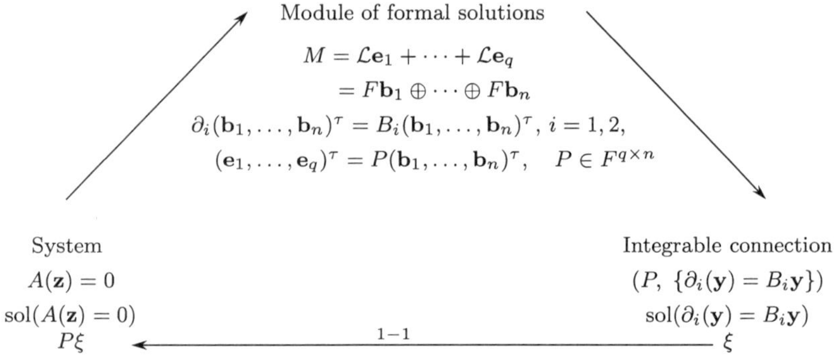


Figure 1 Relationships among Systems, Modules and Solutions

**Definition 4.13.** *Let* $A(\mathbf{z}) = 0$ *with* $A \in \mathcal{S}^{p\times q}$ *be a* $\partial$*-finite system,* $M$ *be its module of formal solutions,* $\{\mathbf{e}_1, \ldots, \mathbf{e}_q\}$ *be a set of* $\mathcal{L}$*-generators of* $M$ *and* $\mathbf{b}_1, \ldots, \mathbf{b}_n$ *be an* $F$*-basis of* $M$ *such that* $A(\mathbf{e}_1, \ldots, \mathbf{e}_q)^\tau = 0$ *and* $(\mathbf{e}_1, \ldots, \mathbf{e}_q)^\tau = P(\mathbf{b}_1, \ldots, \mathbf{b}_n)^\tau$ *where* $P \in F^{q\times n}$.

*A* $q \times n$ *matrix* $V$ *with entries in a* $\Delta$*-extension* $E$ *of* $F$ *is called a* fundamental matrix *for* $A(\mathbf{z}) = 0$ *if* $V = PU$ *where* $U \in E^{n\times n}$ *is a fundamental matrix of the integrable connection of* $A(\mathbf{z}) = 0$ *with respect to* $\mathbf{b}_1, \ldots, \mathbf{b}_n$.

*A Picard-Vessiot ring for an integrable connection of* $A(\mathbf{z}) = 0$ *is called a* Picard-Vessiot ring *for* $A(\mathbf{z}) = 0$.

As a consequence of Theorem 4.10, we have

**Theorem 4.14.** *Every* $\partial$*-finite system* $A(\mathbf{z}) = 0$ *over* $F$ *has a Picard-Vessiot ring* $E$*. If* $F$ *has characteristic* 0 *and* $C_F$ *is algebraically closed, then* $C_E = C_F$.

Assume that $F$ has characteristic 0 with an algebraically closed field of constants. If $E$ is a Picard-Vessiot ring for the system $A(\mathbf{z}) = 0$ then the dimension of $\mathrm{sol}_E(A(\mathbf{z}) = 0)$ as a $C_F$-vector space equals the linear dimension of $A(\mathbf{z}) = 0$, whenever the latter is finite.

In summary, by associating a $\partial$-finite system to its module of formal solutions, we reduce the system to a fully integrable system, define the notion of Picard-Vessiot extension, and compute the dimension of its solutions space by noncommutative Gröbner basis techniques.

# 5 Determining all submodules of a Laurent-Ore module

A module over a Laurent-Ore algebra that is finite-dimensional over the ground field is called a *Laurent-Ore module.* As seen in Section 4.2, a

$\partial$-finite system $S$ has a module of formal solutions, which is a Laurent-Ore module and denoted by $M_S$. A submodule of $M$ corresponds to a subsystem of $S$. Thus, determining all submodules of $M_S$ is equivalent to factoring $S$. In this section, we outline an algorithm for determining all submodules of a Laurent-Ore module.

## 5.1 Generalized Beke's method

Recall that $\mathcal{L} = F[\partial_1, \ldots, \partial_m, \partial_{\ell+1}^{-1}, \ldots, \partial_m^{-1}]$ is a Laurent-Ore algebra. The $d$-th *exterior power* $\wedge^d M$ of an $\mathcal{L}$-module $M$ is the $F$-vector space $\wedge_F^d M$ provided with the actions given by the formulas $\partial_i(\mathbf{w}_1 \wedge \cdots \wedge \mathbf{w}_d) = \sum_{s=1}^d \mathbf{w}_1 \wedge \cdots \wedge (\partial_i \mathbf{w}_s) \wedge \cdots \wedge \mathbf{w}_d$ for $i \leqslant \ell$ and $\partial_j^\nu(\mathbf{w}_1 \wedge \cdots \wedge \mathbf{w}_d) = \partial_j^\nu(\mathbf{w}_1) \wedge \cdots \wedge \partial_j^\nu(\mathbf{w}_d)$ for $j > \ell$ and $\nu \in \{-1, 1\}$. A decomposable element $\mathbf{w} \in \wedge^d M$ is an exterior product of $d$ elements in $M$, *i.e.*, $\mathbf{w} = \mathbf{w}_1 \wedge \cdots \wedge \mathbf{w}_d$.

The next theorem generalizes Lemma 10 in [15] or the corresponding statement in [39, page 111]:

**Theorem 5.1.** *A Laurent-Ore module $M$ has a $d$-dimensional submodule if and only if $\wedge^d M$ has a one-dimensional submodule generated by a decomposable element.*

Remark that the operators $\partial_j^{-1}$ are indispensable in the proof of Theorem 5.1 (see also [47, Theorem 4.3.1]), and this proof yields a correspondence between $d$-dimensional submodules and one-dimensional submodules generated by decomposable elements: if a $d$-dimensional submodule of $M$ has an $F$-basis $\mathbf{v}_1, \ldots, \mathbf{v}_d$, then the linear subspace generated by $\mathbf{v}_1 \wedge \cdots \wedge \mathbf{v}_d$ in $\wedge^d M$ is a one-dimensional submodule; conversely, if a one-dimensional submodule of $\wedge^d M$ is generated by a decomposable element $\mathbf{v}_1 \wedge \cdots \wedge \mathbf{v}_d$, then the $F$-linear subspace generated by $\mathbf{v}_1, \ldots, \mathbf{v}_d$ in $M$ is a $d$-dimensional submodule.

The original idea of Theorem 5.1 is due to Beke [3]. He shows that finding right factors of a linear ordinary differential operator $L$ is equivalent to finding exponential solutions of some associated equations which are constructed by Wronskian techniques. His method is generalized to factor $\partial$-finite differential systems [31, 32]. Theorem 5.1 may be viewed as a module-theoretic generalization of Beke's method. However, this module-theoretic approach avoids not only constructing complicated Wronskian-like determinants but also guessing leading derivatives of Gröbner bases.

## 5.2 Determining one-dimensional submodules

We outline an algorithm for determining one-dimensional submodules of a Laurent-Ore module in [33]. In a $\Delta$-extension $R$ of $F$, a non-zero element $h$ is said to be *hyperexponential with respect to a map* $\phi$ *in* $\Delta$ if $\phi(h)=rh$ for some $r \in F$. The element $r$ is denoted $\ell\phi(h)$. The element $h$ is said to be *hyperexponential* over $F$ if it is hyperexponential with respect to all the maps in $\Delta$. A non-zero vector $V \in R^n$ is said to be *hyperexponential (with respect to a map* $\phi$*)* if there exist $h \in R$, hyperexponential (with respect to $\phi$), and $W \in F^n$ such that $V = hW$.

Let $M$ be an $\mathcal{L}$-module with a finite basis $\mathbf{b}_1, \ldots, \mathbf{b}_n$ over $F$. The module structure of $M$ is determined by $m$ matrices $A_1, \ldots, A_m$ in $F^{n\times n}$, where

$$\partial_i(\mathbf{b}_1, \ldots, \mathbf{b}_n)^T = A_i(\mathbf{b}_1, \ldots, \mathbf{b}_n)^T, \quad i = 1, \ldots, m. \tag{11}$$

Note that $A_{\ell+1}, \ldots, A_m$ are invertible because $\mathcal{L}$ contains $\partial_{\ell+1}^{-1}, \ldots, \partial_m^{-1}$. We call $A_1, \ldots, A_m$ *the structure matrices* with respect to $\mathbf{b}_1, \ldots, \mathbf{b}_n$. For a column vector $Z = (z_1, \ldots, z_n)^T$ of unknowns,

$$\delta_i(Z) = -A_i^T Z, \; i \leqslant \ell, \quad \sigma_j(Z) = \left(A_j^{-1}\right)^T Z, \; j > \ell \tag{12}$$

is called the system associated to $M$ and the basis $\mathbf{b}_1, \ldots, \mathbf{b}_n$. Systems associated to different bases are equivalent in the sense that the solutions of one system can be transformed to those of another by a matrix in $F^{n\times n}$. The multiplicative rules $\partial_s\partial_t = \partial_t\partial_s$ for all $s, t \in \{1, \ldots, m\}$ imply that (12) is fully integrable [8, Definition 2]. A detailed verification of this assertion is presented in [47, Lemma 4.1.1]. On the other hand, every fully integrable system is associated to its module of formal solutions [8, Example 4], which is an $\mathcal{L}$-module of finite dimension.

The next proposition connects one-dimensional submodules of $M$ with hyperexponential solutions of its associated systems.

**Proposition 5.2.** *Let an* $\mathcal{L}$*-module* $M$ *have a finite* $F$*-basis* $\mathbf{b}_1, \ldots, \mathbf{b}_n$ *with structure matrices given in (11) and the associated system in (12). Let* $\mathbf{u} = \sum_{i=1}^n u_i\mathbf{b}_i$ *with* $u_i \in F$ *not all zero.*

(i) *If there exists a hyperexponential element* $h$ *in some* $\Delta$*-extension such that* $h(u_1, \ldots, u_n)^T$ *is a solution of (12), then* $F\mathbf{u}$ *is a submodule of* $M$ *with*

$$\partial_i(\mathbf{u}) = -\operatorname{ldeg}_i(h)\mathbf{u}, \; i \leqslant \ell \text{ and } \partial_j(\mathbf{u}) = \ell\sigma_j(h)^{-1}\mathbf{u}, \; j > \ell. \tag{13}$$

(ii) *If* $F\mathbf{u}$ *is a submodule of* $M$ *then there exists an invertible hyperexponential element* $h$ *in some* $\Delta$*-extension such that* $h(u_1, \ldots, u_n)^T$ *is a solution of (12).*

By Proposition 5.2 we need only to compute hyperexponential solutions of the fully integrable system (12), which can be done by algorithms for computing exponential (resp. hypergeometric) solutions of ordinary differential (resp. difference) matrix equation [2,26], and a back-substitution process described in [33].

# References

[1] J.M. Aroca, J. Cano, R. Feng and X.S. Gao. Algebraic general solutions of algebraic ODEs. *Proceedings of ISSAC 2005*, 29-36, ACM Press, 2005.

[2] M.A. Barkatou. Rational solutions of matrix difference equations: the problem of equivalence and factorization. *Proceedings of ISSAC 1999*, 277-282, ACM Press, 1999.

[3] E. Beke. Die Irrducibilität der homogenen Differentialgleichungen. *Math. Annal.* **45**, 278-294, 1894.

[4] F. Boulier, D. Lazard, F. Ollivier and M. Petitiot. Representation for the radical of a finitely generated differential ideal, *Proceesings of ISSAC 1995*, 158-166, ACM Press, 1995.

[5] D. Bouziane, A. Kandri Rody and H. Maârouf. Unmixed decomposition of a finitely generated perfect differential ideal. *Journal of Symbolic Computation* **31**, 631-649, 2001.

[6] M. Bronstein. An improved algorithm for factoring linear ordinary differential operators. *Proceedings of ISSAC 1994*, 336-347, ACM Press, 1994.

[7] M. Bronstein. *Symbolic Integration I*, Springer, 1997.

[8] M. Bronstein, Z. Li and M. Wu. Picard-Vessiot extensions for linear functional systems. *Proceedings of ISSAC 2005*, 68-75, ACM Press, 2005.

[9] Y. Chen and X. S. Gao. Involutive characteristic set of partial differential polynomial systems. *Science in China (A)*, 33(2), 97-113, 2003.

[10] S.C. Chou and X.S. Gao. Automated reasoning in differential geometry and mechanics. *Journal of Automated Reasoning* **10**, 161-172, 1993.

[11] S.C. Chou and X.S. Gao. Automated reasoning in geometry. *Handbook of Automated Reasoning*, (eds. A. Robinson and A. Voronkov), 709-749, Elsevier, Amsterdam, 2001.

[12] F. Chyzak and B. Salvy. Non-commutative elimination in Ore algebras proves multivariate identities. *Journal of Symbolic Computation* **26**, 187-228, August 1998.

[13] T. Cluzeau and E. Hubert. Resolvent representation for regular differential ideals. *AAECC* **29**, 395-425, 2003.

[14] R.M. Cohn. *Difference Algebra.* Tracts in Mathematics 17, Interscience, New York, 1965.

[15] E. Compoint and J.A. Weil. Absolute reducibility of differential operators and Galois groups. *J. of Algebra*, 275(1): 77-105, 2003.

[16] E.G. Fan. *Integrable Systems and Computer Algebra* (in Chinese), Science Press, 2004.

[17] R.Y. Feng and X.S. Gao. A polynomial time algorithm to find rational general solutions Of first order autonomous ODEs, *Journal of Symbolic Computation* **41**, 739-762, 2006.

[18] X.S. Gao. Implicitization for differential rational parametric equations, *J. of Symbolic Computation*, 811-824, 36(5), 2003.

[19] X.S. Gao and S.C. Chou. A zero structure theorem for differential parametric systems, *Journal of Symbolic Computation* **16**, 585-595, 1994.

[20] R. Feng, X.S. Gao and Z. Huang. Rational general solutions of algebraic ordinary difference equations. *MM-Preprints*, KLMM, CAS, 2005.

[21] X.S. Gao and Y. Luo. A characteristic set method for difference polynomial systems. *Inter Conf on Poly Sys. Sol.* Nov. 24-26, Paris, 2004.

[22] X.S. Gao and C. Yuan. Resolvent systems of difference polynomial ideals. *Proceedings of ISSAC, 2006*, 101-108, ACM Press, 2006.

[23] X.S. Gao and M. Zhang. Decomposition of differential polynomials with constant coefficients. *Proceedings of ISSAC 2004*, 175-182, ACM Press, New York, 2004.

[24] M. van Hoeij. Formal solutions and factorization of differential operators. *Journal of Symbolic Computation* **24**, 1-30, 1997.

[25] M. van Hoeij. Factorization of differential operators with rational function coefficients. *Journal of Symbolic Computation* **24**, 537-561, 1997.

[26] M. van Hoeij. Finite singularities and hypergeometric solutions of linear recurrence equations. *J. Pure Appl. Algebra* **139**, 109-131 1999.

[27] E. Hubert. Factorization-free decomposition algorithms in differential algebra. *Journal of Symbolic Computation* **29**, 641-662, 2000.

[28] G. Labahn and Z. Li. Hyperexponential solutions of finite-rank ideals in orthogonal Ore algebras. In *Proceedings of ISSAC 2004*, 213-220. ACM Press, 2004.

[29] H. Li and M. Cheng. Clifford algebraic reduction method for mechanical theorem proving in differential geometry. *Journal of Automated Reasoning* **21**, 1-21, 1998.

[30] Z. Li. Mechanical theorem proving of the local theory of surfaces. *Ann. Math. Artif. Intell.* **13**, 25-46, 1995.

[31] Z. Li, F. Schwarz and S. Tsarev. Factoring zero-dimensional ideals of linear partial differential operators. In *Proceedings of ISSAC 2002*, 168-175, ACM Press, 2002.

[32] Z. Li, F. Schwarz and S. Tsarev. Factoring systems of linear PDE's with finite-dimensional solution spaces. *Journal of Symbolic Computation* **36**, 443-471, 2003.

[33] Z. Li, M.F. Singer, M. Wu and D. Zheng. A recursive method for determining the one-dimensional submodules of Laurent-Ore modules. *Proceedings of ISSAC 2006*, 220-227, ACM Press, 2006.

[34] B. Malgrange. Motivations and introduction to the theory of $D$-modules. *Computer Algebra and Differential Equations,* vol. 193, *LMS LNS*, 3-20, Cambridge Univ. Press, 1994.

[35] B. Malgrange. On nolinear differential Galois theory, *Chinese Annals of Mathematics, Sereis B*, 23(2), 219-226, 2002.

[36] F. Pauer and A. Unterkircher. Gröbner bases for ideals in Laurent polynomial rings and their applications to systems of difference equations. *AAECC* **9**, 271-291, 1999.

[37] M. Petkovšek, H. Wilf and D. Zeilberger. *A=B*. A K Peters, Ltd, 1996.

[38] M. van der Put and M.F. Singer. *Galois Theory of Difference Equations, Lecture Notes in Mathematics 1666*. Springer, 1997.

[39] M. van der Put and M.F. Singer. *Galois Theory of Linear Differential Equations, Grundlehren der Mathematischen Wissenschaften* **328**. Springer, Heidelberg, 2003.

[40] G. Reid. Algorithms for reducing a system of PDEs to standard form. *European J. of Appl. Math.* **2**, 293-318, 1991.

[41] D. Richardson. Wu's method and the Khovanskii finiteness theorem. *Journal of Symbolic Computation* **12**, 127-141, 1991.

[42] J.F. Ritt. *Differential Algebra.* Amer. Math. Soc. Colloquium, 1950.

[43] M.F. Singer. Testing reducibility of linear differential operators: a group theoretic perspective. *AAECC* **7**, 77-104, 1996.

[44] W. Sit. The Ritt-Kolchin theory for differential polynomials. *Differential Algbra and related Topics, Proceedings of the International Workshop*, 1-70, World Scientific, 2002.

[45] D. Wang. A method for proving theorems in differential geometry and mechanics. *J. Univ. Comput. Sci.* **9**, 658-673, 1995.

[46] D.K. Wang. *Polynomial Equations Solving and Mechanical Geometric Theorem Proving.* PhD Thesis, KLMM, Academia Sinica, 1993.

[47] M. Wu. *On Solutions of Linear Functional Systems and Factorization of Modules over Laurent-Ore Algebras.* PhD thesis, Chinese Academy of Sciences and Université de Nice, 2005.

[48] W.T. Wu. On the decision problem and the mechanization of theorem-proving in elementary geometry. *Scientia Sinica* **21**, 159-172, 1978.

[49] W.T. Wu. Mechanical theorem proving in elementary differential geometry (in Chinese). *Scientia Sinica*, 94-102, 1979.

[50] W.T. Wu. *Basic Principle of Mechanical Theorem Proving in Geometries.* (in Chinese) Science Press, Beijing, 1984; Springer, Wien, 1994.

[51] W.T. Wu. Mechanical derivation of Newton's Gravitational Laws from Kepler's Laws. *MM-Preprints* **1**, 53-61, 1987.

[52] W.T. Wu. On the Foundation of Algebraic Differential Geometry. *Sys.Sci.& Math.Scis.* **2**, 289-312, 1989.

[53] Z.Y. Yan. *Constructive Methods for Complex Nonlinear Waves* (in Chinese), Science Press, 2006.

[54] S. Zampieri. A solution of the Cauchy problem for multidimensional discrete linear shift-invariant systems. *Linear Algebra and Its Applications* **202**, 143-162, 1994.

[55] M. Zhou and F. Winkler. Gröbner bases in difference-differential modules. *Proceedings of ISSAC 2006*, 353-360, ACM Press, 2006.

# A Global Existence Result in Radiation Hydrodynamics*

Song Jiang, Feng Xie
*LCP, Institute of Applied Physics and Computational Mathematics*
*P.O. Box 8009, Beijing 100088, China*
*Email: jiang@iapcm.ac.cn, tzxief@yahoo.com*

Jianwen Zhang
*School of Mathematical Sciences, Xiamen University*
*Xiamen 361005, China*
*Email: jwzhang@xmu.edu.cn*

**Abstract**

We consider the compressible fluid dynamics with taking into account the radiation effect. First, we present the general model in radiation hydrodynamics which is the compressible Navier-Stokes equations coupled with the radiative transport equation with non-local terms and very difficult to solve both numerically and analytically. Practical simplified models are introduced in some physical regions. From the physical and numerical points of view, these models can approximate the general equations of radiation hydrodynamics very well in some particular physical situations. In particular, the Equilibrium Diffusion Approximation model and the Eddington Approximation or Diffusion Approximation model are mainly studied in the present paper. Then, we briefly review recent mathematical results on the equations of radiation hydrodynamics, in particular, on the simplified models. Some remarks on the non-local thermal equilibrium (non-LTE) case are also given. Finally, for a one-dimensional model in non-LTE of radiation hydrodynamics which describes almost isotropic interaction between a viscous heat-conducting gas and photons, we prove the global existence of a unique classical solution, provided that the initial data are suitably smooth and the heat-conductivity coefficient satisfies a physical growth condition with respect to the temperature.

*Supported by the Special Funds for Major State Basic Research Projects (Grant No. 2005CB321700), the NSFC (Grant No. 10225105) and the Morningside Center of Mathematics.

# 1 Introduction

The radiation hydrodynamics [2, 16, 17] is concerned with the propagation of thermal radiation and the effect of this radiation on the hydrodynamics describing the fluid motion. The importance of thermal radiation in the physical problems increases as the temperature is raised. At the moderate temperature, the role of the radiation is primarily one of the transporting energy by radiative process, while at the higher temperature, the energy and the momentum densities of the radiation field may become comparable to or even dominate the corresponding fluid quantities. In this case, the radiation field significantly affects the hydrodynamics of the fluid. The theory of radiation hydrodynamics finds a wide range of applications, including such diverse astrophysical phenomena as waves and oscillations in stellar atmospheres and envelops, nonlinear stellar pulsation, supernova explosions, stellar winds and many others.

In the term radiation hydrodynamics, it is necessary to include effects of the radiation field in the hydrodynamic equations for this class of problems. The equations of hydrodynamics result from particle, momentum, and energy balances for a differential volume of space. If a significant radiation field is present, one has to include the radiation momentum and energy in these balances. This gives rise to radiation terms in the equations of hydrodynamics. We first introduce the basic concepts needed to describe the radiation field and its interaction with matter. Consider the contributions of the radiation field to the energy and momentum density and flux. At any time, $2n$ variables are required to specify the position of a photon in phase space, namely $n$ position variables and $n$ momentum variables. We denote the $n$ position variables by the vector $x$. In radiative transfer work it is conventional to use, rather than the $n$ momentum variables, $N$ equivalent variables, namely the frequency $\nu$ and the direction of travel of the photon $\Omega$. In terms of these variables, we introduce the distribution function $f(x,t,\nu,\Omega)$, such that $f dx d\nu d\Omega$ is the number of photons (at time $t$) at space point $x$ in a differential volume element $dx$, with frequency $\nu$ in a frequency interval $d\nu$, and travelling in a direction $\Omega$ in a solid angle element $d\Omega$. The specific intensity of radiation $I(x,t,\nu,\Omega)$ is then defined as

$$I(x,t,\nu,\Omega) = ch\nu f(x,t,\nu,\Omega)$$

with the Plank constant $h$ and the light speed $c$.

Under the consideration of the three basic interactions between photons and matter, namely absorption, scattering and emission, we find the equation of transfer in the conventional form (cf. [16, 17])

$$\begin{aligned}
&\frac{1}{c}\frac{\partial I(\nu,\Omega)}{\partial t}+\Omega\cdot\nabla I(\nu,\Omega)=S(\nu)-\sigma_a(\nu)I(\nu,\Omega)\\
&+\int_0^\infty d\nu'\int_{S^{n-1}}\Big[\frac{\nu}{\nu'}\sigma_s(\nu'\to\nu,\Omega'\cdot\Omega)I(\nu',\Omega')\\
&-\sigma_s(\nu\to\nu',\Omega\cdot\Omega')I(\nu,\Omega)\Big]\,d\Omega'.
\end{aligned}\tag{1.1}$$

Here $I(\nu,\Omega)\equiv I(x,t,\nu,\Omega)$, $S^{n-1}$ is the unit ball in $\mathbb{R}^n$, $S(\nu)\equiv S(x,t,\nu)$ is the rate of energy emission due to spontaneous processes. $\sigma_a(\nu)\equiv\sigma_a(x,t,\nu,\varrho,\theta)$ denotes the absorption coefficient that may also depend on the mass density $\varrho$ and the temperature $\theta$ of the matter. The dependence of $\sigma_a$ upon $\varrho$ and $\theta$ can have the form $\sigma_a=O(\varrho^\alpha\theta^{-\beta})$, $\alpha,\beta>0$ (see, for example, [11, 20]).

Similar to absorption, a photon can undergo scattering interactions with matter, and the scattering interaction serves to change the photon's characteristics $\nu'$ and $\Omega'$ to a new set of characteristics $\nu$ and $\Omega$. To quantitatively describe the scattering event, one requires a probabilistic statement concerning this change. This leads to the definition of the "differential scattering coefficient" $\sigma_s(\nu'\to\nu,\Omega'\cdot\Omega)\equiv\sigma_s(\nu'\to\nu,\Omega'\cdot\Omega,\varrho,\theta)$ that may depend on $\varrho$ and $\theta$ (in general, $\sigma_s$ is independent of $\theta$, also the example below), such that the probability of a photon being scattered from $\nu'$ to $\nu$ contained in $d\nu$, and from $\Omega'$ to $\Omega$ contained in $d\Omega$, in travelling a distance $ds$ is given by $\sigma_s(\nu'\to\nu,\Omega'\cdot\Omega)d\nu d\Omega ds$. Therefore,

$$\begin{aligned}
\text{outscattering}&=\int_0^\infty d\nu'\int_{S^{n-1}}\sigma_s(\nu\to\nu',\Omega\cdot\Omega')I(\nu,\Omega)d\Omega',\\
\text{inscattering}&=\int_0^\infty d\nu'\int_{S^{n-1}}\sigma_s(\nu'\to\nu,\Omega'\cdot\Omega)I(\nu',\Omega')d\Omega'.
\end{aligned}$$

We give an example of the absorption coefficient and the scattering kernel in the following which describes the Compton scattering (see [17]):

$$\begin{aligned}
\sigma_a(\nu)&=C_1\varrho\theta^{-1/2}\exp\Big[-\frac{C_2}{\theta^{1/2}}\Big(\frac{\nu-\nu_0}{\nu_0}\Big)^2\Big],\\
\sigma_s(\nu\to\nu',\xi)&=\frac{C_3\varrho(1+\xi^2)}{[1+\gamma(1-\xi)]^2}\times\Big\{1+\frac{\gamma^2(1-\xi)^2}{(1+\xi^2)[1+\gamma(1-\xi)]}\Big\}\\
&\quad\times\delta\Big(\nu'-\frac{\nu}{1+\gamma(1-\xi)}\Big),
\end{aligned}$$

where $\gamma=C_4\nu$, $\xi=\Omega\cdot\Omega'$, $C_i$ $(i=1,\cdots,4)$ are positive constants, $\nu_0$ is the fixed frequency.

In the above, for the sake of simplicity, we have assumed that $S$ and $\sigma_a$ are independent of $\Omega$ and $\sigma_s$ depends only upon $\Omega\cdot\Omega$. This means no inherent preferred direction in the matter. However, the fact that in radiation hydrodynamic problems the material is in general in motion changes the situation. This motion does introduce a preferred direction in the matter, namely the direction of motion of the fluid, and consequently, introduces an $\Omega$ (angular) dependence into $S$ and $\sigma_a$, and separate $\Omega$ and $\Omega'$ dependences into $\sigma_s$. These $\Omega$ (angular) dependences are not inherent properties of the material, but arise only from the relative motion between the fluid and the observer. For simplicity of presentation, in what follows, we will suppress the $x, t, \varrho$ and $\theta$ dependences unless it is stated.

In terms of the specific intensity, we define three quantities, namely, the energy density, the radiative flux and the radiative pressure tensor, by

$$\begin{cases} E_r = \dfrac{1}{c}\displaystyle\int_0^\infty d\nu \int_{S^{n-1}} I(\nu,\Omega)d\Omega, \\ F_r = \displaystyle\int_0^\infty d\nu \int_{S^{n-1}} \Omega I(\nu,\Omega)d\Omega, \\ P_r = \dfrac{1}{c}\displaystyle\int_0^\infty d\nu \int_{S^{n-1}} \Omega\otimes\Omega I(\nu,\Omega)d\Omega. \end{cases} \tag{1.2}$$

Including effects due to the presence of a radiation field, the equations of (nonrelativistic) hydrodynamics for a viscous heat-conducting fluid comprise the conservation laws of mass, momentum and energy, and in Eulerian coordinates, can be written as

$$\rho_t + \operatorname{div}(\rho u) = 0, \tag{1.3}$$

$$(\rho u + \frac{1}{c^2}F_r)_t + \nabla(\rho u \otimes u + P_m + P_r) = \operatorname{div}\mathbb{S}, \tag{1.4}$$

$$\begin{aligned} &\Big[\frac{1}{2}\rho u^2 + E_m + E_r\Big]_t + \nabla\Big[(\frac{1}{2}\rho u^2 + E_m + P_m)u + F_r\Big] \\ &\qquad = \operatorname{div}(\mathbb{S}u + \kappa\nabla\theta), \end{aligned} \tag{1.5}$$

where $\rho, u, P_m = P_m(\rho,\theta), E_m = E_m(\rho,\theta)$ and $\theta$ are the density, velocity, pressure, inertial energy and temperature of the fluid respectively, $\kappa = \kappa(\rho,\theta)$ is the heat-conducting coefficient of the fluid, and the symbol $\mathbb{S}$ stands for the viscous stress tensor

$$\mathbb{S} = \lambda(\operatorname{div}u)\,\mathbb{I} + \mu(\nabla u + (\nabla u)^t),$$

$\lambda$ and $\mu$ are the viscosity coefficients of the fluid satisfying $2\lambda + \mu > 0$.

It is well-known that emission and scattering are enhanced by photons already in the final state following interaction. Suppose $\mathbb{P}$ is basic probability of a photon event, then the actual probability $\mathbb{P}' = \mathbb{P}(1+n), n$ is the number of photons in the final states, $n = c^2 I(x,t,\nu,\Omega)/(2h\nu^3)$.

Then Eq. (1.1) can be reduced as

$$\frac{1}{c}\frac{\partial I(\nu,\Omega)}{\partial t}+\Omega\cdot\nabla I(\nu,\Omega)=S(\nu)\Big(1+\frac{c^2I(\nu,\Omega)}{2h\nu^3}\Big)-\sigma_a(\nu)I(\nu,\Omega)$$
$$+\int_0^\infty d\nu'\int_{S^{N-1}}d\Omega'\Big[\sigma_s(\nu'\to\nu,\Omega'\Omega)I(\nu',\Omega')\Big(1+\frac{c^2I(\nu,\Omega)}{2h\nu^3}\Big)$$
$$-\sigma_s(\nu\to\nu',\Omega'\Omega)I(\nu,\Omega)\Big(1+\frac{c^2I(\nu',\Omega')}{2h\nu'^3}\Big)\Big].$$

A change of variables $S=\sigma_a'B$, $\sigma=\sigma_a'[1+c^2B/(2h\nu^3)]$ gives finally

$$\frac{1}{c}\frac{\partial I(\nu,\Omega)}{\partial t}+\Omega\cdot\nabla I(\nu,\Omega)=\sigma_a'(\nu)[B(\nu)-I(\nu,\Omega)]$$
$$+\int_0^\infty d\nu'\int_{S^{N-1}}d\Omega'\Big[\sigma_s(\nu'\to\nu,\Omega'\Omega)I(\nu',\Omega')\Big(1+\frac{c^2I(\nu,\Omega)}{2h\nu^3}\Big)$$
$$-\sigma_s(\nu\to\nu',\Omega'\Omega)I(\nu,\Omega)\Big(1+\frac{c^2I(\nu',\Omega')}{2h\nu'^3}\Big)\Big]. \tag{1.6}$$

The system (1.3)–(1.5) coupled with (1.6) through (1.2) compose the equations of radiation hydrodynamics in $\mathbb{R}^n$. We remark that when the material is in the local thermal equilibrium (LTE) (an important concept discussed in details below), then $B(\nu)=2h\nu^3c^{-2}(e^{h\nu/k\theta}-1)^{-1}$ is the Planck function where $k$ is the Boltzmann constant. Of course, we can also consider the inviscid fluid case, that is, $\lambda=\mu=k=0$. That is, we can ignore the viscous stress tensor $\mathbb{S}$ and the heat-conducting (1.3)–(1.5). In this case, we obtain the system of the Euler equations coupled with the radiation transport equation (1.1). Such system is often used in the study of the inertial constrained fusion (ICF).

The system of radiation hydrodynamics (1.3)–(1.5), (1.6) are very complicated, it is difficult to solve them both analytically and numerically. So, a number of simplified models arise under the special physical assumptions. These simplified models approximate the original model very well in particular physical regions. We should note that all simplified processes are carried out on the equation of transfer. Below, we will introduce two simplified models used in practice.

## 1.1 Eddington or Diffusion Approximation

The basic assumption underlying the classical diffusion, or Eddington, description of transfer is that the angular dependence of the specific intensity can be represented by the first two terms in a spherical harmonic expansion. That is, it is assumed that

$$I(x,t,\nu,\Omega)=\frac{1}{4\pi}I_0(x,t,\nu)+\frac{3}{4\pi}\Omega\cdot I_1(x,t,\nu), \tag{1.7}$$

where the coefficients $I_0$ and $I_1$ have a physical interpretation. Integration of Eq. (1.7) over all solid angle gives

$$I_0(x,t,\nu) = \int_{4\pi} d\Omega I(x,t,\nu,\Omega),$$

and multiplication of (1.7) by $\Omega$ priori to a similar integration yields

$$I_1(x,t,\nu) = \int_{4\pi} d\Omega\Omega I(x,t,\nu,\Omega).$$

To obtain the desired equations for $I_0$ and $I_1$, we use the assumed representation (1.7) in the integro-differential equation of transfer (1.6). Then multiplication of such equation by $\Omega$ priori to an integration over all angle and moreover, neglecting the term $c^{-1}\partial_t I_1$ and requiring the that the scattering kernel is diagonal yields

$$I_1(x,t,\nu) = -D(x,t,\nu)\nabla I_0(x,t,\nu) \quad \text{(Fick's Law)}.$$

On the other hand, inserting (1.7) into (1.6), integrating such an equation over all angle and using the Fick's Law, we obtain

$$\begin{aligned} \frac{1}{c}\frac{\partial I_0(\nu)}{\partial t} &+ \operatorname{div}(D\nabla I_0(\nu)) = \sigma_a'(\nu)[4\pi B(\nu) - I_0(\nu)] - \sigma_s(\nu) I_0 \\ &+ \frac{c^2 I_0(\nu)}{8\pi h}\int_0^\infty d\nu' \Big[\frac{\sigma_{s0}(\nu'\to\nu)}{\nu^2\nu'} - \frac{\sigma_{s0}(\nu\to\nu')}{\nu'^3}\Big] I_0(\nu'). \end{aligned} \tag{1.8}$$

Equation (1.8) is called the diffusion approximate to (1.6). Taking a further simplification, one has the equilibrium diffusion approximation.

## 1.2 Equilibrium diffusion approximation

Assume that

$$I_0(x,t,\nu) \sim I_0(\nu),$$

and $\sigma_a'$ and $\sigma_{s0}$ are appropriate to LTE. Then Eq. (1.8) can approximate radiation field in complete thermal equilibrium at temperature $\theta$. Then we have

$$I_0(\nu) \equiv 4\pi B = \frac{8\pi h\nu^3}{c^2}(e^{h\nu/k\theta} - 1)^{-1}.$$

Consequently, we have

$$E_r = a\theta^4, \quad F_r = -\frac{ac}{3\sigma_R}\nabla\theta^4, \quad P_r = \frac{a}{3}\theta^4\,\mathrm{id}, \tag{1.9}$$

where $\sigma_R$ ($\sim a\theta^n\rho^{-m}$) is called the Rosseland mean.

Thus, such an approximate model is coupled with a hydrodynamic model through the density $\rho$ and the temperature $\theta$. Consequently, in this case, we can ignore the equation of transfer, and therefore, the equations of radiation hydrodynamics become (1.3)–(1.5), (1.9). We should remark here that although it simplifies the original equations (1.3)–(1.5), (1.6), (1.2) dramatically, the system (1.3)–(1.5), (1.9) is still very complicated and contains the high order nonlinear terms of $\theta$ for example. Of course, from the physical point of view, there are many other ways to simplify the equation of transfer. Not only we can approximate the equation by other expansion forms, but also we can consider the equation in relatively simple geometrical structure. For example, the plane 1D case, the spherical symmetry and so on.

When we consider initial boundary value problems in a bounded domain, we need to impose suitable boundary conditions. In the remainder of this section, we induce some physical boundary conditions for the diffusion approximation problem from the boundary condition for the equation of transfer (1.6).

## 1.3 Boundary conditions for the diffusion approximation problem

If we consider a radiation problem in a bounded domain $A$, then we need to impose suitable boundary conditions on the boundary $\partial A$ to guarantee that Eq.(1.6) is well-posed. On physical grounds we know that it is sufficient to specify the specific intensity at all points on the surface $\partial A$ in the incoming direction. We say a domain is a non-re-entrant surface, provided that once they fly out such domain, photons will never enter the domain again. Since a photon travels in a straight line, so the convex domain is a non-re-entrant surface. If the domain is not convex, we can assume that there is a bigger convex domain containing the considered domain and re-define the absorbtion and scattering coefficients to be zero in the extended parts. For a non-re-entrant surface, it is sufficient to impose the boundary condition as follows

$$I(x_s, \nu, \Omega, t) = \Gamma(x_s, \nu, \Omega, t), \quad \mathbf{n}\cdot\Omega < 0, \tag{1.10}$$

where $\Gamma$ is a specific function of all its arguments, $x_s$ is a point on the surface, and $\mathbf{n}$ is the outer normal vector at this point. An important special case of (1.10) is the so-called "vacuum" of "free surface" boundary condition:

$$I(x_s, \nu, \Omega, t) = 0, \quad \mathbf{n}\cdot\Omega < 0. \tag{1.11}$$

However, it is clear that, because of its simple angular dependence, the Eddington representation (1.7) can not satisfy the integro-differential boundary condition (1.10) for arbitrary incoming distribution $\Gamma(x_s, t,$

$\nu, \Omega)$. The best one can do is to demand that Eq. (1.7) is satisfied in an integral sense. That is, we put (1.7) in Eq. (1.10), multiply the resulting equation by a weight function $w(\Omega)$ and integrate over all incoming directions, to obtain

$$\int_{\mathbf{n}\cdot\Omega<0} d\Omega w(\Omega)\Big[\frac{1}{4\pi}I_0(x_s,t,\nu)+\frac{3}{4\pi}\Omega\cdot I_1(x_s,t,\nu)-\Gamma(x_s,t,\nu,\Omega)\Big]=0.$$

There are mainly two boundary conditions (Marshak and Mark Boundary conditions) by choosing different physical weight functions (see [17, Page 53]). However, from the mathematical point of view, both of them have the same form:

$$C_1I_0 - C_2\mathbf{n}\cdot I_1 = G,$$

where $C_i$ $(i=1,2)$ may vary for different cases. Then, by the Fick's Law, we have the following boundary condition for the diffusion approximation model (1.8):

$$C_1I_0 + C_2\mathbf{n}\cdot\nabla I_0 = G.$$

# 2 Mathematical results on radiation hydrodynamics

Mathematical results on radiation hydrodynamics can be separated into two classes. One is that the material is in LTE which refers to the equilibrium diffusion approximation model. Another is that the material is in non-LTE.

For the LTE case, Ducomet and Zlotnik [7] obtained the global existence of weak solutions to the equations of a 1D radiative and reactive viscous gas with large data under general assumptions on the heat conductivity, and then they studied the large-time behavior of the solutions by constructing a suitable Lyapunov function. Recently, Umehara and Tani [21] established the existence of a global classical solution to the one dimensional equations for a self-gravitating viscous radiative and reactive gas. In [21], however, the authors require that the heat-conductivity coefficient satisfies $\kappa\sim(1+\theta^q)$, $q\geqslant 4$ which unfortunately excludes the physically interesting radiation case ($q=3$). Then, Zhang and Xie [22] improved the result of [21] to the case $q>5/2$ which does include the radiation case $q=3$. Moreover, the effect of the magnetic field has been taken into account in [22]. For the multi-dimensional case, since the radiation hydrodynamic system comprises not only the compressible Navier-Stokes (or Euler) but also the nonlinear terms of higher order (the 4th order terms of $\theta$), its mathematical analysis is very complex. There are only few results by now, Ducomet and Feireisl [5,6] established the existence of so-called variation solutions to gaseous stellar equations, which contain also radiative and magnetic effects.

To our best knowledge, there are few mathematical papers on the non-LTE case. We should mention that Zhong and Jiang considered the general radiation hydrodynamic system without the viscous and heat-conducting effects in [24], and they proved the local in time existence and the finite time blow-up phenomenon for large initial data under some physical assumptions for the coefficients of absorbtion and scattering. For some (quite) simplified non-equilibrium models, there are several other results [13–15]. In [15] the authors considered a coupled system of the 1D Euler and Poisson equations with the radiation effect, and showed the existence of shock profiles for inviscid non-equilibrium gases provided that the initial strength are suitable small. Moreover, the smaller of the initial strength, the smoother of the shock profile solution. In [13, 14], a very simplified 1D quasistatic model was studied for a inviscid gas and the global existence was shown.

In this paper we study the following simplified model:

$$\rho_t + (\rho u)_x = 0, \tag{2.1}$$

$$(\rho u)_t + (\rho u^2 + P)_x = \mu u_{xx}, \tag{2.2}$$

$$\begin{aligned} &\left[\frac{1}{2}\rho u^2 + E\right]_t + \left[\left(\frac{1}{2}\rho u^2 + E + P\right)u\right]_x \\ &\qquad = \sigma w - m\theta^4 + (\kappa\theta_x)_x + (\mu u_x u)_x, \end{aligned} \tag{2.3}$$

$$\frac{1}{c}w_t - (a w_x)_x = m\theta^4 - \sigma w, \tag{2.4}$$

where $\rho$, $u$, $P = R\rho\theta$, $E = c_V\rho\theta$, $w$ are the density, velocity, inertial energy and radiation energy density respectively, $\mu$ and $\kappa = \kappa(\rho, \theta)$ are the constant viscosity and the heat conductivity coefficients respectively, $m = 4\sigma\pi\alpha\frac{1}{\beta^4}\int_0^\infty \frac{y^3}{e^y - 1}dy$, $\sigma$ and $a$ are the integrability coefficient of the Planck distribution, the absorption coefficient and the diffusion radiation coefficient, respectively.

The system (2.1)–(2.4) is the diffusion approximation in 1D where the effect of scattering has been omitted, corresponding to the fact that the photons are isotropic. Moreover, for simplicity we have assumed the absorbtion coefficient to be constant. Referring to [1, 2, 9, 15–17], The system (2.1)–(2.4) is the corresponding viscous, heat-conducting and time-dependent case of the model studied in [15]. To our best knowledge, there is no mathematical analysis result on (2.1)–(2.4) in the literature.

The aim of this paper is to prove a global existence theorem for an initial boundary value problem for the system (2.1)–(2.4). For this purpose, we denote the specific volume by $v = 1/\rho$ and assume that the following assumptions on $\kappa$ hold for some $q > 2$.

$$C^{-1}(1+\theta)^q \leqslant \kappa(\rho,\theta) \leqslant C(1+\theta)^q, \tag{2.5}$$

$$|\kappa_v| + |\kappa_{vv}| \leqslant C(1+\theta)^q \qquad \text{for } v,\theta > 0. \tag{2.6}$$

We will consider the system (2.1)–(2.4) in the domain $(0,1)\times(0,\infty)$ with boundary and initial conditions:

$$(u, \kappa\theta_x, w + \mathbf{n}w_x)\Big|_{x=0,1} = (0,0,0), \quad t > 0; \tag{2.7}$$

$$(\rho, u, \theta, w)\Big|_{t=0} = (\rho_0(x), u_0(x), \theta_0(x), w_0(x)), \quad x \in [0,1], \tag{2.8}$$

where $\mathbf{n}$ is the unit outer normal to the domain, in the unit interval domain $(0,1)$, $n$ is just a scalar, i.e., $\mathbf{n} = -1$ at $x = 0$ and $\mathbf{n} = 1$ at $x = 1$. The above boundary condition for $w$ is called the Marshak condition or the Mark condition derived in Section 1.3.

**Remark 2.1.** *We should also mention that in our model (2.1)–(2.4) we have omitted the pressure of radiation in the momentum equation (i.e., $w_x \approx 0$ in (2.2)). In some physical cases, the effect of the radiation pressure is very small and can be therefore neglected, for example, when the energy transfer of photons to plasmas (with subcritical density) completely dominates a process, the radiation pressure could be neglected in the process. On the other hand, for the case that the radiation pressure is present in the momentum equation, we can still obtain all the a priori estimates below, except that we can not verify the non-negativity of the temperature. The global existence is still an open problem in this case.*

Now, let us compare the LTE case studied in [7, 21] and what we consider here. First, since in [7,21], the term $\theta^4$ is contained in the total energy, by virtue of the total energy conservation, the higher integrality of $\theta$ in $L^\infty(0,T;L^4)$-norm is obtained directly. In our case, however, the total energy does not comprise the term of the temperature with higher order exponential, thus the conservation of total energy can not induce the boundedness of $\theta$ of higher order. So, we have to employ new techniques to achieve the higher integrality of $\theta$. In fact, we will see that a better bound for $\theta$ can be gotten, which is very important for our proof. Second, in [21] the authors required that $4 \leqslant q \leqslant 16$ and and the derivatives of $\kappa$ with respect to $v$ and $\theta$ are bounded by $C\theta$. These conditions unfortunately exclude some basic interesting physical situations. On the other hand, the assumptions on $\kappa$ of this paper are the same general as in [7], thus causing some difficulties in the analysis due to the high nonlinearities in $\theta$. Consequently, elaborate new *a priori* estimates are needed in our proof.

Throughout this paper we denote $\|\cdot\|_{L^p(I,B)}$ the space of all strongly measurable, $p^{\text{th}}$-power integrable (essentially bounded if $p=\infty$) functions from $I$ to $B$ respectively its norm, $I \subset \mathbb{R}$ an interval, $B$ a Banach space. Let $Q_T := (0,1) \times (0,T)$. The standard Hölder spaces in $(0,1)$ and in $Q_T$ are denoted by $\mathcal{C}^{\alpha}(0,1)$ and $\mathcal{C}^{\alpha,\alpha/2}(Q_T)$, respectively. For simplicity, we also use the following abbreviations:

$$\|\cdot\| \equiv \|\cdot\|_{L^2}, \quad \|\cdot\|_{L^\infty} \equiv \|\cdot\|_{L^\infty(0,1)}.$$

The capital letter $C$ (or $C(T)$ to emphasize the dependence of $C$ on $T$) will denote a generic positive constant which may depend on the initial data and given time $T$.

Our main result reads as follows.

**Theorem 2.1.** *Let $\alpha \in (0,1)$ and the conditions (2.6) hold. Assume that the initial data satisfy*

$$\begin{aligned}&(\rho_0, u_0, \theta_0, w_0) \in \mathcal{C}^{1+\alpha}(0,1) \times (\mathcal{C}^{2+\alpha}(0,1))^3;\\ &C^{-1} \leqslant \rho_0(x) \leqslant C, \quad \theta_0(x) \geqslant C, \quad w_0(x) \geqslant 0 \quad \text{for } x \in [0,1]. \end{aligned} \tag{2.9}$$

*Then there exists a unique classical solution $(\rho, u, \theta, w)$ of the initial-boundary value problem (2.1)–(2.4), (2.7), (2.8), such that for any $T > 0$,*

$$\begin{aligned}&C^{-1}(T) \leqslant \rho(x,t) \leqslant C(T), \quad \theta(x,t) \geqslant C^{-1}(T), \quad w(x,t) \geqslant 0,\\ &(\rho, \rho_x, \rho_t) \in (\mathcal{C}^{\alpha,\alpha/2}_{x,t}((0,1)\times(0,T)))^3,\\ &(u, \theta, w) \in (\mathcal{C}^{2+\alpha,1+\alpha/2}_{x,t}((0,1)\times(0,T)))^3.\end{aligned}$$

As is well-known that the global existence is obtained by combing the local existence and the uniform *a priori* estimates. It is easy to derive the local existence for our problem in the same manner as in [18, 19, 21] by applying the Banach contraction mapping principle. Hence, we only need to derive the uniform estimates. Without loss of generality we assume $c_V = R = a = m = n = c = \mu = 1$ in (2.1)–(2.4). Then, the system (2.1)–(2.4) becomes

$$\rho_t + (\rho u)_x = 0, \tag{2.10}$$

$$(\rho u)_t + (\rho u^2 + P)_x = u_{xx}, \tag{2.11}$$

$$\Big[\frac{1}{2}\rho u^2 + \rho\theta\Big]_t + \Big[(\frac{1}{2}\rho u^2 + \rho\theta + P)u\Big]_x = w - \theta^4 + (\kappa\theta_x)_x + (u_x u)_x, \tag{2.12}$$

$$w_t - w_{xx} = \theta^4 - w. \tag{2.13}$$

In the next section we derive the *a priori* estimates for Eqs. (2.10)–(2.13) with boundary and initial conditions (2.7) and (2.8).

# 3 A priori estimates

Suppose that $(\rho, u, \theta, w)$ is a smooth solution to (2.10)–(2.13), (2.7), (2.8) on $\bar{Q}_T$. In this section we derive the uniform *a priori* estimates for $(\rho, u, \theta, w)$. We begin with

**Lemma 3.1.** *We have $w(x,t) \geqslant 0$ for any $(x,t) \in \bar{Q}_T$, and*

$$\sup_{t\in[0,T]} \int_0^1 \Big(\rho + \frac{1}{2}\rho u^2 + w + \rho\theta\Big)(x,t)dx \leqslant C, \tag{3.1}$$

$$\int_0^T \int_0^1 \theta^4 dxdt \leqslant C, \tag{3.2}$$

$$\sup_{t\in[0,T]} \int_0^1 \Big(\rho \ln\rho + \rho|\ln\theta|\Big)(x,t)dx + \int_0^T \int_0^1 \Big(\frac{u_x^2 + w}{\theta} + \frac{\kappa\theta_x^2}{\theta^2}\Big)dxd\tau \leqslant C. \tag{3.3}$$

*Proof.* Applying the strong maximal principle to the parabolic problem (2.13), (2.7), one easily gets $w(x,t) \geqslant 0$. Integrating Eq. (2.13) in $(0,1)\times(0,t)$, and using the boundary condition $w + nw_x = 0$, we deduce

$$\int_0^1 wdx + \int_0^t (w(1,\tau) + w(0,\tau))d\tau + \int_0^1\int_0^t wdxd\tau = \int_0^1 w_0(x)dx + \int_0^1\int_0^t \theta^4 dxd\tau,$$

which, by virtue of the positivity of $w$, gives

$$\int_0^1 wdx \leqslant \int_0^1\int_0^t \theta^4 dxd\tau - \int_0^1\int_0^t wdxd\tau + \int_0^1 w_0(x)dx. \tag{3.4}$$

Adding (2.10) to (2.12) and integrating the resulting equation over $(0,1)$, recalling the boundary conditions (2.7), and combining (3.4), we obtain (3.1) immediately. An integration of (2.12) and use of (3.4) gives (3.2) easily.

Rewriting the energy equation in the following form

$$\rho\theta_t + \rho u\theta_x + Pu_x = w - \theta^4 + (\kappa\theta_x)_x + u_x^2, \tag{3.5}$$

multiplying then (3.5) by $-1/\theta$ and integrating over $(0,1)\times(0,t)$, we obtain by (2.7) that

$$\int_0^1 \rho\ln\theta + \rho\ln\rho dx + \int_0^t\int_0^1 \Big(\frac{u_x^2 + w}{\theta} + \frac{\kappa\theta_x^2}{\theta^2}\Big)dxd\tau$$

$$= -\int_0^t \int_0^1 \theta^3 dx d\tau + \int_0^1 \rho_0 \ln \theta_0 - \rho_0 \ln \rho_0 dx,$$

which yields (3.3). □

Integrability of the $L^\infty$-norm of the temperature is given in the following:

**Lemma 3.2.** *Assume that heat conductivity $\kappa$ satisfies (2.6) for some $r < q+2$. Then*

$$\int_0^T \|\theta(t)\|_{L^\infty}^r dt \leqslant C. \tag{3.6}$$

*Proof.* Multiplying (3.5) by $\theta^{-\alpha}$ ($\alpha \in (0,1)$), integrating by parts and applying (2.7), we get

$$\int_0^T \int_0^1 \frac{\kappa \theta_x^2}{\theta^{1+\alpha}} dx dt \leqslant C,$$

whence,

$$\begin{aligned}
\int_0^T \|\theta\|_{L^\infty}^r dt &\leqslant C \int_0^T \int_0^1 \theta^r dx dt + C \int_0^T \int_0^1 \theta^{r-1} \theta_x dx dt \\
&\leqslant C \int_0^T \|\theta\|_{L^\infty}^{r-1} dt + C \int_0^T \int_0^1 \left( \frac{\theta^{2r-1+\alpha}}{k} + \frac{\kappa \theta_x^2}{\theta^{1+\alpha}} \right) dx dt \\
&\leqslant C + C \int_0^T \|\theta\|_{L^\infty}^{r-1} dt + \int_0^T \|\theta\|_{L^\infty}^{2r-2+\alpha-q} dt,
\end{aligned}$$

which, by recalling the Young inequality and the fact that $2r-2+\alpha-q < r$, gives

$$\int_0^T \|\theta\|_{L^\infty}^r dt \leqslant C \quad \text{for any } 1 \leqslant r < q+2.$$

□

The following lemma gives us pointwise uniform upper and lower bounds of the density, the proof of which is the similar as in [8] and is therefore omitted.

**Lemma 3.3.**

$$C^{-1} \leqslant \rho(x,t) \leqslant C \quad \textit{for all } (x,t) \in \bar{Q}_T.$$

For the simplicity of presentation, we introduce the Lagrangian coordinates. Without causing any confusion, we still denote the Lagrange coordinates by $(x,t)$ in what follows:

$$x \mapsto x = \int_0^x \rho(s,t) ds, \quad t = t. \tag{3.7}$$

Due to the fact that $\rho(x,t)$ has positive lower and upper bounds, this coordinate transformation is invertible. Since the mass is conserved, for simplicity we normalize the mass $\int_0^1 \rho(x,t)dx = 1$. In this way, the transformation (3.7) maps $[0,1]$ onto $[0,1]$. Denoting $v = 1/\rho$, the system (2.10)–(2.13) in Lagrangian coordinates turns to

$$v_t = u_x, \tag{3.8}$$

$$u_t = \left(\frac{u_x}{v} - P\right)_x, \tag{3.9}$$

$$\theta_t + Pu_x = \left(w - \theta^4\right)v + \left(\frac{\kappa\theta_x}{v}\right)_x + \frac{u_x^2}{v}, \tag{3.10}$$

$$vw_t - uw_x - \left(\frac{w_x}{v}\right)_x = (\theta^4 - w)v. \tag{3.11}$$

With the corresponding boundary condition,

$$(u, \kappa\theta_x, vw + \mathbf{n}w_x)\Big|_{x=0,1} = (0,0,0). \tag{3.12}$$

Thanks to the fact that density $\rho$ has lower and upper bound, so ignore the factor of $\rho$ in the integration. We now can transform all the estimates from Lemmas 3.1–3.2 to the ones in Lagrangian coordinates and obtain therefore

$$\sup_{t\in[0,T]} \int_0^1 \left(\frac{1}{2}u^2 + w + \theta\right)(x,t)dx \leqslant C, \quad C^{-1} \leqslant v(x,t) \leqslant C, \tag{3.13}$$

$$\int_0^T \int_0^1 \left(\theta^4 + \frac{u_x^2 + w}{\theta} + \frac{k\theta_x^2}{\theta^2}\right)dxdt \leqslant C, \quad \int_0^T \|\theta\|_{L^\infty}^r dt \leqslant C, \tag{3.14}$$

for $1 \leqslant r < q+2$. Next, we derive more uniform *a priori* estimates in Lagrangian coordinates.

**Lemma 3.4.** *We have*

$$\sup_{t\in[0,T]} \|v_x(t)\| + \int_0^T \int_0^1 \theta v_x^2 dxdt \leqslant C.$$

*Proof.* Equations (3.8) and (3.9) give $(u - v_x/v)_t = -P_x$. We multiply this equality by $(u - v_x/v)$ and integrate it over $(0,1)$ to infer

$$\frac{1}{2}\frac{d}{dt}\int_0^1 \left(u - \frac{v_x}{v}\right)^2 dx + \int_0^1 \frac{v_x^2\theta}{v^3}dx = \int_0^1 \left(\frac{v_x\theta u}{v^2} - \frac{\theta_x}{v}(u - \frac{v_x}{v})\right)dx \equiv H, \tag{3.15}$$

where the right hand side $H$ can be bounded as follows, using Cauchy-Schwarz's inequality and recalling $q > 2$.

$$H \leqslant \frac{1}{2}\int_0^1 \frac{v_x^2\theta}{v^3}dx + C\|\theta\|_{L^\infty}\int_0^1 u^2 dx + C\int_0^1 \frac{\theta_x^2}{v^2}dx + C\int_0^1 \left(u - \frac{v_x}{v}\right)^2 dx$$

$$\leqslant \frac{1}{2}\int_0^1 \frac{v_x^2\theta}{v^3}dx + C\left(\|\theta\|_{L^\infty} + \int_0^1 \frac{\kappa\theta_x^2}{\theta^2}dx + \int_0^1 \left(u - \frac{v_x}{v}\right)^2 dx\right).$$

Inserting the above inequality into (3.15), applying Gronwall's Lemma, (3.13) and Lemma 3.3, we conclude

$$\int_0^1 \left(u - \frac{v_x}{v}\right)^2 (x,t)dx + \int_0^t \int_0^1 \frac{v_x^2\theta}{v^3}dxd\tau \leqslant C,$$

which proves the lemma. □

Now, we multiply Eq. (3.9) by $u^3$ and integrate by parts and utilize Lemma 3.3 to find that

$$\frac{1}{4}\int_0^1 u^4 dx + \int_0^t \int_0^1 \frac{3u^2u_x^2}{v}dxd\tau = \int_0^t \int_0^1 3Pu^2u_x dxd\tau + \frac{1}{4}\int_0^1 u_0^4 dx$$
$$\leqslant \frac{1}{2}\int_0^t \int_0^1 \frac{3u^2u_x^2}{v}dxd\tau + C\int_0^t \|\theta^2\|_{L^\infty}\int_0^1 u^2 dxd\tau + \frac{1}{4}\int_0^1 u_0^4 dx,$$

from which, (3.13) and (3.6), the following lemma follows.

**Lemma 3.5.**

$$\sup_{t\in[0,T]}\int_0^1 u^4(x,t)dx + \int_0^T \int_0^1 u^2u_x^2 dxdt \leqslant C.$$

The following lemma is devoted to the integration of the derivatives of velocity $u$.

**Lemma 3.6.**

$$\sup_{t\in[0,T]}\int_0^1 u_x^2(x,t)dx + \int_0^T \int_0^1 (u_t^2 + u_{xx}^2)dxdt \leqslant C.$$

*Proof.* From Eq. (3.9) we get

$$\left(\sqrt{v}u_t - \frac{u_{xx}}{\sqrt{v}}\right)^2 = \left(\sqrt{v}P_x + \frac{u_xv_x}{v^{3/2}}\right)^2.$$

An integration of this equality over $(0,1)\times(0,t)$ yields

$$\int_0^1 u_x^2 dx + \int_0^t \int_0^1 \left(vu_t^2 + \frac{u_{xx}^2}{v}\right)dxd\tau$$
$$= \int_0^1 (u_0)_x^2 dx + \int_0^t \int_0^1 \left(vP_x^2 + \frac{u_x^2v_x^2}{v^3} + 2\frac{P_xu_xv_x}{v}\right)dxd\tau$$
$$\equiv \int_0^1 (u_0)_x^2 dx + \sum_{j=1}^3 I_j, \tag{3.16}$$

where $I_j$ can be bounded as follows, using (3.13), (3.14), Lemma 3.4 and Sobolev's imbedding theorem.

$$\begin{aligned}I_1 &= \int_0^t\int_0^1 v\Big(\frac{\theta_x}{v}-\frac{v_x\theta}{v^2}\Big)^2 dxd\tau \leqslant C\int_0^t\int_0^1\Big(\frac{\theta_x^2}{v^2}+\frac{v_x^2\theta^2}{v^2}\Big)dxd\tau\\ &\leqslant C\int_0^t\int_0^1\frac{k\theta_x^2}{\theta^2}dxd\tau + C\int_0^t\|\theta^2\|_{L^\infty}d\tau\int_0^1 v_x^2dx \leqslant C;\end{aligned}$$

$$\begin{aligned}I_2 &\leqslant C\int_0^t\|u_x^2\|_{L^\infty}d\tau\int_0^1 v_x^2dx\\ &\leqslant C\int_0^t\Big(\int_0^1 u_x^2dx + 2\int_0^1|u_x|\,|u_{xx}|dx\Big)d\tau\\ &\leqslant C\int_0^t\Big(\int_0^1 u_x^2dx + \frac{1}{4C}\int_0^1\frac{u_{xx}^2}{v}dx\Big)d\tau,\end{aligned}$$

and similarly

$$\begin{aligned}I_3 &= \int_0^t\int_0^1\frac{\theta_x v u_x v_x - (v_x)^2\theta u_x}{v^2}dxd\tau\\ &\leqslant C\int_0^t\int_0^1\Big(\theta_x^2 + v_x^2\|u_x^2\|_{L^\infty} + v_x^2\|\theta^2\|_{L^\infty} + \|u_x^2\|_{L^\infty}v_x^2\Big)dxd\tau\\ &\leqslant C\int_0^t\int_0^1\Big(\frac{\kappa\theta_x^2}{\theta^2} + u_x^2 + \frac{1}{2C}\frac{u_{xx}^2}{v}\Big)dxd\tau.\end{aligned}$$

Substituting the above estimates for $I_j$ into (3.16) and applying Gronwall's Lemma, one obtains Lemma 3.6. □

To derive higher integrability for the temperature, we multiply Eq. (3.10) by $\theta^7$ and integrate over $(0,1)\times(0,t)$ to deduce

$$\begin{aligned}\frac{1}{8}\int_0^1\theta^8dx &+ \int_0^t\int_0^1\Big(v\theta^{11} + \frac{7\kappa(\theta_x\theta^3)^2}{v}\Big)dxd\tau\\ &= \int_0^t\int_0^1\Big(-\theta^7Pu_x + wv\theta^7 + \frac{u_x^2\theta^7}{v}\Big)dxd\tau, \quad (3.17)\end{aligned}$$

where the right hand side can be estimated as follows, using (3.13), (3.14), Sobolev's imbedding theorem and Lemma 3.6.

$$\begin{aligned}&\Big|\int_0^t\int_0^1\theta^7Pu_xdxd\tau\Big| \leqslant C\int_0^t\int_0^1\theta^{11/2}\theta^{5/2}|u_x|dxd\tau\\ &\leqslant \frac{1}{4}\int_0^t\int_0^1 v\theta^{11}dxd\tau + C\int_0^t\|\theta^5\|_{L^\infty}\int_0^1 u_x^2dxd\tau\\ &\leqslant \frac{1}{4}\int_0^t\int_0^1 v\theta^{11}dxd\tau + C\int_0^t\int_0^1\theta^5dxd\tau + C\int_0^t\int_0^1\theta^4|\theta_x|dxd\tau\end{aligned}$$

$$\leqslant C+\frac{1}{4}\int_0^t\int_0^1 v\theta^{11}dxd\tau+\frac{1}{4}\int_0^t\int_0^1\frac{\kappa(\theta_x\theta^3)^2}{v}dxd\tau$$

and

$$\begin{aligned}&\int_0^t\int_0^1\Big(|vw\theta^7|+\frac{u_x^2\theta^7}{v}\Big)dxd\tau\\&\leqslant C\int_0^t\int_0^1(\theta^7+|7\theta^6\theta_x|)dx\int_0^1 wdxd\tau+\int_0^t\|\theta^7\|_{L^\infty}d\tau\int_0^1 u_x^2dx\\&\leqslant C+C\int_0^t\int_0^1\theta^8dxd\tau+\frac{1}{4}\int_0^t\int_0^1\frac{\kappa(\theta_x\theta^3)^2}{v}dxd\tau.\end{aligned}$$

Inserting the above two inequalities into (3.17) and applying Gronwall's Lemma, we obtain

**Lemma 3.7.**

$$\sup_{t\in[0,T]}\int_0^1\theta^8(x,t)dx+\int_0^T\int_0^1\Big(v\theta^{11}+\frac{\kappa(\theta^3\theta_x)^2}{v}\Big)dxdt\leqslant C.$$

As a consequence of Lemma 3.7, we can easily deduce the following lemma.

**Lemma 3.8.**

$$\sup_{t\in[0,T]}\int_0^1 w_x^2(x,t)dx+\int_0^T\int_0^1\Big(v^2w_t^2+\frac{w_{xx}^2}{v^2}\Big)dxdt\leqslant C.$$

*Proof.* Multiplying (3.11) by $2w$, integrating over $(0,1)\times(0,t)$, we find that

$$\int_0^1 vw^2dx+2\int_0^t\int_0^1\Big(\frac{w_x^2}{v}+vw^2\Big)dxd\tau\leqslant C.$$

On the other hand, we can also rewrite Eq. (3.11) in the following form:

$$vw_t-\frac{w_{xx}}{v}=(\theta^4-w)v+uw_x-\frac{w_xv_x}{v^2}.$$

Thus, following a process very similar to the one used for the proof of Lemma 3.6, the proof is complete. □

With the help of Lemma 3.8, it is easy to see that $w\in L^\infty((0,1)\times(0,T))$. Next, we will show the uniform estimates of the derivatives of $\theta$. As in [10, 12, 21], we introduce the function

$$K(v,\theta):=\int_0^\theta\frac{\kappa(v,s)}{v}ds.$$

Then, we have

**Lemma 3.9.**

$$\sup_{t\in[0,T]}\int_0^1\Big(\frac{\kappa}{v}\theta_x\Big)^2(x,t)dx+\int_0^T\int_0^1\frac{\kappa\theta_t^2}{v}dxdt\leqslant C. \tag{3.18}$$

We multiply Eq. (3.10) by $K_t$ and integrate the resulting equation over $(0,1)\times(0,t)$ to obtain

$$\int_0^t\int_0^1\Big(\theta_t+Pu_x-\frac{u_x^2}{v}-(w-\theta^4)v\Big)K_tdxd\tau+\int_0^t\int_0^1\frac{\kappa\theta_x}{v}K_{xt}dxd\tau=0. \tag{3.19}$$

Here

$$\begin{cases} K_t=\dfrac{\kappa}{v}\theta_t+K_vu_x,\;\; K_{xt}=\Big(\dfrac{\kappa}{v}\theta_x\Big)_t+K_{vv}v_xu_x+K_vu_{xx}+\Big(\dfrac{\kappa}{v}\Big)_v v_x\theta_t,\\ |K_v|,\;|K_{vv}|\leqslant C(1+\theta^{q+1}). \end{cases}$$

After rearranging (3.19), we arrive at

$$\frac12\int_0^1\Big(\frac{\kappa}{v}\theta_x\Big)^2dx+\int_0^t\int_0^1\frac{\kappa\theta_t^2}{v}dxd\tau=\frac12\int_0^1\Big(\frac{\kappa(v_0,\theta_0)}{v_0}\theta_{0x}\Big)^2dx+\text{RHS}, \tag{3.20}$$

where

$$\begin{aligned}\text{RHS}=&-\int_0^t\int_0^1\Big\{\Big(\theta_t+Pu_x-(w-\theta^4)v-\frac{u_x^2}{v}\Big)K_vu_x\\ &-\Big(Pu_x-(w-\theta^4)v-\frac{u_x^2}{v}\Big)\frac{\kappa}{v}\theta_t\Big\}dxd\tau\\ &-\int_0^t\int_0^1\Big(K_{vv}v_xu_x+K_vu_{xx}+\Big(\frac{\kappa}{v}\Big)_v v_x\theta_t\Big)\frac{\kappa}{v}\theta_xdxd\tau.\end{aligned}$$

Before we proceed, we need the following estimate.

$$\begin{aligned}\|(1+\theta)^r\|_{L^\infty(Q_T)}&\leqslant\sup_{t\in[0,T]}\Big[\int_0^1(1+\theta)dx\Big]^r+\sup_{t\in[0,T]}\int_0^1(1+\theta)^{r-1}|\theta_x|dx\\ &\leqslant C+C\sup_{t\in[0,T]}\Big(\int_0^1\Big(\frac{\kappa}{v}\theta_x\Big)^2dx\Big)^{1/2}\Big(\int_0^1(1+\theta)^{2r-2q-2}dx\Big)^{1/2}\\ &\leqslant C+C\sup_{t\in[0,T]}\Big(\int_0^1\Big(\frac{\kappa}{v}\theta_x\Big)^2dx\Big)^{1/2},\quad\text{provided } r\leqslant q+5.\end{aligned} \tag{3.21}$$

With the help of all the bounds established so far, the inequality (3.21) and Young's inequality (for $(1+\theta)^{q+1}$), we can estimate each term in RHS as follows.

$$\Big|\int_0^t\int_0^1\theta_tK_vu_xdxd\tau\Big|$$

$$\leqslant \frac{1}{16}\int_0^t\int_0^1 \frac{\kappa\theta_t^2}{v}dxd\tau + C\|(1+\theta)\|_{L^\infty(Q_T)}^{q+2}\int_0^t\int_0^1 u_x^2 dxd\tau$$
$$\leqslant \frac{1}{16}\int_0^t\int_0^1 \frac{\kappa\theta_t^2}{v}dxd\tau + C + \frac{1}{32}\sup_{t\in[0,T]}\int_0^1 \Big(\frac{\kappa}{v}\theta_x\Big)^2 dx;$$

and similarly,

$$\Big|\int_0^t\int_0^1 (Pu_x - wv)K_v u_x dxdt\Big|, \Big|\int_0^t\int_0^1 (Pu_x + \theta^4 v - wv)\frac{\kappa}{v}\theta_t dxd\tau\Big|,$$
$$\Big|\int_0^t\int_0^1 \theta^4 vK_v u_x dxd\tau\Big|, \Big|\int_0^t\int_0^1 \frac{u_x^2}{v}K_v u_x dxd\tau\Big|, \Big|\int_0^t\int_0^1 \frac{u_x^2}{v}\frac{\kappa}{v}\theta_t dxd\tau\Big|$$
$$\leqslant \frac{1}{16}\int_0^t\int_0^1 \frac{\kappa\theta_t^2}{v}dxd\tau + C + \sup_{t\in[0,T]}\frac{1}{32}\int_0^1 \Big(\frac{\kappa}{v}\theta_x\Big)^2 dx;$$

and

$$\Big|\int_0^t\int_0^1 \Big(\frac{\kappa}{v}\Big)_v v_x\theta_t\frac{\kappa}{v}\theta_x dxd\tau\Big|$$
$$\leqslant \frac{1}{16}\int_0^t\int_0^1 \frac{\kappa\theta_t^2}{v}dxd\tau + C\|(1+\theta)^q\|_{L^\infty(Q_T)}\int_0^t \|\frac{\kappa}{v}\theta_x\|_{L^\infty}^2 d\tau, \quad (3.22)$$

where we have to bound the last term on the right hand side of (3.22) as follows: keeping in mind that $\theta_x|_{x=0} = 0$, one has

$$\int_0^t \|\frac{\kappa}{v}\theta_x\|_{L^\infty}^2 d\tau \leqslant 2\int_0^t\int_0^1 \Big|\frac{\kappa}{v}\theta_x\Big|\Big|\Big(\frac{\kappa}{v}\theta_x\Big)_x\Big|dxd\tau$$
$$\leqslant C + C\Big[\int_0^t\int_0^1 (1+\theta)^{q+2}\Big(\frac{\kappa\theta_x}{v}\Big)_x^2 dxd\tau\Big]^{1/2}$$
$$\leqslant C + C\Big[\int_0^t\int_0^1 (1+\theta)^{q+2}\Big(\theta_t + Pu_x - (w-\theta^4)v - \frac{u_x^2}{v}\Big)^2 dxd\tau\Big]^{1/2}$$
$$\leqslant C + C\|(1+\theta)\|_{L^\infty(Q_T)}\Big[\int_0^t\int_0^1 \frac{\kappa\theta_t^2}{v}dxd\tau\Big]^{1/2} + C\|(1+\theta)^{q+4}\|_{L^\infty(Q_T)}.$$

Substituting the above inequality into (3.22), one finds that

$$\Big|\int_0^t\int_0^1 \Big(\frac{\kappa}{v}\Big)_v v_x\theta_t\frac{\kappa}{v}\theta_x dxd\tau\Big|$$
$$\leqslant \frac{1}{16}\int_0^t\int_0^1 \frac{\kappa\theta_t^2}{v}dxd\tau + C + C\|(1+\theta)^{2q+4}\|_{L^\infty(Q_T)}$$
$$+C\|(1+\theta)^{q+1}\|_{L^\infty(Q_T)}\Big(\int_0^t\int_0^1 \frac{\kappa\theta_t^2}{v}dxd\tau\Big)^{1/2}$$

$$\leqslant \frac{1}{8}\int_0^t\int_0^1 \frac{\kappa\theta_t^2}{v}dxd\tau + C + C\sup_{t\in[0,T]}\left[\int_0^1\left(\frac{\kappa\theta_x}{v}\right)^2dx\right]^{(2q+4)/(2q+10)}$$
$$\leqslant \frac{1}{8}\int_0^t\int_0^1 \frac{\kappa\theta_t^2}{v}dxd\tau + C + \frac{1}{32}\sup_{t\in[0,T]}\int_0^1\left(\frac{\kappa}{v}\theta_x\right)^2dx.$$

We proceed to derive bounds on the rest terms in RHS. Similarly to the above, we deduce

$$\left|\int_0^t\int_0^1 K_v u_{xx}\frac{\kappa}{v}\theta_x dxd\tau\right|, \qquad \left|\int_0^t\int_0^1 K_{vv}v_x u_x\frac{\kappa}{v}\theta_x dxd\tau\right|$$
$$\leqslant C + \frac{1}{32}\sup_{[0,T]}\int_0^1\left(\frac{\kappa}{v}\theta_x\right)^2dx.$$

Now, combining all the above estimates concerning the terms in RHS together, applying Young's inequality, we obtain finally (3.18). This completes the proof of Lemma 3.9. □

As a direct consequence of Lemmas 3.9 and 3.7, one sees that

$$\|\theta\|_{L^\infty(Q_T)} \leqslant C. \tag{3.23}$$

Next, we estimate $u_t, u_{xx}$ in $L^\infty(0,T;L^2)$ in the following lemma.

**Lemma 3.10.**

$$\sup_{t\in[0,T]}\int_0^1(u_t^2+u_{xx}^2)(x,t)dx + \|u_x\|_{L^\infty(Q_T)} + \int_0^T\int_0^1 u_{xt}^2dxdt \leqslant C.$$

*Proof.* We start with the following inequality, which is obtained by differentiating (3.9) with respect to $t$, multiplying then by $u_t$ in $L^2((0,1)\times(0,t))$, and making use of Eq.(2.10), Lemmas 3.6 and 3.9, and (2.5), (3.13), Sobolev's imbedding theorem and (3.23).

$$\frac{1}{2}\int_0^1 u_t^2(x,t)dx + \int_0^t\int_0^1\frac{u_{xt}^2}{v}dxd\tau$$
$$= \int_0^t\int_0^1\left(\frac{u_x v_t}{v^2} - P_t\right)u_{xt}dx + \frac{1}{2}\int_0^1 u_t^2(x,0)dxd\tau$$
$$\leqslant C + \frac{1}{2}\int_0^t\int_0^1\frac{u_{xt}^2}{v}dxd\tau$$
$$+C\int_0^t(\|u_x^2\|_{L^\infty}\|u_x\|^2 + \|\theta_t\|^2 + \|\theta^2\|_{L^\infty}\|u_x\|)d\tau$$
$$\leqslant C + \frac{1}{2}\int_0^t\int_0^1\frac{u_{xt}^2}{v}dxd\tau + C\sup_{t\in[0,T]}\|u_{xx}\|,$$

whence,

$$\sup_{t\in[0,T]}\int_0^1 u_t^2(x,t)dx+\int_0^T\int_0^1\frac{u_{xt}^2}{v}dxd\tau\leqslant C+C\sup_{t\in[0,T]}\|u_{xx}\|. \quad (3.24)$$

On the other hand, from Eq. (3.9) and (2.5) we easily get

$$\|\frac{u_{xx}}{v}\|\leqslant C\Big(\|u_t\|+\|\frac{\theta_x}{v}\|+\|\frac{(u_x-\theta)v_x}{v^2}\|\Big)\leqslant C\Big(1+\|\frac{k\theta_x}{v}\|+\sup_{t\in[0,T]}\|u_{xx}\|^{\frac{1}{2}}\Big),$$

which, by applying Young's inequality and Lemma 3.9, results in

$$\sup_{t\in[0,T]}\|u_{xx}(t)\|\leqslant C.$$

Inserting this inequality into (3.24), we immediately obtain the lemma, noting that the boundedness of $\|u_x\|_{L^\infty(Q_T)}$ follows from Sobolev's imbedding theorem and the boundedness of $\|u_{xx}\|$. □

Finally, we show pointwise boundedness of the temperature and derive bounds for its higher derivatives.

**Lemma 3.11.**

$$\theta(x,t)\geqslant C \qquad \textit{for any } (x,t)\in\bar{Q}_T, \quad (3.25)$$

$$\sup_{t\in[0,T]}\int_0^1(\theta_t^2+\theta_{xx}^2)(x,t)dx+\int_0^T\int_0^1\theta_{xt}dxdt\leqslant C.$$

*Proof.* Utilizing the *a priori* estimates established so far and the standard comparison principle, arguing in the same manner as in [21, Lemma 11], we obtain (3.25) easily.

To prove the rest of the lemma, we differentiate Eq.(3.11) with respect to $t$ and carefully use the boundary condition (3.12) and then deduce by arguments similar to those used in the proof of Lemma 3.10 that

$$\int_0^1 w_t^2(x,t)dx+\int_0^t\int_0^1\frac{w_{xt}^2}{v^2}dxd\tau\leqslant C, \qquad 0\leqslant t\leqslant T. \quad (3.26)$$

Now, we differentiate Eq. (3.10) with respect to $t$ and multiply the resulting equation by $\theta_t$ in $L^2(0,1)$ to arrive at

$$\frac{1}{2}\frac{d}{dt}\int_0^1\theta_t^2dx+\int_0^1\frac{\kappa}{v}\theta_{xt}^2dx$$
$$=\int_0^1\Big\{\Big(\frac{\kappa}{v}\Big)_t\theta_x\theta_{xt}+\big[(w-\theta^4)v+\frac{u_x^2}{v}-(Pu_x)\big]_t\theta_t\Big\}dx, \quad (3.27)$$

where the terms on the right hand side can be bounded as follows, using (3.13), (3.23), (3.26), Lemma 3.6

$$\begin{aligned}\left|\int_0^1 [(w-\theta^4)v]_t\theta_t dx\right| &= \left|\int_0^1 (w_t v\theta_t + wu_x\theta_t + \theta^4 u_x\theta_t + 4\theta^3\theta_t^2 v)dx\right| \\ &\leqslant C\int_0^1 \frac{\kappa\theta_t^2}{v}dx + C\int_0^1 w_t^2 dx + C\int_0^1 u_x^2 dx \\ &\leqslant C + C\int_0^1 \frac{\kappa\theta_t^2}{v}dx;\end{aligned}$$

$$\left|\int_0^1 \left(\frac{\kappa}{v}\right)_t \theta_x\theta_{xt}dx\right| \leqslant \frac{1}{8}\int_0^1 \frac{\kappa}{v}\theta_{xt}^2 dx + C + C\|\theta_t^2\|_{L^\infty};$$

and

$$\left|\int_0^1 (Pu_x)_t\theta_t dx\right|, \left|\int_0^1 \left(\frac{u_x^2}{v}\right)_t \theta_t dx\right| \leqslant C + C\int_0^1 \frac{\kappa\theta_t^2}{v}dx.$$

Thus, substituting the above three inequalities into (3.27), integrating the resulting inequality over $(0,t)$ and using Lemma 3.9, we conclude

$$\|\theta_t(t)\|^2 + \int_0^t \|\theta_{xt}\|^2 d\tau \leqslant C + C\int_0^t \|\theta_t^2\|_{L^\infty} d\tau, \qquad t\in[0,T]. \tag{3.28}$$

On the other hand, we have by Sobolev's imbedding theorem that

$$C\|\theta_t^2(t)\|_{L^\infty} \leqslant C\int_0^1 (\theta_t^2 dx + |\theta_t|\,|\theta_{xt}|)dx \leqslant \frac{1}{2}\int_0^1 \theta_{xt}^2 dx + C\int_0^1 \theta_t^2 dx.$$

Inserting the above estimate into (3.28) and applying Gronwall's Lemma, we obtain

$$\sup_{t\in[0,T]} \|\theta_t(t)\|^2 + \int_0^T \|\theta_{xt}\|^2 dt \leqslant C,$$

which, together with Eq.(3.10) and the *a priori* estimates established so far, implies

$$\|\theta_{xx}(t)\| \leqslant C + C\|\theta_x\|_{L^\infty} \leqslant C + C\|\theta_x\|^{1/2}\|\theta_{xx}\|^{1/2} \leqslant C + \frac{1}{2}\|\theta_{xx}(t)\|,$$

which yields

$$\sup_{t\in[0,T]} \|\theta_{xx}(t)\| \leqslant C.$$

Thus, we have proved Lemma 3.11. □

With the help of Lemmas 3.1–3.11 together, we are able to employ the same arguments as in [21, 22] to deduce the following Hölder-estimates:

$$\|(v,v_x,v_t)\|_{C^{\alpha,\alpha/2}(Q_T)} + \|(u,\theta,w)\|_{C^{2+\alpha,1+\alpha/2}(Q_T)} \leqslant C, \tag{3.29}$$

$$C^{-1} \leqslant v(x,t), \quad \theta(x,t) \leqslant C, \quad w(x,t) \geqslant 0, \quad \forall\,(x,t) \in \bar{Q}_T. \tag{3.30}$$

We can transform the estimates (3.29) and (3.30) to the corresponding ones in Eulerian coordinates and obtain therefore the same uniform *a priori* estimates for $(\rho, u, \theta, w)$. With the help of these uniform a priori estimates in Eulerian coordinates, we can continue a local smooth solution globally in time and obtain a unique global smooth solution. The proof of Theorem 2.1 is complete.

# References


[1] C. Buet and B. Després. Asymptotic analysis of fluid models for coupling of radiation and hydrodynamics. *J. of Quantative Spectroscopy and Radiative Transfer* **85**, 385-418, 2004.

[2] J.W. Bond, K.M. Watson and J.A. Welch. *Atomic Theory of Gas Dynamics.* Addison-Wesley Publishing Company, INC. 1965.

[3] B. Ducomet. Hydrodynamical models of gsaeous stars. *Reviews of Math. Phys.* **8**, 957-1000, 1996.

[4] B. Ducomet. A model of thermal dissipation for a one-dimensional viscous reactive and radiative gas. *Math. Meth. Appl. Sci.* **22**, 1323-1349, 1999.

[5] B. Ducomet and E. Feireisl. On the dynamics of gaseous stars, Arch. *Rational Mech. Anal.* **174**, 221-266, 2004.

[6] B. Ducomet and E. Feireisl. The equations of magnetohydrodynamics: on the interation between matter and radiation in the evlution of gaseous stars. *Comm. Math. Phys.* **266**, 595-629, 2006.

[7] B. Ducomet and A. Zlotnik. Lyapunov functional methods for 1D radiative and ractive viscous gas dynamics. *Arch. Rational Mech. Anal.* **177**, 185-229, 2005.

[8] J. Fan, S. Jiang and G. Nakamura. Vanishing shear viscosity limit in the magnetohydrodynamic equations. *Comm. Math. Phys.* **270**, 691-708, 2007.

[9] Th. Goudon and P. Lafitte. A coupled model for radiative transfer: Doppler effects, equilibrium and non equilibrium diffusin asympotics. *SIAM Multiscale Model Simul.* **4**, 1245-1279, 2005 (electronic).

[10] S. Jiang. On initial boundary value problems for a viscous heat-conducting one-dimensional real gas. *J. Diff. Eqns.* **110**, 157-181, 1994.

[11] N. Kaiser, J. Meyer-ter-Vehn and R. Siegel. The x-ray-driven heating wave. *Phys. Fluids B* **8**, 1747-1752, 1989.

[12] B. Kawohl. Global existence of large solution to a initial boundary value problems for a viscous heat-conducting one-dimensional real gases. *J. Diff. Eqns.* **58**, 76-103, 1985.

[13] S. Kawashima and S. Nishibata. Cauchy problem for a model system of radiating gas: weak solutions with a jump and classical solutions. *Math. Models Meth. Appl. Sci.* **9**, 69-91, 1999.

[14] S. Kawashima and S. Nishibata. Shock waves for a model system of radiating gas. *SIAM J. Math. Anal.* **30**, 95-117, 1999.

[15] C. Lin, J.F. Coulomrel and T. Goudon. *Shock profiles for non equilibrium radiating gases*, Preprint, 2006.

[16] D. Mihalas and B. Weibel-Mihalas. *Foundation of Radiation Hydrodynamics*, Oxford University Press, 1984.

[17] G.C. Pamraning. *The Equations of Radiation Hydrodynamics*, Pergamon Press, 1973.

[18] P. Secchi. On the motion of gaseous stars in the presence of radiation. *Commu. PDE* **15**, 185-204, 1990.

[19] V.A. Solonnikov and A.V. Kazhikhov. Existence theorems for the equations of motion of a compressible viscous fluid. *Ann. Rev. Fluid Mech.* **13**, 79-95, 1981.

[20] G.D. Tsakiris and K. Eidmann. An approximate method for calculating Plank and Rosseland mean opacities in hot, dense plasmas. *J. Quant. Spectrosc. Radiat. Transfer* **38**, 353-368, 1987.

[21] M. Umehara and A. Tani. Global solution to the one-dimensional equations for a self-gravitating viscous radiative and ractive gas. *J. Diff. Eqns.* **234**, 439-463, 2007.

[22] J.W. Zhang and F. Xie. *Global solution for a one-dimensional model problem in thermally radiative magnetohydrodynamics*, Preprint, 2007.

[23] Y.B. Zeldovich and Y.P. Raizer. *Phsics of Shock Waves and High-Temperature Hydrodynamic Phenomenon*, Academic Press, 1966.

[24] X. Zhong and S. Jiang. Local existence and finite time blow-up in multidimensional radiation hydrodynamics. *J. Math. Fluid Mech.* (Online).

# Recent Computational Methods for High Frequency Waves in Heterogeneous Media*

Shi Jin
*Department of Mathematics, University of Wisconsin-Madison*
*Madison, WI 53706, USA*
*Email: jin@math.wisc.edu*

**Abstract**

In this note, we review our recent results on the Eulerian computation of high frequency waves in heterogeneous media. We cover three recent methods: the moment method, the level set method, and the computational methods for interface problems in high frequency waves. These approaches are all based on high frequency asymptotic limits.

## 1 Introduction

High frequency wave computation is a classical field of applied mathematics, with many important applications in acoustic waves, elastic waves, optics, and electromagnetism, etc.. The main computational challenge in these problems is that one cannot afford to numerically resolve the small wave length. Approximate models based on asymptotic methods are often used.

One of the most important computational methods for high frequency waves uses geometric optics. A classical way of solving geometric optics is the Lagrangian framework, which uses *ray tracing* that traces the trajectory of particles. This method is easy to implement, since one just needs to solve a system of ODEs—which is a Hamiltonian system. The disadvantage of it is that the method loses accuracy when the rays diverge, in which case a complicated regridding is needed.

The Eulerian methods, based on solving partial differential equations (PDEs) on *fixed grids*, provide uniformly accurate numerical solutions regardless of the ray behavior, thus have many attractive advantages when compared with the traditional Lagrangian method by ray tracing.

*This work was supported by NSF grant DMS-0608720, and a Van Vleck Distinguished Research Prize from the University of Wisconsin-Madison.

In this note, we will review several of our recently introduced Eulerian computational methods for high frequency waves. Specifically, we will review the moment methods, the level set methods and the computational methods for interface problems in high frequency waves.

For recent comprehensive reviews on high frequency wave computations, see [18, 52].

# 2 The high frequency limit

As an example, consider the linear Schrödinger equation with the high frequency initial data,

$$i\epsilon\,\psi_t + \frac{\epsilon^2}{2}\Delta\psi - V(\mathbf{x})\psi = 0, \qquad \mathbf{x} \in \Re^n, \tag{2.1}$$

$$\psi(\mathbf{x}, 0) = A_0(\mathbf{x})e^{i\frac{S_0(\mathbf{x})}{\epsilon}}. \tag{2.2}$$

In (2.1)–(2.2), $\psi(\mathbf{x}, t)$ is the complex-valued wave function, $\epsilon$ is the rescaled Planck constant, and $V(\mathbf{x})$ denotes the potential. In the semiclassical regime, where the Planck constant $\epsilon$ is small, the wave function $\psi$ and the related physical observables become oscillatory of wave length $O(\epsilon)$. Mathematically, the rapid oscillations will forbid any strong convergence, and the limits have to be defined in the weak sense.

A related problem is the wave equation:

$$u_{tt} - c(\mathbf{x})^2\Delta u = 0, \tag{2.3}$$

where $c(\mathbf{x})$ is the local wave speed of the medium ($c_0/c(\mathbf{x})$, with $c_0$ a reference sound speed, is the index of refraction). When the essential frequencies of the wave field are relatively high, the wavelengths are small compared to the overall size of the physical domain.

In a direct numerical simulation of these problems, one needs a few grid points per wavelength in order to guarantee the numerical convergence [4, 45]. For sufficiently high frequencies, such a direct simulation in not feasible, especially in high space dimension, thus methods based on approximations of these equations are needed. Geometrical optics studies the high frequency limit, $\epsilon \to 0$, of solution to (2.3) in the form $u(\mathbf{x}, t) \sim A(\mathbf{x}, t)e^{i\phi(\mathbf{x},t)/\epsilon}$, where $A$ is the amplitude of the wave while $S$ is the phase. The similar limit for (2.1) is referred to as the semiclassical limit of the Schrödinger equation.

A classical approach for an Eulerian computation is the WKB (Wentzel-Kramers-Brillouin) method, which, by assuming the form of solution of (2.1)–(2.2) to be $\psi^\epsilon(\mathbf{x}, t) = A(\mathbf{x}, t)e^{iS(\mathbf{x},t)/\epsilon}$, yields, to the leading order, an eikonal equation for the phase $S$ and a linear transport equation for the position density $|A|^2$:

$$\partial_t S + \frac{1}{2}|\nabla S|^2 + V(\mathbf{x}) = 0, \tag{2.4}$$

$$\partial_t(|A|^2) + \nabla \cdot (|A|^2 \nabla S) = 0. \tag{2.5}$$

The eikonal equation is a nonlinear Hamilton-Jacobi equation. Even for smooth initial data, its solution may become singular in finite time, which corresponds to the formation of caustics (in the context of hyperbolic conservation law, which is the gradient of the Hamilton-Jacobi equation (2.4), it corresponds to the formation of shocks). Beyond this singularity, modern 'shock-capturing' numerical methods for the eikonal equation (2.4) will select the very stable *viscosity solution* [13,16], which is *not* the *dispersive* semiclassical limit of the Schrödinger equation, since it violates the superposition principle, an essential property of the linear Schrödinger equation. In fact, beyond the caustics, the solution becomes multivalued or multiphased, as can be studied by the classical stationary phase method [15].

A mathematically convenient tool to study the semiclassical limit, beyond the caustics, is the Wigner transform [59]:

$$W[\phi,\psi](\mathbf{x},\mathbf{p},t) = \frac{1}{(2\pi)^n}\int_{\Re^n} e^{i\mathbf{p}\cdot\mathbf{y}}\phi\Big(\mathbf{x}-\frac{\epsilon\mathbf{y}}{2}\Big)\overline{\psi}\Big(\mathbf{x}+\frac{\epsilon\mathbf{y}}{2}\Big)d\mathbf{y}. \tag{2.6}$$

The moments of the Wigner function $W$ give the physical observables, such as

$$\text{position density} \quad |\psi|^2 = \int W[\psi,\psi]\,d\mathbf{p}, \tag{2.7}$$

$$\text{current} \quad \epsilon\,\mathrm{Im}(\overline{\psi}\nabla\psi) = \int \mathbf{p}W[\psi,\psi]\,d\mathbf{p}, \tag{2.8}$$

$$\text{energy} \quad -\frac{\epsilon^2}{2}\mathrm{Re}(\overline{\psi}\Delta\psi) + \frac{\epsilon^2}{2}|\nabla\psi|^2 = \int |\mathbf{p}|^2 W[\psi,\psi]\,d\mathbf{p}\,, \tag{2.9}$$

where $\overline{\psi}$ is the complex conjugate of $\psi$. For $\psi^\epsilon$ satisfying (2.1)–(2.2), and a *smooth potential* $V(\mathbf{x})$, $W^\epsilon = W[\psi^\epsilon,\psi^\epsilon]$ can be shown to converge weakly towards a measure-valued solution of the Liouville or Vlasov equation in classical mechanics [21,42]:

$$\partial_t w + \mathbf{p}\cdot\nabla_{\mathbf{x}} w - \nabla V \cdot \nabla_{\mathbf{p}} w = 0\,, \tag{2.10}$$

$$w(\mathbf{x},\mathbf{p},0) = |A_0(x)|\delta(\mathbf{p}-\nabla S_0(\mathbf{x}))\,. \tag{2.11}$$

The Liouville equation (2.10) naturally unfolds the caustics, and is the correct semiclassical limit *globally in time.* If one uses the ansatz

$$w(\mathbf{x},\mathbf{p},t) = \rho(\mathbf{x},t)\delta(\mathbf{p}-\mathbf{u}(\mathbf{x},t)) \tag{2.12}$$

in (2.10) and takes the first two moments, one obtains the *pressureless gas* equations

$$\rho_t + \nabla \cdot \rho \mathbf{u} = 0 \,, \tag{2.13}$$

$$\mathbf{u}_t + \mathbf{u} \cdot \nabla \mathbf{u} + \nabla V = 0 \,, \tag{2.14}$$

which are equivalent to (2.4)–(2.5) with $\mathbf{u} = \nabla S$ for smooth solutions. The ansatz (2.12) is no longer good after the formation of caustics. In fact, the correct solution is multivalued, which is a superposition of the (smooth) solution to (2.13)–(2.14), in the physical space (see [26, 53]).

The initial value problem (2.10)–(2.11) is the starting point of the numerical methods to be described below. Most of recent computational methods are derived from, or related to, this equation. The main advantage here is that (2.10)–(2.11) filters out the $O(\epsilon)$ oscillations, thus allows a numerical mesh size *independent of* $\epsilon$. However, there are several major difficulties in its numerical approximation:

- High dimensionality. The Liouville equation is defined in the phase space, thus the memory requirement exceeds the current computational capability.
- Singularity. The initial data (2.11) is a delta function. The solution at later time remains a delta function (for single valued solution) or a sum of delta functions (for multi-valued solutions) beyond caustics [26, 53], which is poorly resolved numerically.
- Potential barrier. If $V(\mathbf{x})$ is discontinuous, corresponding to a potential barrier, there are subtle analytical and numerical issues with respect to (2.10) since it is a linear hyperbolic equation with a *measure-valued* coefficient $\nabla V$.

In the past few years, several new numerical methods have been introduced to overcome these difficulties. Below we will review the moment methods, the level set methods and methods for discontinuous potentials.

# 3 The moment method

A classical approach in kinetic theory to reduce the dimension of the Boltzmann equation is to use moment closure. This can be done using a local Maxwellian, which yields the compressible Euler equations defined in the physical space, or some other *ad-hoc* density distributions [25, 40, 48] which yield higher order moment equations. For multivalued solution of (2.4)–(2.5) or (2.13)–(2.14), with $N < \infty$ phases, as shown in [26, 53], the semiclassical limit of (2.1)–(2.2), away from the caustics, takes the form

$$w(\mathbf{x},\mathbf{p},t)=\sum_{k=1}^{N}\rho_k(\mathbf{x},t)\delta(\mathbf{p}-\mathbf{u}_k(\mathbf{x},t))\,, \tag{3.1}$$

where each $(\rho_k, \mathbf{u}_k)$ satisfies the pressureless gas equations (2.13)–(2.14). Using distribution (3.1) one can close the Liouville equation (2.10) in the *physical* space, resulting a system of $(d+1)N$ weakly hyperbolic equations for a $d$-dimensional problem [26].

For example, in one space dimension, define the moment variables

$$m_l(x,t)=\int p^l w(x,p,t)\,dp\,,\quad l=0,1,\cdots,2N. \tag{3.2}$$

Multiplying the Liouville equation (2.10) by $p^l$ (for $l=0,1,\cdots,2N-1$), and integrating over $p$, one gets the following moment system

$$\partial_t m_0+\partial_x m_1=0, \tag{3.3}$$
$$\partial_t m_1+\partial_x m_2=-m_0\partial_x V, \tag{3.4}$$
$$\cdots\cdots\cdots \tag{3.5}$$
$$\partial_t m_{2N-1}+\partial_x m_{2N}=-(2N-1)m_{2N-2}\partial_x V. \tag{3.6}$$

With the special distribution function (3.1), one can express the last moment $m_{2N}$ as a function of the first $2N$ moments:

$$m_{2N}=g(m_0,m_1,\cdots,m_{2N-1}). \tag{3.7}$$

Thus the above moment system is closed. Moreover, it was shown in [26] that this $2N\times 2N$-system is *weakly* hyperbolic, in the sense that the Jacobian matrix of the flux is a Jordan Block, with only $N$-distinct eigenvalues $u_1,u_2,\cdots,u_N$.

By solving the moment system numerically, one produces the multi-valued solution to (2.13)–(2.14). In [26] explicit flux function $g$ in (3.7) was given for $N\leqslant 5$. For larger $N$ a numerical procedure was proposed in [26] for evaluating $g$.

The moment method for multivalued solution of Burger's equation was first introduced by Brenier and Corrias [6, 7], and used computationally by Engquist and Ronborg [17] and Gosse [22] for multivalued solutions in geometrical optics, which is the high frequency limit of the wave equation (2.3).

Since the moment system is weakly hyperbolic, with phase jumps which are *undercompressive shocks* [23], standard shock capturing schemes such as the Lax-Friedrichs scheme and the Godunov scheme face severe numerical difficulties as in pressureless gas equations [8, 17]. Following our work for pressureless gas system [8], a kinetic scheme derived from the Liouville equation (2.10), with the closure (3.1), was used in [26] for this moment system, which outperforms both the Lax-Friedrichs and Godunov schemes.

The multivalued solution also arises in the high frequency approximation of nonlinear waves, for example, in the modeling of electron transport in vacuum electronic devices [24]. There the underlying equations are the Euler-Poisson equations, which is a coupled nonlinear hyperbolic-elliptic system. A similar moment method was introduced in [41] which uses the moment closure ansatz (3.7) for the Vlasov-Poisson system. See also [57]. The validity of the semiclassical limit from the Schrödinger-Poisson system to the Vlasov-Poisson system remains a theoretical challenge, although it was studied numerically [29].

The moment systems lead to an Eulerian method defined in the physical space, thus offers greater efficiency compared with the computation in the phase space. However, when the number of phases becomes very large, or in high space dimensions, the moment systems become very complex. It is also hard to estimate, *a priori*, the total number of phases in high space dimension, which is needed to construct the moment equations. Moreover, the caustics for the moment system are *undercompressive shocks* [23], which are difficult to analyze and hard to compute accurately. These provide very interesting yet challenging numerical problems for the future.

# 4 The level set methods

One of the recently introduced numerical methods for multivalued solution in the high frequency limit is the level set method. This method is rather general, applicable to the computation of multivalued solutions of any (scalar) multi-dimensional quasilinear hyperbolic equations and Hamilton-Jacobi equations.

We now review the level set method, following the derivation of [34]. See also [10]. The original mathematical formulation is classical, see for example [12].

Let $u(\mathbf{x}, t) \in \Re$ be a scalar satisfying an initial value problem of an $n$-dimensional first order hyperbolic PDE with source term:

$$\partial_t u + \mathbf{F}(u) \cdot \nabla_{\mathbf{x}} u + q(\mathbf{x}, u) = 0 \,, \tag{4.1}$$

$$u(\mathbf{x}, 0) = u_0(\mathbf{x}) \,. \tag{4.2}$$

Here $\mathbf{F}(u) : \Re \to \Re^n$ is a vector, and $q : \Re^{n+1} \to \Re$, $B$ is the source term. Introduce a level set function $\phi(\mathbf{x}, p, t)$ in $n+1$ dimension, whose zero level set is the solution $u$:

$$\phi(\mathbf{x}, p, t) = 0 \qquad \text{at} \quad p = u(\mathbf{x}, t) \,. \tag{4.3}$$

Therefore one evolves the entire solution $u$ as the zero level set of $\phi$. Simple calculation gives

$$\partial_t \phi + \mathbf{F}(p) \cdot \nabla_{\mathbf{x}} \phi - q(\mathbf{x}, p)\, \partial_p \phi = 0\,. \tag{4.4}$$

This is the level set equation. It resembles a Liouville equation, which is linear hyperbolic with variable coefficients, with the solution governed by the characteristics, even beyond the singularity of $u$. By solving this linear transport equation, and then finding the zero level set of $\phi$, we generate the multivalued solution to $u$.

For smooth initial data $u_0(x)$, the initial condition for $\phi$ can be chosen simply as

$$\phi(\mathbf{x}, p, 0) = p - u_0(\mathbf{x})\,. \tag{4.5}$$

However, if the initial data are discontinuous, such as in the Riemann problem, such a choice of the initial level set will miss the line that connects the two constant states, thus forming a vacuum. In this case, a good choice for the initial level set function is the *signed distance function* [54].

Similar idea can also be applied to Hamilton-Jacobi equations. Consider the time dependent, $n$-dimensional Hamilton-Jacobi equation

$$\partial_t S + H(\mathbf{x}, \nabla_{\mathbf{x}} S) = 0\,, \tag{4.6}$$

$$S(0, \mathbf{x}) = S_0(\mathbf{x})\,. \tag{4.7}$$

Introduce $\mathbf{u} = (u_1, \cdots, u_n) = \nabla_{\mathbf{x}} S$. Taking the gradient on (4.6), one gets an equivalent (at least for smooth solutions) form of the Hamilton-Jacobi equation

$$\partial_t \mathbf{u} + \nabla_{\mathbf{x}} H(\mathbf{x}, \mathbf{u}) = 0\,, \tag{4.8}$$

$$\mathbf{u}(\mathbf{x}, 0) \equiv \mathbf{u}_0(\mathbf{x}) = \nabla_{\mathbf{x}} S_0(\mathbf{x})\,. \tag{4.9}$$

We use $n$ level set functions $\phi_i = \phi_i(\mathbf{x}, \mathbf{p}, t)$, $i = 1, \cdots, d$, where $\mathbf{p} = (p_1, \cdots, p_d) \in \Re^d$, such that the intersection of their zero level sets yields $\mathbf{u}$, namely,

$$\phi_i(\mathbf{x}, \mathbf{p}, t) = 0 \qquad \text{at} \quad \mathbf{p} = \mathbf{u}(\mathbf{x}, t)\,, \qquad i = 1, \cdots, n\,. \tag{4.10}$$

Then $\phi_i$ solves the following initial value problem of the Liouville equation for Hamiltonian $H(\mathbf{x}, \mathbf{p})$:

$$\partial_t \phi_i + \nabla_{\mathbf{p}} H \cdot \nabla_{\mathbf{x}} \phi_i - \nabla_{\mathbf{x}} H \cdot \nabla_{\mathbf{p}} \phi_i = 0\,, \quad i = 1, \cdots, n, \tag{4.11}$$

$$\phi_i(\mathbf{x}, \mathbf{p}, 0) = p_i - u_i(\mathbf{x}, 0). \tag{4.12}$$

It is the Liouville equation. When $H = \frac{1}{2}|\mathbf{p}|^2 + V(\mathbf{x})$, it corresponds to the semiclassical limit (2.10) of the linear Schrödinger equation (2.1), while for geometrical optics limit of the wave equation (2.3) $H = c(\mathbf{x})|\mathbf{p}|$. The intersection of the zero level sets of all $\phi_i$ give the multivalued solution of $\mathbf{u}$.

While the eiconal (Hamilton-Jacobi) equation gives the multivalued velocity $\mathbf{u}$, it is desirable to also compute multivalued density, energy, etc. A simple idea was introduced in [30, 31]. This method is equivalent to a decomposition of the measure-valued initial data (2.11), namely, we solve $\phi(\mathbf{x}, \mathbf{p}, t)$ satisfying the Liouville equation (2.10) with initial data

$$\phi(\mathbf{x}, \mathbf{p}, 0) = \rho_0(\mathbf{x}) \tag{4.13}$$

and $\psi_i(\mathbf{x}, \mathbf{p}, t) \in \Re^n \quad (i = 1, \cdots, n)$ satisfying the same Liouville equation, with initial data (4.12). A simple mathematical argument shows that the solution to (2.10)–(2.11) is simply

$$w(\mathbf{x}, \mathbf{p}, t) = \phi(\mathbf{x}, \mathbf{p}, t)\Pi_{i=1}^{n}\delta(\psi_i(\mathbf{x}, \mathbf{p}, t)), \tag{4.14}$$

while the moments can be recovered through

$$\rho(\mathbf{x}, t) = \int \phi(\mathbf{x}, \mathbf{p}, t)\Pi_{i=1}^{n}\delta(\psi_i(\mathbf{x}, \mathbf{p}, t))d\mathbf{p}, \tag{4.15}$$

$$\mathbf{u}(\mathbf{x}, t) = \frac{1}{\rho(\mathbf{x}, t)} \int \mathbf{p}\phi(\mathbf{x}, \mathbf{p}, t)\Pi_{i=1}^{d}\delta(\psi_i(\mathbf{x}, \mathbf{p}, t))d\mathbf{p}. \tag{4.16}$$

Thus the only time we have to deal with the delta-function is at the output, while during the evolution we solve $\phi$ and $\psi_i$ which are $l^\infty$ functions! This avoids the singularity problem mentioned earlier, and gives numerical methods with much better resolution than the one based directly on (2.10)–(2.11) by approximating the initial delta-function numerically and then marching on time. This idea has been successfully applied to the semiclassical limit of Schrödinger equation [30], and to general linear symmetric hyperbolic systems (including the geometrical optics) in [31].

Another advantage of this level set approach is that we only need to care about the zero level sets, thus the technique of local level set methods [1, 11, 50], which restricts the computation to a narrow band around the zero level set, can be used to reduce the computational cost to $O(N \ln N)$ for $N$ computational points in the physical space. This is an nice alternative for dimension reduction of the Liouville equation.

The Liouville-based methods were also proposed earlier but for the computation of only the *wave fronts*, see [19, 20, 49]. Here it was shown that it can actually be used to construct the *entire solution.* When solution with many initial data need to be computed, fast algorithms can be used, see [20, 60].

So far the level set methods have not formulated for nonlinear hyperbolic systems (not the type of (4.8) which is the gradient of the Hamilton-Jacobi equations), except for 1-d Euler-Poisson equations [44] where a three dimensional Liouville equation has to be used for a 1-d calculation of multivalued solutions.

For a recent review on these level set methods see also [43].

# 5 Computation of high frequency waves through potential barriers or interfaces

When the medium is heterogeneous, the potential $V$ or the local wave speed $c$ could be *discontinuous*, creating a sharp potential barrier or interface where waves can be partially reflected and partially transmitted as in Snell-Descartes' Law of Refraction. This gives rise to new numerical challenges not faced in the smooth potential case. Clearly, the semiclassical limit (2.10)–(2.11) does not hold at the barrier. Analytical study of the semiclassical limit with interface was carried out in [3, 47]. When $V$ or $c$ is discontinuous, the Liouville equation (4.11) contains characteristics that are discontinuous and even measure-valued. Its bicharacteristics, given by the Hamiltonian system:

$$\partial_t \mathbf{x} = \nabla_{\mathbf{p}} H\,, \tag{5.1}$$

$$\partial_t \mathbf{p} = -\nabla_{\mathbf{x}} H \tag{5.2}$$

is a system of ODEs with the right hand side that are not Lipschitz (for which the classical well-posedness theory was established). It does not even have a bounded variation, for which the renormalized solution was introduced by DiPerna and Lions [14] (see also [2]).

## 5.1 Notion of the solution

One first needs to introduce a notion of solution to such singular Liouville equation (2.10) and the underlying singular Hamiltonian system (5.1)–(5.2). One can then design robust numerical methods for such problems that capture such solutions. The solution so constructed will be physically relevant, namely, it should give the correct transmission and reflection of waves through the barrier, obeying Snell's Law of Refraction.

In [37], we provide an interface condition to connect the Liouville equations at both sides of the interface. Let us concentrate in one space dimension. Consider a particle moving with velocity $p > 0$ to the barrier. The interface condition is

$$w(x^+, p^+) = \alpha_T w(x^-, p^-) + \alpha_R w(x^+, -p^+). \tag{5.3}$$

Here the superscripts "$\pm$" represent the right and left limits of the quantities, $\alpha_T \in [0, 1]$ and $\alpha_R \in [0, 1]$ are the transmission and reflection coefficients respectively, satisfying $\alpha_R + \alpha_T = 1$. $x^+ = x^-$ (for a sharp interface), while $p^+$ and $p^-$ are connected by the *Hamiltonian preserving* condition:

$$H(x^+, p^+) = H(x^-, p^-). \tag{5.4}$$

We remark that in classical mechanics, the Hamiltonian $H = \frac{1}{2}p^2 + V(x)$ is conserved along the particle trajectory, even across the barrier. In this case, $\alpha_T, \alpha_R = 0$ or $1$, namely, a particle can be either transmitted or reflected.

In geometric optics, condition (5.4) is equivalent to Snell's Law of Refraction for a flat interface [36]. The coefficients $\alpha_T$ and $\alpha_R$ are between 0 and 1, namely, waves can be partially transmitted or reflected. They can be determined from the original wave equation (2.3) before the geometric optics limit is taken. Thus (5.3) is a *multiscale* coupling between the (more macroscopic) Liouville equation and the (microscopic) wave equation.

The well-posedness of the initial value problem to the singular Liouville equation with the interface condition (5.3) was established in [37], using the method of characteristics. To determine a solution at $(x, p, t)$ one traces back along the characteristics determined by the Hamiltonian system (5.1)–(5.2) until hitting the interface. At the interface, the solution bifurcates with the interface condition (5.3), one corresponds to the transmission and the other reflection, and this process continues until one arrives the line of $t = 0$. The interface condition (5.3) thus provides a generalized characteristic method.

We will also introduce a notion of the solution to the Hamiltonian system (5.1)–(5.2), using a *probability* interpretation. Basically, one solves the system using a standard ODE or Hamiltonian solver, but at the interface, we introduce the following *Monte-Carlo* solution (we give the solution in the case of $p^- > 0$; the other case is similar):

- With probability $\alpha_R$, the particle (wave) is reflected with

$$x \to x, \quad p^- \to -p^-. \tag{5.5}$$

- With probability $\alpha_T$, the particle (wave) is transmitted, with

$$x \to x, \quad p^+ \text{ is obtained from } p^- \text{ using (5.4).} \tag{5.6}$$

Although the original problem is deterministic, this probability solution allows us to go beyond the interface with the new value of $(x, p)$ defined in (5.5)–(5.6). This is clearly the Lagrangian picture of the Eulerian solution determined by using the interface condition (5.3). This solution also motivates a (Monte-Carlo) particle method for thin quantum barriers, see [33].

## 5.2 Numerical flux at the interface

While the Liouville equation (4.11) can be solved by a standard finite difference or finite volume shock capturing methods, such schemes face difficulties when the Hamiltonian is discontinuous, since ignoring the discontinuity of the Hamiltonian during the computation will result in solutions inconsistent with the notion of the (physically relevant) solution defined in the preceding subsection. Even with a smoothed Hamiltonian, it is usually impossible—at least in the case of *partial* transmission and reflection–to obtain transmission and reflection with the correct transmission and reflection coefficients. A smoothed Hamiltonian will also give a severe time step constraint like $\Delta t \sim O(\Delta x \Delta p)$, where $\Delta t, \Delta x$ and $\delta p$ are time step, mesh sizes in the $x$- and $p$-directions respectively. This is a parabolic type CFL condition, despite that we are solving a hyperbolic problem!

Our idea of approximating the Liouville equation (4.11) at the interface in [35,37] is to *build the interface condition (5.3) into the numerical flux*. This is in the spirit of the Immersed interface method [39,46]. It was also motived by an idea of Perthame and Simeoni for a well-balanced kinetic scheme for shallow water equations with bottom topography [51]. Our new numerical schemes overcome the aforementioned analytic and numerical difficulties. In particular, they have the following important properties:

- They produce the solution crossing the interface defined by the mathematical solution introduced in the previous subsection, thus obtain physically relevant solution of particle/wave transmission and reflection at the interfaces. In particular, in the case of geometric optics, this solution is consistent to Snell-Descartes' Law of Refraction at the interface. The Snell's Law was built into the numerical flux!
- It allows a *hyperbolic* CFL condition $\Delta t = O(\Delta x, \Delta p)$.

This idea has been applied successfully to compute the semiclassical limit of the linear Schrödinger equation with potential barriers [35] and the geometrical optics with complete transmission/reflection [36] or partial transmission/reflection [37]. Positivity, and both $l^1$ and $l^\infty$ stabilities were also established, under the "good" (hyperbolic) CFL condition. For piecewise constant Hamiltonians, an $l^1$-error estimate of the first order finite difference of the type introduced in [35] was established in [56], following [55].

This is the *first* Eulerian numerical methods for high frequency waves that are able to capture correctly the transmission and reflection of waves through the barriers or interfaces. It has also been extended to high frequency elastic waves [27], and high frequency waves in random media [28] with diffusive interfaces.

## 5.3 Thin quantum barriers

A correct modeling of electron transport in nanostructures, such as resonant tunneling diodes, superlattices or quantum dots, require the treatment of quantum phenomena in localized regions of the devices, while the rest of the device is governed by classical mechanics. The quantum barrier that separates the quantum and classical regions differ from a classical barrier, in that a quantum wave can transmit through any barrier, a phenomenon known as tunneling. While solving the Schrödinger equation in the entire physical domain is too expensive, it is rather attractive to use a *multiscale* approach, namely, solve the quantum mechanics in the quantum well, and classical mechanics outside the well [5]. It is highly desirable to have a semiclassical computational model for quantum barriers, with a cost slightly higher than a classical approach, but much less than a quantum approach. In [32], we introduced the following semiclassical model:

- solve the time-independent Schrödinger equation—either analytically if possible, or numerically—for the local barrier/well to determine the scattering data (transmission and reflection coefficients)
- solve the classical Liouville equation elsewhere, using the scattering data at the barrier for the interface condition (5.3) and the numerical method of [35] for a classical barrier.

Our 1d [32] and 2d [33] results indicate the success of this approach when the well is very thin (a few $\epsilon$'s) and well-separated. It can correctly capture both transmitted and reflected waves that a classical Liouville equation cannot, and the results agree (in the sense of weak convergence) with the solution obtained by solving directly the Schrödinger equation with small $\epsilon$ with a much less cost.

Currently, more study is underway, in particular, for highly resonant wells, time delay, phase information, and higher dimensional problems.

# References

[1] D. Adelsteinsson and J. Sethian. A fast level set method for propagating interfaces. *J. Comp. Phys.* **118**, 269-277, 1995.

[2] L. Ambrosio. Transport equation and Cauchy problem for BV vector fields. *Invent. math.* **158**, 227-260, 2004.

[3] G. Bal, J.B. Keller, G. Papanicolaou and L. Ryzhik. Transport theory for acoustic waves with reflection and transmission at interfaces. *Wave Motion* **30**, 303-327, 1999.

[4] W. Bao, S. Jin and P. Markowich. On Time-Splitting Spectral Approximations for the Schrodinger Equation in the Semiclassical Regime. *J. Comp. Phys.* **175**, 487-524, 2002.

[5] N. Ben Abdalla, I. Gamba and P. Degond. Coupling one-dimensional time-dependent classical and quantum transport models. *J. Math. Phys.* **43**, 1-24, 2002.

[6] Y. Brenier and L. Corrias. A kinetic formulation for multibranch entropy solutions of scalar conservation laws. *Ann. Inst. Henry Poincaré* **15**, 169-190, 1998.

[7] Y. Brenier and L. Corrias. *Capturing multi-valued solutions*, preprint.

[8] F. Bouchut, S. Jin and X.T. Li. Numerical approximations of pressureless and isothermal gas dynamics. *SIAM J. Num. Anal.* **41**, 135-158, 2003.

[9] M. Crandall and P.L. Lions. Viscosity solutions of Hamilton-Jacobi equations. *Trans. AMS* **277**, 1-42, 1983.

[10] L.-T. Cheng, H.L. Liu and S. Osher. Computational high frequency wave propagation using the level set method with applications to the semiclassical limit of the Schrödinger equation. *Comm. Math. Sci.* 1(3), 593-612, 2003.

[11] D. Chopp. Computing minimal surfaces via level set curvature flow. *J. Comp. Phys.* **106**, 77-91, 1993.

[12] R. Courant and D. Hilbert. *Methods of Mathematical Physics.* vol. 2, New York, Interscience Publishers, 1953-1962.

[13] M. Crandall, P.L. Lions. Viscosity solutions of Hamilton-Jacobi equations. *Trans. Amer. Math. Soc.* **282**, 487-502, 1984.

[14] R.J. DiPerna and P.-L. Lions. Ordinary differential equations, transport theory and Sobolev spaces. *Invent. Math.* 98(3), 511-547, 1989.

[15] J.J. Duistermaat. *Fourier Integral Operators.* Birkhäuser, 1995.

[16] B. Engquist, E. Fatemi and S. Osher. Numerical solution of the high frequency asymptotic expansion for the scalar wave equation. *J. Comput. Phys.* 120(1): 145-155, 1995.

[17] B. Engquist and O. Runborg. Multi-Phase computations in geometrical optics. *J. Comp. Appl. Math.* **74**, 175-192, 1996.

[18] B. Engquist and O. Runborg. Computational high frequency wave propagation. *Acta Numerica* **12**, 181-266, 2003.

[19] B. Engquist, O. Runborg and A.-K. Tornberg. High frequency wave propagation by the segment projection methods. *J. Comp. Phys.* **178**, 373-390, 2002.

[20] S. Fomel and J.A. Sethian. Fast-phase space computation of multiple arrivals. *Proc. Nat. Acad. Sci.* **99**, 7329-7334, 2002.

[21] P. Gerard, P.A. Markowich, N.J. Mauser and F. Poupaud. Homogenization limits and Wigner transforms. *Comm. Pure Appl. Math.* **50**, 323-379, 1997.

[22] L. Gosse. Using K-branch entropy solutions for multivalued geometric optics computation. *J. Comp. Phys.* **180**, 155-182, 2002.

[23] L. Gosse, S. Jin and X.T. Li. On two moment systems for computing multiphase semiclassical limits of the Schrodinger equation. *Math. Model Methods Appl. Sci.* **13**, 1689-1723, 2003.

[24] V.L. Granatstein, R.K. Parker and C.M. Armstrong. Proc. IEEE 87, 702, 1999.

[25] H. Grad. *Comm. Pure Appl. Math.* **2**, 331-407, 1949.

[26] S. Jin and X.T. Li. Multi-phase computations of the semiclassical limit of the schrodinger equation and related problems: Whitham vs. Wigner. *Physica D* **182**, 46-85, 2003.

[27] S. Jin and X. Liao. A Hamiltonian-preserving scheme for high frequency elastic waves in heterogeneous media. *J. Hyperbolic Diff. Eqn.* 3(4), 741-777, 2006.

[28] S. Jin, X. Liao and X. Yang. Computation of interface reflection and regular or diffuse transmission of the planar symmetric radiative transfer equation with isotropic scattering and its diffusion limit. *SIAM J. Sci. Comp.*, to appear.

[29] S. Jin, X. Liao and X. Yang. The Vlasov-Poisson equations as the semiclassical Limit of the Schrodinger-Poisson Equations: a numerical study. *J. Hyperbolic Diff. Eqn.*, to appear.

[30] S. Jin, H.L. Liu, S. Osher and R. Tsai. Computing multivalued physical observables for the semiclassical limit of the Schrodinger equations. *J. Comp. Phys.* **205**, 222-241, 2005.

[31] S. Jin, H.L. Liu, S. Osher and R. Tsai. Computing multi-valued physical observables for high frequency limit of symmetric hyperbolic systems. *J. Comp. Phys.* **210**, 497-518, 2005.

[32] S. Jin and K. Novak. A semiclassical transport model for thin quantum barriers. *Multiscale Modeling and Simulation* **5**, 1063-1086, 2006.

[33] S. Jin and K. Novak. A semiclassical transport model for two-dimensional thin quantum barriers. *J. Comp. Phys.* **226**, 1623-1644, 2007.

[34] S. Jin and S. Osher. A level set method for the computation of multivalued solutions to quasi-linear hyperbolic PDEs and Hamilton-Jacobi equations. *Comm. Math. Sci.* 1(3), 575-591, 2003.

[35] S. Jin and X. Wen. Hamiltonian-preserving schemes for the Liouville equation with discontinuous potentials. *Comm. Math. Sci.* **3**, 285-315, 2005.

[36] S. Jin and X. Wen. Hamiltonian-preserving schemes for the Liouville equation of geometrical optics with discontinuous local wave speeds. *J. Comp. Phys.* **214**, 672-697, 2006.

[37] S. Jin and X. Wen. Hamiltonian-preserving schemes for the Liouville equation of geometrical optics with partial transmissions and reflections. *SIAM J. Num. Anal.* **44**, 1801-1828, 2006.

[38] S. Jin and D. Yin. Computational high frequency waves through curved interfaces via the Loiuville equation and Geometric Theory of Diffraction. *J. Comp. Phys.*, submitted.

[39] R.J. LeVeque and Z.L. Li. Immersed interface methods for Stokes flow with elastic boundaries. *SIAM J. Sci. Comp.* **18**, 709-735, 1997.

[40] C.D. Levermore. Moment closure hierarchies for kinetic theories. *J. Stat. Phys.* **83**, 1021-1065, 1996.

[41] X.T. Li, J.G. Wohlbier, S. Jin and J.H. Booske. An Eulerian method for computing multi-valued solutions of the Euler-Poisson equations and applications to wave breaking in klystrons. *Phys Rev E.* **70**, 016502, 2004.

[42] P.L. Lions and T. Paul. Sur les measures de Wigner. *Revista. Mat. Iberoamericana* **9**, 553-618, 1993.

[43] H. Liu, S. Osher and R. Tsai. Multi-valued solution and level set methods in computational high frequency wave propagation. *Commun. Comput. Phys.* **1**, 765-804, 2006.

[44] H. Liu and Z.M. Wang. A field space-based level set method for computing multi-valued solutions to 1D Euler-Poisson equations. *J. Comp. Phys.*, to appear.

[45] P.A. Markowich, P. Pietra and C. Pohl. Numerical approximation of quadratic observables of Schrödinger-type equations in the semi-classical limit. *Numer. Math.* **81**, 595-630, 1999.

[46] A. Mayo. The fast solution of Poisson'sand the biharmonic equations on irregular regions. *SIAM J. Sci. Comp.* **21**, 285-299, 1984.

[47] L. Miller. Refraction of high frequency waves density by sharp interfaces and semiclassical measures at the boundary. *J. Math. Pures Appl.* IX(79), 227-269, 2000.

[48] I. Müller and T. Ruggeri. *Rational Extended Thermodynamics*, 2nd ed., Springer, 1998.

[49] S. Osher, L.T. Cheng, M. Kang, H. Shim and Y.H. Tsai. Geometric optics in a phase space based level set and Eulerian framework. *J. Comp. Phys.* **79**, 622-648, 2002.

[50] D. Peng, B. Merriman, S. Osher, H.K. Zhao and M. Kang. A PDE-based fast local level set method. *J. Comp. Phys.* **155**, 410-438, 1999.

[51] B. Perthame and C. Simeoni. A kinetic scheme for the Saint-Venant system with a source term. *Calcolo* **38**, 201-231, 2001.

[52] O. Runborg. Mathematical models and numerical methods for high frequency waves. *Commun. Comput. Phys.* **2**, 827-880, 2007.

[53] C. Sparber, M. Markowich, N. Mauser. Wigner functions versus WKB-methods in multivalued geometrical optics. *Asymptot. Anal.* **33**, 153-187, 2003.

[54] Y.-H. R. Tsai, Y. Giga and S. Osher. A level set approach for computing discontinuous solutions of Hamilton-Jacobi equations. *Math. Comp.* **72**, 159-181, 2003.

[55] X. Wen and S. Jin. Convergence of an immersed interface upwind scheme for linear advection equations with piecewise constant coefficients I: $l^1$-error estimates. *J. Comp. Math.*, to appear.

[56] X. Wen and S. Jin. The $l^1$-error estimates for a Hamiltonian-preserving scheme to the Liouville equation with piecewise constant coefficients. *SIAM J. Num. Anal.*, submitted.

[57] J.G. Wohlbier, S. Jin and S. Sengele. Eulerian calculations of electron overtaking and multi-valued solutions in a traveling wave tube. *Physics of Plasmas* **12**, 023106, 2005.

[58] G.B. Whitham. *Linear and Nonlinear Waves*, Wiley, New York, 1974.

[59] E. Wigner. On the quantum correction for thermodynamic equilibrium. *Phys. Rev.* **40**, 749-759, 1932.

[60] L. Ying and E.J. Candés. The phase flow method. *J. Comput. Phys.* **220**, 184-215, 2006.

# Some Recent Results on Ranking Webpages and Websites

Ying Bao[1,2], Zhiming Ma[1], Yanhong Shang[3]
[1] *Academy of Mathematics and Systems Science*
*Chinese Academy of Sciences, Beijing 100080, China*
[2] *Graduate University of the Chinese Academy of Sciences*
[3] *Department of Mathematics*
*Beijing Jiaotong University, Beijing 100044, China*
*Email: ybao@amss.ac.cn, mazm@amt.ac.cn, 05121740@bjtu.edu.cn*

**Abstract**

In this paper we briefly review some of our recent results on the research of the design and analysis of search engine algorithms. The contents include: the limiting behavior of PageRank when the damping factor tends to 1; comparison of the convergence rate of maximal and minimal irreducible Markov chains on the Internet; a new proposal of $N$-step PageRank algorithm; a new proposal of ranking Websites-AggregateRank algorithm.

As is well known that in recent years Web search engines have been more and more important in modern science and technology, and more and more popular in civil daily life. The design of Web search engines has been becoming a focus of the research on the Web search and mining. One popular aspect is to calculate Static Rank by exploiting the hyperlink structure of the Web. Researchers have made great progress on link analysis models and algorithms since 1998, such as HITS and PageRank ([9,17]). In nowadays, PageRank has emerged a popular link analysis model, mostly due to its query-independence, using only the web graph structure, and Google's huge business success.

We are grateful to the colleagues in MSRA (Microsoft Research Asia) who brought our attention to the research area of the design and analysis of search engine algorithms. The story began in October 2004, when a colleague of MSRA presented a talk at our regular RGCN (Random Graph and Complex Networks) seminar about the subject of PageRank. Since then we have had frequent discussions with the colleagues of MSRA, which stimulated our research in this direction and yielded some joint research work. The present paper is a brief review of some recent results on ranking Webpages and Websites made by the RGCN group in AMSS, of which some results are collaborated with MSRA.

# 1 Ranking Webpages

In this section, we present some results about ranking Webpages. For the convenience of the reader we recall first the basic PageRank model in Subsection 1.1. Then in Subsection 1.2 we discuss the limiting behavior of PageRank when the damping factor tends to 1. In Subsection 1.3 we compare the convergence rate of maximal and minimal irreducible Markov chains on the Internet. In Subsection 1.4 we briefly introduce our recent result of $N$-step PageRank algorithm.

## 1.1 Basic PageRank model

Consider the hyperlink structure of the webpages on a network as a directed graph $G = (V(G), E(G))$. A vertex $i \in V(G)$ of the graph represents a webpage and a directed edge $\overrightarrow{ij} \in E(G)$ represents a hyperlink from the webpage $i$ to $j$. Let $d_i$ be the outdegree of vertex $i$ and $|V(G)|$ be the cardinal number of $V(G)$. Suppose that $|V(G)| = n$. We can construct a matrix $W = [w_{ij}]_{n\times n}$, the normalized adjacent matrix of $G$, by setting $w_{ij} = \begin{cases} \frac{1}{d_i}, & \text{if } \overrightarrow{ij} \in E(G), \\ 0, & \text{otherwise.} \end{cases}$

In the real web there always exists page $i$ which does not link to any other webpage, then $d_i = 0$ and the entries of the $i$th row in $W$ are all 0. Hence the matrix $W$ is not a (conservative) transition matrix (nonnegative matrix and the sum of every row is 1 ). There are two methods to change $W$ into a transition matrix. One is to replace all $(0, 0, \cdots, 0)$ in $W$ by $(\frac{1}{n}, \frac{1}{n}, \cdots, \frac{1}{n})$ to get a new matrix $P$. The other is to add a new vertex $\beta_{(n+1)}$ into the graph and then to construct a corresponding matrix $\widetilde{P} =: \begin{pmatrix} W & R \\ 0 & 1 \end{pmatrix}$, where $W, R, 0, 1$ are $n \times n$, $n \times 1$, $1 \times n$, $1 \times 1$ submatrixes respectively, $R(i) = \begin{cases} 0, & \text{if } d_i \neq 0, \\ 1, & \text{if } d_i = 0. \end{cases}$

Let $\{X_t\}_{t\geqslant 0}$ be a Markov chain associated with the above transition matrix $P$. Then $\{X_t\}$ can be intuitively interpreted as a surfer surfs at the Internet: When $X_t = i$, if $d_i \neq 0$, then he chooses the next page by randomly clicking on one of the outlinks of $i$ ; if $d_i = 0$, then he chooses the next page randomly in the whole Web. The transition matrix $\widetilde{P}$ is intuitively interpreted similarly: if $d_i \neq 0$, then the surfer behaviors in the same manner as above; however, if $d_i = 0$, then the surfer will choose the vertex $\beta_{(n+1)}$ in the $t + 1$ step. With the above interpretations, it is reasonable to think that the average clicking ratio can be interpreted as a measurement of the relative importance of webpages. If the corresponding Markov chain has a unique stationary distribution $\pi = (\pi_1, \pi_2, \cdots, \pi_n)$, then by the ergodic theory we will have:

$$\lim_{m\to\infty} E\Big[\frac{1}{m}\sum_{k=0}^{m-1} I_{\{\text{visiting } i \text{ at } k\text{th step}\}}\Big] = \lim_{m\to\infty}\frac{1}{m}\sum_{k=0}^{m-1} p_{ji}^k := \pi_i, \quad \text{a.s..}$$

Thus the stationary distribution $\pi = (\pi_1, \pi_2, \cdots, \pi_n)$ is a suitable candidate of the ranking of the webpages.

However, the adjacent matrix of the real Web is always very large and is sparse [8], and hence $P$ and $\widetilde{P}$ are most likely reducible, which means that most likely their stationary distributions are not unique. There are several methods to modify $P$ or $\widetilde{P}$ for forcing it irreducible [18], among them the most popular one is the famous PageRank model.

In the PageRank model, the Markov chain is forced irreducible by making every state directly reachable from every other state. This is achieved by adding a perturbation matrix $E = \frac{1}{n}\cdot \mathbf{1}^T\mathbf{1}$ to $P$. Then, the mathematical algorithm of PageRank is formulated as

$$\mathbf{P}(\alpha) = \alpha\mathbf{P} + (1-\alpha)\cdot\frac{1}{n}\cdot\mathbf{1}^T\mathbf{1}, \tag{1}$$

$$\pi(\alpha) = \lim_{l\to\infty}\frac{1}{n}\cdot\mathbf{1}\cdot\mathbf{P}(\alpha)^l, \tag{2}$$

where $\mathbf{1} := (1, 1, \cdots, 1)$ represents a row vector with all entries equal to 1, $0 < \alpha < 1$ is a constant called the damping factor in the literature. By the theory of Markov chains, $\pi(\alpha)$ is the stationary distribution of $\mathbf{P}(\alpha)$, which is called the PageRank vector and is utilized to measure the relative importance of webpages (see e.g. [1, 7, 18, 19]).

The above perturbation has a reasonable intuitive explanation. We may imagine that a surfer on the Internet goes along the hyperlinks with probability $\alpha$, and he may also open a new Webpage randomly with probability $1-\alpha$.

PageRank was originally proposed by Google founders Larry Page and Sergey Brin in 1998 ( [9]). Later the PageRank model adopted a slightly more general perturbation by using a "personalized" distribution $\mu$ replacing the uniform distribution $\frac{1}{n}\cdot\mathbf{1}$ (see (3) in next subsection). In nowadays, PageRank has emerged a popular link analysis model, mostly due to its query-independence, using only the web graph structure, and Google's huge business success.

## 1.2 Limit of PageRank

It was reported in the literature that the most common choice of the damping factor in practical algorithms is taking $\alpha = 0.85$ (cf. [18]).

Then, one would naturally think about the relations between $\alpha$ and $\pi(\alpha)$, e.g. how $\alpha$ affects the rank $\pi(\alpha)$ and affects the convergence rate of (2), what is the limit of $\pi(\alpha)$ as $\alpha \to 1$, and if the limit exists is it better than $\pi(\alpha)$ with $\alpha = 0.85$, etc. In the literature there are various discussions along the above questions (see e.g. [1, 7, 18]). In this subsection we report a recent work of Bao and Liu [1]. They discussed the limit of the PageRank $\pi(\alpha)$ as $\alpha \to 1$. Their result verifies a conjecture proposed by Boldi et al. [7] in the 14th International World Wide Web Conference (2005).

To state the result of [1] let us work on the following PageRank model with personalized distribution $\mu$:

$$\mathbf{P}(\alpha) = \alpha\mathbf{P} + (1-\alpha)\cdot\mathbf{1}^T\cdot\mu, \tag{3}$$

$$\pi(\alpha) = \lim_{l\to\infty}\frac{1}{n}\cdot\mathbf{1}\cdot\mathbf{P}(\alpha)^l, \tag{4}$$

where $\mu = (\mu_1, \cdots, \mu_n)$ is an arbitrary probability vector with $\mu_i > 0$ for all $i$, which means that when a surfer surfs to the next step by opening a new page, he will choose the next page not randomly but according to his personal favor.

In practice $\mathbf{P}$ is most likely aperiodic, so $\mathbf{P}(\alpha)$ is an irreducible and aperiodic Markov transition probability matrix on a finite state space, its invariant probability $\pi(\alpha)$ exists and is unique. By the tightness property, as $\alpha \to 1^-$ (i.e. $\alpha$ tends to 1 from below), the limit point of $\pi(\alpha)$ always exists. But the limit is in general not unique. The theorem below shows that the limit is unique and gives an analytic representation of the limit.

**Theorem 1.** ([1] Theorem 1): *Assume that* $\mathbf{P}$ *is an aperiodic n-dimensional transition matrix,* $\mu = (\mu_1, \mu_2, \cdots, \mu_n)$ *is a probability vector satisfying* $\mu_i > 0, i = 1, \cdots, n$; $\mathbf{P}(\alpha)$, $\pi(\alpha)$ *are defined in (3) and (4), respectively. Then,* $\pi^* = \lim_{\alpha\to 1^-}\pi(\alpha)$ *exists and is unique. Moreover,* $\pi^* = \mu\cdot\mathbf{V}$, *where* $\mathbf{V} = \lim_{l\to\infty}\mathbf{P}^l$.

Note that when $\mu = \frac{1}{n}\cdot\mathbf{1}$, the above result confirms the corresponding conjecture proposed in [7]. The above result shows also that $\pi^*$ is not suitable to be used for ranking Webpages. This is because that if a page $i$ is in a transient state of the Markov chain, then by Theorem 1 we will have $\pi^*_i = (\mu\cdot\mathbf{V})_i = 0$. However, in the real Web structure, more than half pages are transient states, and these transient states are often interesting [8]. So, it is unreasonable to choose $\mu\cdot\mathbf{V}$ or $\pi(\alpha)$ with very large $\alpha$ as the PageRank vector. This conclusion is agreeable with the opinion in [7].

## 1.3 Comparison of different irreducible Markov chains

In the above PageRank model the irreducibility of the Markov chain is obtained by adding direct connections between each vertex, which we will refer to as the maximal irreducible chain. Some authors suspected that this approach might be overkill and proposed alternative approaches to force irreducibility, among them a practical one forces irreducibility in a minimal sense and is refer to as the minimal irreducible chain ([6,18,26]).

The minimal irreducible chain is constructed by adding a new vertex $\nu_{(n+1)}$ to $V(G)$ and revising the matrix accordingly. The revised transition matrix is $A(\alpha) = \begin{pmatrix} \alpha P & (1-\alpha)\mathbf{1}^T \\ \mu & 0 \end{pmatrix}$, where $\alpha, \mathbf{1}, \mu$ are the same as in the above subsection, $\alpha P, (1-\alpha)\mathbf{1}^T, \mu, 0$ are $n \times n, n \times 1, 1 \times n, 1 \times 1$ sub matrixes respectively. The behavior of Markov chain with the transition matrix $A(\alpha)$ can be interpreted as follows: when the current state is $i$, in the next step the surfer will either choose a webpage from pages pointed by $i$ with probability $\alpha$, or choose the new vertex $\nu_{(n+1)}$ with probability $(1-\alpha)$, and then from $\nu_{(n+1)}$ he will choose a webpage from $V(G)$ with distribution $\mu$ in the further next step. All states of the Markov chain determined by $A(\alpha)$ are reachable from each other, so this Markov chain is irreducible, which we will refer to as the minimal irreducible chain. It is then interesting to make comparison between the maximal and minimal irreducible chains ([19]). The paper ([18]) contains some discussion in this aspect. Recently Bao and Zhu ([2, 29, 30]) made some further comparison between different irreducible Markov chains. Their results concerning the comparison of the stationary distributions, the convergence rates, and the Maclaurin series of stationary distributions. Below we present part of the results obtained in [2, 29, 30].

At first, we compare the stationary distributions of the maximal and minimal Markov chains. The stationary distribution of the maximal irreducible Markov chain is

$$\pi(\alpha) = (1-\alpha)\mu \cdot \sum_{k=0}^{\infty} (\alpha P)^k. \tag{5}$$

Denote the stationary distribution of the minimal irreducible Markov chain as $\widehat{\pi}(\alpha) = \big(\widehat{\pi}_n(\alpha), \widehat{\pi}_{(n+1)}(\alpha)\big)$, where $\widehat{\pi}_n(\alpha)$ is the distribution values of the $n$ vertexes of $V(G)$, and $\widehat{\pi}_{(n+1)}(\alpha)$ is the distribution value of the new vertex $\nu_{(n+1)}$ . In [29] (see also [2]) it is calculated that

$$\widehat{\pi}_n(\alpha) = \frac{1-\alpha}{2-\alpha}\mu \cdot \sum_{k=0}^{\infty} (\alpha P)^k. \tag{6}$$

Comparing (5) with (6), we have $\widehat{\pi}_n(\alpha) = \dfrac{1}{2-\alpha}\pi(\alpha)$, which means that $\widehat{\pi}_n(\alpha)$ is identified with $\pi(\alpha)$ after being normalized. Consequently the minimal and maximal Markov chains yield the same ranking of the webpages.

To compare the convergence rates of the above chains, we already know that he convergence rate of the maximal irreducible Markov chain is ([18])

$$\|\pi(\alpha) - \mu \cdot [P(\alpha)]^k\|_\infty < 2\alpha^{k+1}.$$

Therefore we need only to calculate the convergence rate of the minimal irreducible Markov chain, to this end we let

$$A^k(\alpha) := \begin{pmatrix} A^k(1,1) & A^k(1,2) \\ A^k(2,1) & A^k(2,2) \end{pmatrix},$$

where $A^k(1,1)$, $A^k(1,2)$, $A^k(2,1)$, $A^k(2,2)$ are $n \times n, n \times 1, 1 \times n, 1 \times 1$ submatrixes respectively. Then the convergence rate of $A^k(\alpha) \overset{k\to\infty}{\longrightarrow} \mathbf{1}_{n+1}^T \cdot \widehat{\pi}(\alpha)$ can be calculated separately for each sub-matrix's. In [2] it is calculated as follows:

$$\|A^k(2,2) - \widehat{\pi}_{(n+1)}(\alpha)\|_\infty = \frac{(1-\alpha)^{k-1}}{2-\alpha},$$

$$\|A^k(1,2) - \widehat{\pi}_{(n+1)}(\alpha) \cdot \mathbf{1}^T\|_\infty = \frac{(1-\alpha)^k}{2-\alpha},$$

$$\|\mu \cdot A^k(1,1) - \widehat{\pi}_n(\alpha) \cdot \mathbf{1}^T\|_\infty < \left|\frac{2\alpha}{(2\alpha-1)(2-\alpha)}\right|\alpha^k,$$

$$\|A^k(2,1) - \widehat{\pi}_n(\alpha)\|_\infty < \left|\frac{2\alpha}{(2\alpha-1)(2-\alpha)}\right|\alpha^k.$$

From the above results we see that in $\|\cdot\|_\infty$ norm the convergence rate of the two matrixes are almost the same. But if we use the $\|\cdot\|_1$ norm, we may have a slightly finer result as follows:

**Theorem 2.** ([2] Theorem 10): *If* $\alpha > \dfrac{1}{\sqrt{2}}$, *then the convergence rate of the minimal Markov chain has the same order as* $\alpha^k$, *more precisely we have*

$$\lim_{k\to\infty} \frac{\|\mu \cdot A^k(1,1) - \widehat{\pi}_n(\alpha) \cdot \mathbf{1}^T\|_1}{\alpha^k} = \frac{(2\alpha^2-1)}{(2-\alpha)(2\alpha-1)}.$$

By the above discussion we see that when $\alpha > \frac{1}{\sqrt{2}}$ (note that in practice $\alpha = 0.85$), the maximal Markov chain converges faster than the minimal Markov chain.

The maximal and minimal irreducible chains are both based on the transition matrix $P$, in [2] the authors introduced another irreducible Markov chain based on the transition matrix $\widetilde{P}$, which is referred as the middle irreducible Markov chain. The interested reader may refer [2] for the comparison of the three (maximal, minimal and middle) Markov chains.

## 1.4 $N$-step PageRank

Although PageRank is an eminent search engine algorithm, researchers still continuously make effort to improve it or invent new algorithms for seeking the better accuracy and/or speed. In this subsection we present shortly an improved algorithm to boost the search accuracy of the classical PageRank. This new algorithm was proposed very recently in [28] and is named as $N$-step PageRank.

The motivation of the $N$-step PageRank comes from the strategy used in computer chess. The key to the winning of computer "Deep Blue" [15] over human is that it can predict all the situations within much more steps than a human being can do at the same time.

We use this idea into the design of the search engine algorithm. In the classical PageRank algorithm, when the surfer chooses the next webpage, he uses only information of 1-step out-links of the current webpage, and chooses every out-link page with equal probability. That is, PageRank assumes that each out-link has the same importance. In fact, the surfer can estimate the importance of the different links according to the knowledge he has, and the webpage which contains more information will have more opportunity to be chosen. So, we assume that the out-link number of a webpage can represent its information capacity. That is, after $n$ steps the more links there are, the more information the user can get.

According to this principle, we can compute a new transition matrix $P^{(N)} = [p_{ij}^{(N)}]_{n\times n}$ as follows: for two arbitrary vertexes $i$, $j$,

$$p_{ij}^{(N)} = \frac{d_j^{(N-1)} \cdot 1_{\vec{ij}\in E(G)}}{\sum_{\vec{ik}\in E(G)} d_k^{(N-1)}},$$

where $d_j^{(N)}$ is the vertex number after vertex $j$ jump $N$ steps, and $d^{(0)} = (1, 1, \cdots, 1)^T$.

Replace the transition matrix $P$ in the PageRank algorithm by $P^{(N)}$, we obtain the $N$-step PageRank algorithm. It was shown in [28] that both $P^{(N)}$ and the stationary distribution of $P^{(N)}(\alpha)$ can be easily calculated.

Some experiments in comparing the $N$-step PageRank algorithm with the classical PageRank algorithm are reported in [28]. The experiments are based on the dataset of TREC Web track. The results show that the $N$-step PageRank algorithm can boost the search accuracy of PageRank by more than 15% in terms of mean average precision.

# 2 Ranking Websites

In this section we present our recent work about ranking Websites ([3,14]). In Subsection 2.1, we describe the traditional approaches to the Websites-ranking and discuss their weakness. We then present our approach of ranking Websites, the AggregateRank algorithm, in Subsection 2.2.

## 2.1 Traditional approaches of Websites-ranking

In the literature of Website ranking, people used to apply those technologies proposed for ranking Webpages to the ranking of Websites. For example, the PageRank algorithm was used to rank Websites in [13,27]. In order to apply PageRank to the ranking of Websites, they constructed a HostGraph. In the HostGraph, the nodes denote Websites and there is an edge between two nodes if there are hyperlinks from the Webpages in one Website to the Webpages in the other. According to the different definitions of the edge weights, two categories of HostGraphs were used in the literature. In the first category, the weight of an edge between two Websites was defined by the number of hyperlinks between the two sets of Webpages in these sites [5]. In the second category, the weight of any edge was simply set to 1 [12]. For the sake of clarity, we refer to the two categories as weighted HostGraph and naïve HostGraph respectively.

After constructing the HostGraph, the similar random walk was conducted. That is, a random surfer was supposed to jump between Websites following the edges with a probability of $\alpha$, or jump to a random Website with a probability of $1 - \alpha$. In such a way, one can obtain the HostRank, which is the importance measure of Websites.

At first glance, the above random walk model over the HostGraph seems to be a natural extension of the PageRank algorithm. However, in [3,14] we point out that it is actually not as reasonable as PageRank because it is not in accordance with the browsing behaviors of the Web surfers. As we know, real world Web surfers usually have two basic ways to access the Web. One is to type URL in the address edit of the Web browser (using favorite folder can also be considered as a shortcut of typing URL). And the other one is to click any hyperlink in the current loaded Webpage. These two manners can be well described by the parameter $\alpha$ used in PageRank. That is, with a probability of $1-\alpha$, the Web users visit a random Webpage by inputting its URL, and with a probability of $\alpha$, they visit a Webpage by clicking a hyperlink.

Nevertheless, as for the random walk in the HostGraph, we can hardly find the same evident correlation between the random walk model and real world user behaviors. For example, even if there is a edge between two Websites A and B in the HostGraph, when a Web surfer visits a Webpage in Website A, he may not be able to jump to Website B because the hyperlink to Website B may exist in another Webpage in Website A which is even unreachable from the Webpage that he is currently visiting. In other words, the HostGraph is only a kind of approximation to the Web graph: it loses much transition information, especially as for the naïve HostGraph. So we propose a new algorithm to rank websites in accordance with the browsing behaviors of the web surfers [3, 14]. Our new algorithm is named AggregateRank algorithm, which will be discussed in the next subsection.

## 2.2 AggregateRank algorithm

The basic idea of the AggregateRank algorithm is that the importance of a Website should be measured by the mean frequency of a surfer's visiting.

Suppose there are totally $N$ Websites in the Web. As each Webpage belongs to some determinate Website, we rearrange the transition matrix $P(\alpha)$ and partition it into $N \times N$ blocks according to the $N$ Websites. Then it has the following form

$$P(\alpha) = \begin{pmatrix} P_{11}(\alpha) & P_{12}(\alpha) & \cdots & P_{1N}(\alpha) \\ P_{21}(\alpha) & P_{22}(\alpha) & \cdots & P_{2N}(\alpha) \\ \vdots & \vdots & \ddots & \vdots \\ P_{N1}(\alpha) & P_{N2}(\alpha) & \cdots & P_{NN}(\alpha) \end{pmatrix},$$

where the elements in each diagonal block denote the transition probabilities between Webpages in the same Website, and the elements of each off-diagonal block denote the transition probabilities between Webpages in different Websites. The diagonal blocks $P_{ii}(\alpha)$ are square and of order $n_i$, for $i = 1, 2, \cdots, N$, and $n = \sum_{i=1}^{N} n_i$. The stationary distribution $\pi(\alpha)$, known as the PageRank vector, is given by

$$\pi(\alpha)P(\alpha) = \pi(\alpha) \quad \text{with} \quad \pi(\alpha)e = 1.$$

Let $\pi(\alpha)$ be partitioned conformally with $P(\alpha)$, i.e.,

$$\pi(\alpha) = (\pi_1(\alpha), \pi_2(\alpha), \cdots, \pi_N(\alpha)),$$

and $\pi_i(\alpha)$ is a row vector of length $n_i$.

We assume that $\pi(\alpha)$ is the initial distribution of the Webpage surfing Markov chain $\{X_k\}_{k\geqslant 0}$, and a surfer is browsing on some Website $S_i$ at time $m$. We calculate the number of visiting the Website $S_j$ from now on. Let $N_j(l)$ denote the number of $\{X_k\}_{k\geqslant 0}$ visiting the Website $S_j$ during the $l$ times $\{m+1, m+2, \cdots, m+l\}$. Then we can get the following conclusion [3].

**Theorem 3.** ([3] Theorem 3.3): $\|\pi_j(\alpha)\|_1 = E\big(\lim_{l\to\infty} \frac{N_j(l)}{l}\big)$.

From this theorem, we know that $\|\pi_j(\alpha)\|_1$ is the mean frequency of visiting the Website $S_j$. Hence the probability vector $(\|\pi_1(\alpha)\|_1, \|\pi_2(\alpha)\|_1, \cdots, \|\pi_N(\alpha)\|_1)$, is a suitable candidate for ranking the importance of Websites.

It seems that the direct approach of computing the measurement of the website importance is to accumulate PageRank values (denoted by PageRankSum). However, this approach is unfeasible because the computation of PageRank is not a trivial task when the number of Web pages is as large as several billions. Therefore, we propose an approximate algorithm named AggregateRank, which can be much more efficient than PageRankSum with very little accuracy loss.

In the AggregateRank model, we define

$$c_{ij}(\alpha) = Pr_{\pi(\alpha)}\{X_{m+1} \in S_j \mid X_m \in S_i\}$$

as the one-step transition probability from the Website $S_i$ to the Website $S_j$. Then, the $N \times N$ matrix $C(\alpha) = (c_{ij}(\alpha))$ is the transition matrix between Websites [22]. It can be proved that $C(\alpha)$ is an irreducible stochastic matrix, and the unique stationary probability vector of it is

$$\xi(\alpha) = (\|\pi_1(\alpha)\|_1, \|\pi_2(\alpha)\|_1, \cdots, \|\pi_N(\alpha)\|_1).$$

Therefore, we can get the measurement of the website importance by calculating the stationary distribution of $C(\alpha)$.

By the theory of stochastic complement [22] and some further approximations, we design the AggregateRank algorithm as follows:

1. Divide the $n \times n$ matrix $P(\alpha)$ into $N \times N$ blocks according to the $N$ sites.

2. Construct the stochastic matrix $P^*_{ii}(\alpha)$ for $P_{ii}(\alpha)$ by changing the diagonal elements of $P_{ii}(\alpha)$ to make each row sum up to 1.

3. Determine $u_i(\alpha)$ from

$$u_i(\alpha)P^*_{ii}(\alpha) = u_i(\alpha) \quad \text{with} \quad u_i(\alpha)e = 1. \tag{1}$$

4. Form an approximation $C^*(\alpha)$ to the coupling matrix $C(\alpha)$, by evaluating

$$(C^*(\alpha))_{ij} = u_i(\alpha)P_{ij}(\alpha)e. \tag{2}$$

5. Determine the stationary distribution of $C^*(\alpha)$ and denote it $\xi^*(\alpha)$, i.e.,

$$\xi^*(\alpha)C^*(\alpha) = \xi^*(\alpha) \quad \text{with} \quad \xi^*(\alpha)e = 1. \tag{3}$$

Through the error analysis, we conclude that the error bound of $\xi(\alpha) - \xi^*(\alpha)$ can be well controlled [22]. Therefore, $\xi^*(\alpha)$ is a good measurement for the website importance.

We did some experiments on the dataset of TREC Web track [14], the results show that the AggregateRank algorithm has the best performance and it is the best approximation to PageRankSum. Moreover the AggregateRank algorithm is faster than PageRankSum, while a little more complex than the HostRank algorithms. So, by taking the effectiveness and efficiency into consideration at the same time, we consider the AggregateRank algorithm as a better solution to website-ranking.

# 3 Final remarks

In this paper we have focused our discussion on Static Rank by exploiting the hyperlink structure of the Web. The notable advantages of this technology are its query independence and content independence. We should remark that there are also other interesting research subjects to be further studied. For example, the queries submitted by the users and the contents of Webpages contain also very important information which should not be ignored. In fact there are many researchers have already paid their attention to the methods of ranking Webpages which depend also on the queries, such as RankNet and RankSVM. In practice search engine companies compose their rank of Webpages not only based the PageRank, which is of course a very important feature of Webpages, but also based on many other features of Webpages, such as the relevance of Webpages to the query, the contents of Webpages, and other features. From this point of view, "learning to rank" is becoming an important subject of the research on the Web search and mining. Note that different ranking algorithms possess different merits, no one performs absolutely better than all of the others in all circumstances. Therefore, some researchers have made their efforts on the research subject of "aggregating algorithms", which aims to compose a better ranking function by making use of the results obtained from different ranking functions, and which possesses potential applications in the meta-search, similarity search and other areas of the research on the Web search and mining. All these research subjects are interesting and are worth to be further studied.

## Acknowledgements

We thank Tieyan Liu and Yuting Liu for valuable discussions in preparing this paper. We thank all our co-authors for permitting us to present our joint results here. This work is partly supported by NSFC, 973 Project and by Science and Practice Fund for Graduate of CAS.

## References

[1] Y. Bao and Y. Liu. Limit of PageRank with Damping Factor. *Dynamics of Continuous, Discrete and Impulsive Systems*, Series B, 13(3), 497-504, 2006.

[2] Y. Bao and Z.H. Zhu. Comparison of Three Web Search Algorithms. *Acta Mathematicae Applicatae Sinica*, English Series, 22(3), 517-528, 2006.

[3] Y. Bao, G. Feng, T.Y. Liu, Z.M. Ma, and Y. Wang. Ranking WebSites, a Probabilistic View, to appear in *Internet Mathematics.*

[4] H. Bavulcu, S. Vadrevu and S. Nagarajan. OntoMiner: Bootstrapping Ontologies From Overlapping Domain Specific Web site. *Proceedings of the Thirteenth International World Wide Web Conference*, New York, USA, May 2004.

[5] K. Bharat, B.W. Chang, M. Henzinger and M. Ruhl. Who links to whom: Mining linkage between Websites. *Proceedings of the IEEE International Conference on Data Mining (ICDM'01)*, San Jose, USA, November 2001.

[6] M. Bianchini, M. Gori, F. Scarselli. Inside PageRank. *ACM Transactions on Internet Technology*, 2(5): 92-128, 2005.

[7] P. Boldi, M. Santini, S. Vigna. PageRank as a function of the damping factor. 14th International World Wide Web Conference 2005, http://www2005.org/cdrom/docs/p557.pdf.

[8] A. Border, R. Kumar, F. Maghoul, P. Raghavan, S. Rajagopalan, R. Stata, A. Tomkins, J. Wiener. Graph structure in the Web. 9th International World Wide Web Conference, May, 2000.

[9] S. Brin, L. Page, R. Motwami and T. Winograd. The PageRank Citation Ranking: Bringing Order to the Web. *Technical Report*, 1999-0120, Computer Science Department, Stanford University, Stanford, CA, 1999.

[10] G.E. Cho and C.D. Meyer. Aggregation/Disaggregation Methods of Nearly Uncoupled Markov Chains. http://meyer.math.ncsu.edu/Meyer/PS_Files/Numcad.ps.

[11] T. Despeyroux. Practical Semantic Analysis of Web Sites and Documents. *Proceedings of the Thirteenth International World Wide Web Conference*, New York, USA, May 2004.

[12] S. Dill, R. Kumar, K. McCurley, S. Rajagopalan, D. Sivakumar and A. Tomkins. Self-similarity in the Web. *Proceedings of International Conference on Very Large Data Bases*, 69-78, Rome, 2001.

[13] N. Eiron, K.S. McCurley and J.A. Tomlin. Ranking the Web frontier. *Proceedings of the 13th International World Wide Web Conference (WWW)*, 309-318, ACM Press, New York, USA, 2004.

[14] G. Feng, T.Y. Liu, Y. Wang, Y. Bao, Z.M. Ma, X.D. Zhang and W.Y. Ma.AggrerateRank: Bringing Order to Web Sites. *Proceedings of the 29th ACM Conference on Research and Development on Information Retrieval*, 75-82, Scattlc, 2006.

[15] Feng-hsiung Hsu. Behind Deep Blue. Princeton University Press, Princeton, NJ, 2002.

[16] O. Kallenberg. Foundations of Modern Probability, Springer, 152, 2001.

[17] J. Kleinberg. Authoritative sources in a hyperlinked environment. *J. ACM* **46**, 604-632, 1999.

[18] A.N. Langville and C.D. Meyer. Deeper inside PageRank. *Internet Mathematics*, 1(3): 355-400, 2004.

[19] A.N. Langville and C.D. Meyer. A Survey of Eigenvector Methods of Web Information Retrieval. *SIAM Review*, 47(1): 135-161, 2005.

[20] Y. Lei, E. Motta and J. Domingue. Modelling Data-Intensive Web Sites with OntoWeaver. *Proceedings of International Workshop on Web Information Systems Modeling*, Riga, Latvia, June 2004.

[21] K. Lerman, L. Getoor, S. Minton and C. Knoblock. Using the Structure of Web Sites for Automatic Segmentation of Tables. *Proceedings of the ACM International Conference on Management of Data*, Paris, France, June 2004.

[22] C.D. Meyer. Stochastic complementation, uncoupling markov chains, and the theory of nearly reducible systems. *SIAM Review*, 31(2): 240-272, 1989.

[23] M.E.J. Newman. The structure and function of complex networks. arXiv: cond-mat/0303516 v1 25 Mar 2003.

[24] T. Qin, T.Y. Liu, X.D. Zhang, G. Feng and W.Y. Ma. Subsite Retrieval: A Novel Concept for Topic Distillation. *Proceedings of 2nd Asia Information Retrieval Symposium*, Jeju Island, Kotea, October 2005.

[25] G. Salton and M.J. McGill. Introduction to Modern Information Retrieval. McGraw-Hill, 1983.

[26] J.A. Tomlin. A New paradigm for ranking pages on the world wide web. Twelth International World Wide Web, 2003.

[27] J. Wu and K. Aberer. Using SiteRank for P2P Web Retrieval. EPFL Technical Report ID: IC/2004/31, 2004.

[28] L. Zhang, Q. Tao, T.Y. Liu, Y. Bao and H. Li. N-Step PageRank for Web Search. *Proceedings of 29th European Conference on Information Retrieval*, 2007.

[29] Z.H. Zhu and Y. Bao. Comparison of Two Web Search Algorithms. Acta Mathematicae Applicatae Sinica, Chinese Series, 28(4), 577-586, 2005.

[30] Z.H. Zhu. Research on Web Search Algorithms (in Chinese). Master Degree Thesis, 2006.

# Report on Testing and Finding the Generating Functions $g$ of an Option Pricing Mechanism through Market Data

Lifeng Chen, Shige Peng
*Institute of Mathematics, Institute of Finance*
*Shandong University, Jinan 250100, China*
*Email: chenlifeng80@126.com, peng@sdu.edu.cn*

**Abstract**

We study dynamic pricing mechanisms of financial derivatives. A typical model of such pricing mechanism is the so-called $g$-expectation defined by solutions of a backward stochastic differential equation with $g$ as its generating function. Black-Scholes pricing model is a special linear case of this pricing mechanism. We are mainly concerned with two types of pricing mechanisms in an option market: the market pricing mechanism through which the market prices of options are produced, and the ask-bid pricing mechanism operated through the system of market makers. The later one is a typical nonlinear pricing mechanism. Data of prices produced by these two pricing mechanisms are usually quoted in an option market.

We introduce a criteria, i.e., the domination condition (A5) in (4.1) to test if a dynamic pricing mechanism under investigation is a $g$-pricing mechanism. This domination condition was statistically tested using CME data documents. The result of test is significantly positive. We also provide some useful characterizations of a pricing mechanism by its generating function.

# 1 Introduction

Our research group of Shandong University, collaborated with other research groups in mathematical finance, such as Ecole Polytechnique de Paris, Universite de Rennes I, ETH, Zurich, have struggled, for a quite longtime, a similar problem of finding the generating function $g$ of a pricing mechanism of financial derivatives, including European call and put options, American options, Asian options, and some other exotic options.

The history of mathematical finance can be traced from Louis Bachelier's 1900 Thesis on option pricing. The theoretical and practical breakthrough of option pricing was Black-Scholes formula. But till then, the central problem of mathematical finance is still and, maybe always, the pricing mechanism of traders, market makers, small and large investors, and of a market.

A pricing mechanism of a financial institution is a black box. Its input is a derivative product and its output is the price of this, or the prices—since a market maker has the privilege to offer two prices: the ask price and the bid price. It is this dynamic black box of derivative pricing mechanism with, usually a huge quantity of input-output data that attract us to find if there is a generating function hidden behind us.

To quantitatively describe the pricing mechanism of a market of derivatives is a very interesting problem. A model of dynamic pricing mechanism of derivatives is formulated (see (A1)–(A4) in the next section) to characterize this pricing behavior.

We are mainly concerned with two types of pricing mechanisms in an option market: the market pricing mechanism which outputs the trading prices of options and the bid-ask pricing mechanism operated according the system of market makers. We stress here that, in our point of view, the ask prices and the bid prices quoted in a market are determined by a single pricing mechanism. The difference of a ask price and the corresponding bid price, called bid-ask spread, reflects the nonlinearity of this mechanism. The data of prices of above mentioned two pricing systems is usually systematically quoted in the internet thus the models under our investigation can be statistically tested. We hope that our modeling can also be applied to describe the pricing mechanism of some other financial institutions.

The well-known Black-Scholes formula is a typical model of dynamic pricing mechanism of derivatives. It is a linear pricing mechanism. In fact, the prices produced by this mechanism is solved by a linear Backward Stochastic Differential Equation (BSDE). This means that the corresponding generating function $g$ of the BSDE is a linear function. A nonlinear pricing mechanism by BSDE was originally proposed in [9]. In this paper we show that each well-defined BSDE with a fixed generating function $g$ forms a dynamic pricing mechanism, called $g$-expectation and that the behaviors of this mechanism are perfectly characterized by the behaviors of $g$. Several conditions of equivalence provided in this paper will be very helpful to characterize and to find the generating function, or in some other circumstances, to regulate or to design a pricing mechanism.

A very interesting problem is how to design a test procedure to verify whether an existing pricing mechanism of derivatives is a $g$-expectation. We will present the following result: if a dynamic pricing mechanism is uniformly dominated by a $g_\mu$-expectation with a sufficiently large number $\mu$ for the function $g_\mu = \mu(|y| + |z|)$, then it is a $g$-expectation. This domination inequality (4.1) has been applied as a testing criteria in our data analysis. The results strongly support that both the market pricing mechanism and the bid-ask pricing mechanism under our investigation can be modelled as $g$-expectations, and that the bid-ask prices are then produced by this single mechanism.

In this paper we present the notion of $g$-pricing mechanism and show that, for each well-defined function $g$ it satisfies the basic conditions (A1)–(A4) of a dynamic pricing mechanism of derivatives. We then show that, a dynamic pricing mechanism dominated by a $g_\mu$-expectation, i.e., (4.1) is satisfied, is a $g$-expectation. In Section 3, we will present some equivalent conditions to show that the behaviors of a $g$-expectation are perfectly reflected by its generating function $g$. We also provide some examples and explain how to statistically find the function $g$ by testing the input-output data of prices.

We apply the crucial domination inequality (4.1) to test the market pricing mechanisms and the bid-ask pricing mechanisms of S&P500 index future options and S&P500 index options, using data of parameter files provided by CME and CBOE. The result supports that they are $g$-expectations.

The main references of this paper are [25] and [28].

# 2 A pricing mechanism of derivatives as an input-output black box

Let us consider a $d$-stock market with price $S(t) = (S^1(t), \cdots, S^d(t))$. A derivative $X$ underlying on the stock $S$ with maturity $T$ is a contract of which the value $X$ is determined by the prices of $S$ before $T$. A typical example is a European call option $X = \max\{0, S^1(T) - k)$ where the strike price $k$ is fixed. In this case $X$ depends only on $S$ at the time $T$. An Asian option $X = \frac{1}{T}\int_0^T S^1(t)dt$ depends on the whole prices of $S^1$ before $T$. The central problem how, at a time $t < T$, a financial institute make a price of a derivative $X$ with maturity $T$?

A method quite different from the Black-Scholes's one is to formulate a pricing mechanism as an input-output box. This input-output system is significantly different from a traditional dynamic system e.g. a control system or a regulator in the sense that the input data, the option contract will be realized at its maturity time $T$ whereas the output of this mechanism is the price of this option at the present time $t < T$. This means that the input data is realized after the output price. If we denote $X_T$ to be the option contract whose data will be realized at the time $T$, the output of this pricing mechanism is the price of $X_T$ at time $t$, denoted by $\mathbb{O}_{t,T}[X_T]$. However we make a basic and reasonable assumption that the input $X_T$ depends on the prices of the underlying stocks $(S_s)_{s\leqslant T}$, before the maturity $T$, and the output $\mathbb{O}_{t,T}[X_T]$ depends only on $(S_s)_{s\leqslant t}$.

Our main assumption on this dynamic blackbox $\{\mathbb{O}_{t,T}[X_T]\}_{t\leqslant T}$ is as follows: for each $s \leqslant t \leqslant T < \infty$, derivative contracts $X$, $\bar{X}$ depending on the price of $S$ before $t$, the price of $\mathbb{O}_{t,T}[X]$ (respectively the price of $\mathbb{O}_{s,T}[X]$) depends on the price of $S$ before $t$ (respectively, before $s$) and we have

**(A1)** $\mathbb{O}_{t,T}[X] \geqslant \mathbb{O}_{t,T}[\bar{X}]$, if $X \geqslant \bar{X}$;

**(A2)** $\mathbb{O}_{T,T}[X] = X$;

**(A3)** $\mathbb{O}_{s,t}[\mathbb{O}_{t,T}[X]] = \mathbb{O}_{s,T}[X]$ for $s \leqslant t$;

**(A4)** $1_A\mathbb{O}_{t,T}[X] = \mathbb{O}_{t,T}[1_A X]$, where $1_A$ is valued on 0 and 1 and it depends on the prices of $S$ before $t$.

**Remark 2.1.** *(A1) and (A2) are economically obvious conditions for a rational derivative pricing mechanism. Condition (A3) means that, at the time $s$, the value $\mathbb{O}^g_{t,T}[X]$ depends on the price of $S$ before $t$, thus $\mathbb{O}^g_{t,T}[X]$ can be regarded as a derivative contract with maturity $t$. The price of this derivative at $s$ is $\mathbb{O}^g_{s,t}[\mathbb{O}^g_{t,T}[X]]$. This price must be the same as $\mathbb{O}^g_{s,T}[X]$.*

**Remark 2.2.** *The meaning of condition (A4) is that, since at time $t$, the agent knows already the value of $I_A$ which either 1 or 0. When $I_A$ is 1, then $1_AX = X$ thus the price $\mathbb{O}^g_{t,T}[1_AX]$ must be the same as $\mathbb{O}^g_{t,T}[X]$, otherwise $1_AX = 0$, so it worthies $\mathbb{O}^g_{t,T}[1_AX] = \mathbb{O}^g_{t,T}[0] = 0 = 1_A\mathbb{O}^g_{t,T}[X]$.*

# 3 BSDE pricing mechanism model and Black-Scholes formula

Let us consider a market of financial derivatives in which the price $(S(t))_{t\geqslant 0}$ of the underlying assets is driven by a $d$-dimensional Brownian motion $(B_t)_{t\geqslant 0}$. We assume that the past information $\mathcal{F}^S_t$ of the price $S$ before $t$ depends on the values of $B$ before $t$. A derivative $X$

with maturity $T$ is an $\mathcal{F}_T$-measurable and random value called maturity value. $X$ is considered as an input. The output is the price $Y_t$ of $X$ at the time $t < T$ of a given pricing mechanism.

Here we make a basic technique requirement that each process $\eta_t$ is $\mathcal{F}_t$-adapted, namely, the value of $\eta_t$ depends on the values of the Brownian motion $B$ before $t$. Our BSDE pricing mechanism is to solve $Y_t$ by the following backward stochastic differential equation (BSDE):

$$Y_t = X + \int_t^T g(s, Y_s, Z_s)ds - \int_t^T Z_s dB_s. \tag{3.1}$$

Here $(Y, Z)$ a pair of the $\mathcal{F}_t$-adapted processes to be solved, $g$ is a given function

$$g : (\omega, t, y, z) \in \Omega \times [0, \infty) \times R \times R^d \to R.$$

We call $g$ the generating function of the BSDE. It satisfies the following basic assumptions for each $\forall y,\ \bar{y} \in R$ and $z,\ \bar{z} \in R^d$,

$$|g(t, y, z) - g(T, \bar{y}, \bar{z})| \leqslant \mu(|y - \bar{y}| + |z - \bar{z}|). \tag{3.2}$$

It is important to consider the following special situation:

$$\begin{cases} \text{(a)} \quad g(\cdot, 0, 0) \equiv 0, \\ \text{(b)} \quad g(\cdot, y, 0) \equiv 0,\ \forall y \in R. \end{cases} \tag{3.3}$$

Obviously (b) implies (a). This BSDE (3.1) was introduced by Bismut [1] for the case where $g$ is a linear function of $(y, z)$. [17, Pardoux-Peng, 1990] obtained the following basic result: for each $X \in L^2(\mathcal{F}_T)$, there exists a unique square-integrable adapted solution $(Y, Z)$ of the BSDE (3.1). The following notion of $g$-expectations was introduced by [20, Peng 1997a] and [21, Peng 1997].

**Definition 3.1.** *We denote by* $\mathbb{O}^g_{t,T}[X] := Y_t$*:*

$$\mathbb{O}^g_{t,T}[\cdot] : L^2(\mathcal{F}_T) \to L^2(\mathcal{F}_t), \quad 0 \leqslant t \leqslant T < \infty. \tag{3.4}$$

$(\mathbb{O}^g_{t,T}[\cdot])_{0 \leqslant t \leqslant T < \infty}$ *is called* $g$*-pricing mechanism.*

As an example of BSDE pricing mechanism, we consider the following Black-Scholes pricing mechanism:

**Example 3.2.** ***(Black-Scholes is a*** $g$***-pricing mechanism)*** *Consider a financial market consisting of* 2 *underlying assets: one bond and one stock. We denote by* $S_0(t)$ *the price of the bond and by* $S(t)$ *the price of the stock at time* $t$*. We assume that* $S_0(t)$ *satisfies an ordinary differential equation:* $dS_0(t) = r_t S_0(t)dt$*, and* $S(t)$ *is the solution of* *the following stochastic differential equation (SDE) with* 1*-dimensional Brownian motion $B$ (i.e., $d = 1$) as driven noise:*

$$dS(t) = S(t)(b_t dt + \sigma_t dB_t), \quad S(0) = p.$$

*Here $r_t$ is the interest rate, $b_t$ the rate of the expected return and $\sigma_t$ the volatility of the stock at the time $t$. $r_t$, $b_t$, $\sigma_t$ and $\sigma^{-1}$ are assumed to be $\mathcal{F}_t$-measurable and uniformly bounded. Black and Scholes have solved the problem of the market pricing mechanism of a European call option $X_{call} = \max\{0, S_T - k\}$ and put option $X_{call} = \max\{0, k - S_T\}$, where $k$ is the strike price, under the assumption that $r$, $b$ and $\sigma$ are constant. Their main idea can be easily adapted to our slightly more general situation for a derivative $X \in L^2(\mathcal{F}_T)$ with maturity $T$. Consider an investor with the following investment portfolio at a time $t \leqslant T$: he has $n_0(t)$ bonds and $n(t)$ stock, i.e., he invests $n_0(t)S_0(t)$ in bond and $\pi(t) = n(t)S(t)$ in the stock. We define by $Y_t$ the investor's wealth invested in the market at time $t$:*

$$Y_t = n_0(t)P_0(t) + n(t)P(t).$$

*We make the so called "self-financing assumption":*

$$dY_t = n_0(t)dS_0(t) + n(t)dS(t)$$

*or*

$$dY_t = [r_t Y_t + (b_t - r_t)\pi(t)]dt + \sigma_t \pi_t dB_t.$$

*We denote $g(t, y, z) := -r_t y - (b_t - r_t)\sigma_t^{-1} z$. Then, by denoting $Z_t = \sigma_t \pi(t)$, the above equation is*

$$-dY_t = g(t, Y_t, Z_t)dt - Z_t dB_t.$$

*We observe that the above function $g$ satisfies (3.2). It follows from the existence and uniqueness theorem of BSDE that for each derivative $X \in L^2(\mathcal{F}_T)$, there exists a unique adapted solution $(Y, Z)$ with the terminal condition $Y_T = X$. This result of existence and uniqueness is economically meaningful: in order to replicate the derivative $X$ at the maturity $T$, the investor needs and only needs to invest the $Y_t$ at the present time $t$ and then, during the time interval $s \in [t, T]$, to perform the portfolio strategy $\pi(s) = \sigma_s^{-1} Z_s$. Furthermore, by Comparison Theorem of BSDE, if he wants to replicate a derivative $\bar{X}$ with the same maturity $T$ which is bigger than $X$ (i.e., $\bar{X} \geqslant X$ and $P(\bar{X} \geqslant X) > 0$) then he must invest more than $Y_t$ at the time $t$. This means there this no arbitrage opportunity. In this situation $Y_t = \mathbb{O}^g_{t,T}[X]$ is called the Black–Scholes price, and $(\mathbb{O}^g_{t,T}[\cdot])_{0 \leqslant t \leqslant T < \infty}$ is called the corresponding Black–Scholes pricing mechanism. We observe that the generating function $g$ satisfies (a) of condition (3.3).*

**Example 3.3.** $X_{call} = \max\{0, S_T - K\}$, $X_{put} = \max\{0, K - S_T\}$,

$$\mathbb{O}^g_{t,T}[X_{call}] = S_t N(d_1) - K\exp\left(-r(T-t)\right) N(d_2),$$
$$\mathbb{O}^g_{t,T}[X_{put}] = K\exp(-r(T)N(-d_2) - S_0 N(-d_1),$$

*where*

$$d_1 = \frac{\ln(S_0/K) + (r + \sigma^2/2)(T-t)}{\sigma\sqrt{T-t}},$$
$$d_2 = \frac{\ln(S_0/K) + (r - \sigma^2/2)(T-t)}{\sigma\sqrt{T-t}} = d_1 - \sigma\sqrt{T-t},$$
$$N(x) = \frac{1}{\sqrt{2\pi}}\int_{-\infty}^{x} \exp(-\frac{\eta^2}{2})d\eta.$$

**Example 3.4.** *The following problem was considered in [5] and [9]: the investor is allowed to borrow money at time t at an interest rate $R_t > r_t$. The amount borrowed at time t is equal to $(Y_t - \pi(t))^-$. In this case the wealth process $Y$ still satisfies BSDE:*

$$-dY_t = g(t, Y_t, Z_t)dt \;-\; Z_t dW_t$$

*with $g(t, y, z) := -r_t y - (b_t - r_t)\sigma_t^{-1} z + (R_t - r_t)(y - \sigma_t^{-1} z)^-$. This derives a g-pricing mechanism with a sub-additive generating function $g$.*

Similar equations appear in continuous trading with short sales constraints with different risk premium for long and short positions (cf. [15], [11] and [9]). In this case $g(t, y, z) := -r_t y - (b_t - r_t)\sigma_t^{-1} z + k_t z^-$. We observe that in each of the above three examples, $g$ is sub-additive in $(y, z)$.

## 3.1 Numerical calculation of solutions of BSDEs

As the method of binormal tree in the option pricing model, our main idea of the calculation of BSDE is to replace the above Brownian motion $B$ by a random walk. We refer to [29] for the details of our numerical approach. We make a standard time-partition of the interval $[0, 1]$: $0 = t_0 < t_1 < \cdots < t_n = 1$, $\delta := t_j - t_{j-1} = \frac{1}{n}$, for $1 \leqslant j \leqslant n$. Consider $\{(\varepsilon_m)_{1 \leqslant m \leqslant n}\}$, a Bernoulli sequence, with $\varepsilon_0 = 0$, which are i.i.d. random variable satisfying

$$\varepsilon_m = \begin{cases} +1, & p = 0.5, \\ -1, & p = 0.5. \end{cases}$$

Now we define the scaled random walk $\{B^n_\cdot\}$, by setting $B^n_0 = 0$,

$$B^n_t = \sqrt{\delta}\sum_{m=0}^{[t/\delta]} \varepsilon_m, \quad 0 \leqslant t \leqslant T \tag{3.5}$$

and denote $B_j^n = B_{t_j}^n$, i.e., $B_j^n = \sqrt{\delta}\sum_{m=1}^{j}\varepsilon_m$. And we define the discrete filtration $\mathcal{F}_j^n := \sigma\{\varepsilon_m; 0 \leqslant m \leqslant j\} = \sigma\{B_t^n; 0 \leqslant t \leqslant t_j\}$, for $1 \leqslant j \leqslant n$.

Then on the small interval $[t_j, t_{j+1}]$, the equation

$$y_{t_j} = y_{t_{j+1}} + \int_{t_j}^{t_{j+1}} f(s, y_s, z_s)ds - \int_{t_j}^{t_{j+1}} z_s dB_s \tag{3.6}$$

can be approximated by the discrete equation

$$y_j^n = y_{j+1}^n + f(t_j, y_j^n, z_j^n)\delta - z_j^n\sqrt{\delta}(B_{j+1}^n - B_j^n). \tag{3.7}$$

If $f(t, y, z)$ satisfies the Lipschitz condition with constant $k$, for $\delta < 1/k$, there exists a unique couple $(y_j^n, z_j^n)$ satisfying equation (3.7). In fact with $B_{j+1}^n - B_j^n = \varepsilon_{j+1}$, and $E[\varepsilon_{j+1}|\mathcal{F}_j^n] = 0$, we get immediately

$$z_j^n = \frac{1}{2\sqrt{\delta}}E[y_{j+1}^n\varepsilon_{j+1}|\mathcal{F}_j^n]. \tag{3.8}$$

Then taking conditional expectation on (3.7), it follows

$$y_j^n = E[y_{j+1}^n|\mathcal{F}_j^n] + f(t_j, y_j^n, z_j^n)\delta. \tag{3.9}$$

Consider the mapping $\Theta(y) = y - f(t_j, y, z_j^n)\delta$, from the Lipschitz property of $f$, we obtain

$$\langle\Theta(y) - \Theta(y'), y - y'\rangle \geqslant (1 - \delta k)\,|y - y'|^2 > 0,$$

which implies that the mapping $\Theta(y)$ is a monotonic mapping. So there exists a unique value $y$ s.t. $\Theta(y) = E[y_{j+1}^n|\mathcal{F}_j^n]$ holds, i.e. $y_j^n = \Theta^{-1}(E[y_{j+1}^n|\mathcal{F}_j^n])$.

**Remark 3.5.** *The existence of the solution of discrete BSDE only depends on the Lipschitz condition of $f$ on $y$. In fact, if $f$ does not depend $y$, we can easily get $\Theta^{-1}(y) = y + f(t_j, z_j^n)\delta$.*

**Remark 3.6.** *In general, if $f$ nonlinearly depends on $y$, then $\Theta^{-1}$ can not be solved explicitly, so sometimes we use $(\overline{y}_j^n, \overline{z}_j^n)$, where*

$$\begin{aligned} \overline{y}_j^n &= E[y_{j+1}^n|\mathcal{F}_j^n] + f(t_j, E[y_{j+1}^n|\mathcal{F}_j^n], \overline{z}_j^n)\delta, \\ \overline{z}_j^n &= \frac{1}{2\sqrt{\delta}}E[y_{j+1}^n\varepsilon_{j+1}|\mathcal{F}_j^n] \end{aligned} \tag{3.10}$$

*to approximate the solution for of $\Theta(y) = E[y_{j+1}^n|\mathcal{F}_j^n]$. (3.10) is called the explicit scheme for BSDE, while (3.9) is called the implicit scheme for BSDE.*

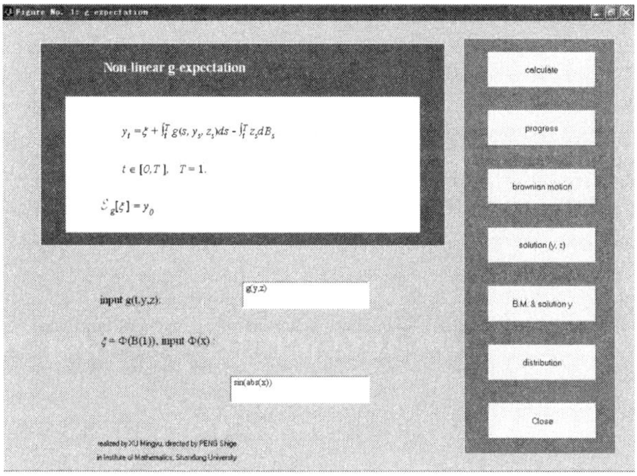


Figure 1 To calculate the output $y_t$, we input the derivative $\xi$ toolbox of $g$-pricing calculations

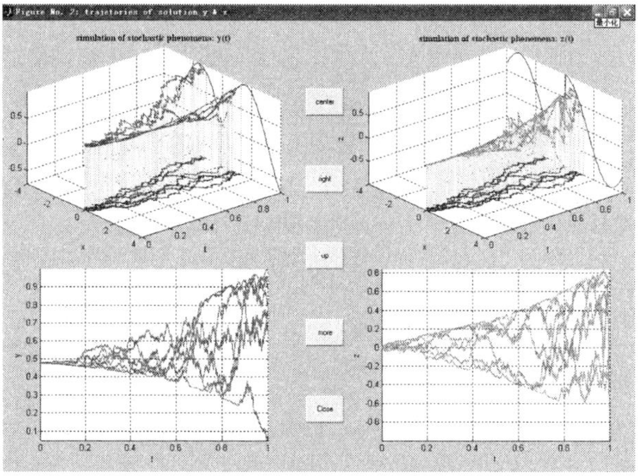


Figure 2

# 4 BSDE as a pricing mechanism of derivatives

The following result, obtained in [25]-Theorem 3.4, explains why this $g$-expectation is a good candidate to model a dynamic pricing mechanism of derivatives:

**Proposition 4.1.** *Let the generating function g satisfies (3.2) and (3.3)-(a). Then the above defined g-expectation* $\mathbb{O}^g[\cdot]$ *is a dynamic pricing mechanism of derivatives, i.e., it satisfies, for each* $t \leqslant T < \infty$, $X$,

$\bar{X} \in L^2(\mathcal{F}_T)$,
***(A1)*** $\mathbb{O}^g_{t,T}[X] \geqslant \mathbb{O}^g_{t,T}[\bar{X}]$, *a.s., if* $X \geqslant \bar{X}$;
***(A2)*** $\mathbb{O}^g_{T,T}[X] = X$;
***(A3)*** $\mathbb{O}^g_{s,t}[\mathbb{O}^g_{t,T}[X]] = \mathbb{O}^g_{s,T}[X]$, *for* $s \leqslant t$;
***(A4)*** $1_A \mathbb{O}^g_{t,T}[X] = \mathbb{O}^g_{t,T}[1_A X]$, $\forall A \in \mathcal{F}_t$,
*where* $I_A$ *is the indicator function of* $A$, *i.e.,* $I_A(\omega)$ *equals to* 1, *when* $\omega \in A$ *and* 0 *otherwise.*

**Remark 4.2.** *(A1) and (A2) are economically obvious conditions for a pricing mechanism. Condition (A3) means that, at the time* $s$, *the random value* $\mathbb{O}^g_{t,T}[X]$ *can be regarded as a maturity value with maturity* $t$. *The price of this derivative at* $s$ *is* $\mathbb{O}^g_{s,t}[\mathbb{O}^g_{t,T}[X]]$. *It must be the same as the price* $\mathbb{O}^g_{s,T}[X]$ *of* $X$ *at* $s$.

**Remark 4.3.** *The meaning of condition (A4) is that, since at time* $t$, *the agent knows the value of whether* $I_A$ *is* 1 *or* 0. *When* $I_A$ *is* 1, *then the price* $\mathbb{O}^g_{t,T}[1_A X]$ *of* $1_A X$ *must be the same as* $\mathbb{O}^g_{t,T}[X]$, *otherwise* $1_A X = 0$, *so it worthies* 0.

From the above results we see that $\mathbb{O}^g$ is a good candidate to be a dynamic pricing mechanism. The following result provides a criteria to test if a dynamic pricing mechanism is a $g$-expectation. The proof can be found in [27].

**Definition 4.4.** *A system of mappings* $(\mathbb{O}_{t,T}[\cdot])_{0 \leqslant t \leqslant T < \infty}$

$$\mathbb{O}_{t,T}[X] : X \in L^2(\mathcal{F}_T) \mapsto L^2(\mathcal{F}_t)$$

*is called a dynamic pricing mechanism of derivatives if it satisfies (A1)–(A4) (with* $\mathbb{O}[\cdot]$ *in the place of* $\mathbb{O}^g[\cdot]$*).*

**Theorem 4.5.** *Let* $\mathbb{O}_{t,T}[\cdot]_{0 \leqslant t \leqslant T < \infty}$ *be a dynamic pricing mechanism. If there exists a sufficiently large constant* $\mu > 0$, *such that the following domination criteria is satisfied*
***(A5)***

$$\mathbb{O}_{t,T}[X] - \mathbb{O}_{t,T}[\bar{X}] \leqslant \mathbb{O}^{g_\mu}_{t,T}[X - \bar{X}]. \tag{4.1}$$

$\mathbb{O}^{g_\mu}$ *is a* $g$*-expectation with the generating function* $g_\mu$ *defined by*

$$g_\mu(y, z) := \mu|y| + \mu|z|, \quad (y, z) \in R \times R^d. \tag{4.2}$$

*Then there exists a unique generating function* $g(\omega, t, y, z)$ *satisfying (3.2) and (3.3)-(a) such that, for each* $t \leqslant T$ *and for each derivative* $X \in L^2(\mathcal{F}_T)$, *we have*

$$\mathbb{O}_{t,T}[X] = \mathbb{O}^g_{t,T}[X], \tag{4.3}$$

*namely* $\mathbb{O}$ *is a* $g$*-expectation.*

**Remark 4.6.** *This theorem also implies that, for a generating function $g$ satisfying (3.2) and (3.3)-(a), the corresponding $g$-expectation $\mathbb{O}^g$ is also dominated by $\mathbb{O}^{g_\mu}$, i.e., (A5) is satisfied. This can be also directly proved by using the comparison theorem of BSDE.*

**Remark 4.7.** *It turns out that the domination condition (4.1) becomes a crucial criteria to test whether a dynamic pricing mechanism of derivatives is a $g$-expectation. We provide a test in Appendix 4.2 to use market data to check the inequality (4.1).*

**Remark 4.8.** *This deep result has non-trivially generalized the main result of [4], theoretically and practically, where a special case $g = g(t, z)$ with $g(s, 0) \equiv 0$ is considered. The $g$-expectation originally introduced in [21] corresponds such situation of "zero interest rate"(cf. [25]).*

# 5 Testing the criteria (A5) by market data

With Chen L. and Sun P. of our research group, we proceed a data test for the criteria (A5), i.e., the domination inequality (4.1), to check if a specific pricing mechanism is a $g$-expectation, or $g$-pricing mechanism $\mathbb{O}^g$.

We have firstly tested the CME's (Chicago Mercantile Exchange) market pricing mechanism of derivatives by taking the daily closing prices of options with S&P500 index futures as the underlying asset. The data is obtained from parameter files published from CME's ftp site, named cmeMMDDs.par (MM for month, DD for day) of call and put prices, from 05 January 2000 to November 2003, of totally 960 trading days. The corresponding S&P500 future's prices is obtained from the parameter files of SPAN (Standard Portfolio Analysis of Risk) system downloaded from CME's ftp site.

We denote by $X_T^i = (S_T - k_i)^+$ (resp. $Y_T^i = (S_T - k_i)^-$), the market maturity value of the call (resp. put) option with maturity $T$ and strike price $k_i$. The corresponding values of the short positions are $-X_T^i$ and $-Y_T^i$. We denote the market price of the corresponding prices of options at time $t < T$ by $\mathbb{O}_{t,T}^m[X_T^i]$, $\mathbb{O}_{t,T}^m[Y_T^i]$, $\mathbb{O}_{t,T}^m[-X_T^i]$ and $\mathbb{O}_{t,T}^m[-Y_T^i]$ respectively. The inequalities we need to put to the test are, according to (4.1), in the following different combinations (Figure 3), with different $(t, T)$ and different strike prices

$$\begin{cases} \text{Call-Call:} \ \mathbb{O}_{t,T}^m[X_T^i] - \mathbb{O}_{t,T}^m[X_T^j] \leqslant \mathbb{O}_{t,T}^{g_\mu}[X_T^i - X_T^j], \\ \text{Put-Put:} \ \mathbb{O}_{t,T}^m[Y_T^i] - \mathbb{O}_{t,T}^m[-Y_T^j] \leqslant \mathbb{O}_{t,T}^{g_\mu}[Y_T^i - Y_T^j], \\ \text{Call-Put:} \ \mathbb{O}_{t,T}^m[X_T^i] - \mathbb{O}_{t,T}^m[Y_T^j] \leqslant \mathbb{O}_{t,T}^{g_\mu}[X_T^i - Y_T^j], \\ \text{Put-Call:} \ \mathbb{O}_{t,T}^m[Y_T^i] - \mathbb{O}_{t,T}^m[X_T^j] \leqslant \mathbb{O}_{t,T}^{g_\mu}[Y_T^i - X_T^j], \end{cases} \tag{5.1}$$

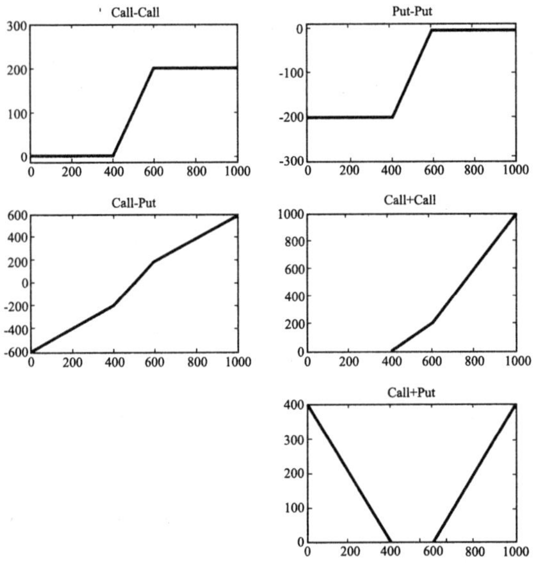


Figure 3 Seven kinds of combination

and

$$\begin{cases} \text{Call-ShortCall:} & \mathbb{O}^m_{t,T}[X^i_T] - \mathbb{O}^m_{t,T}[-X^j_T] \leqslant \mathbb{O}^{g_\mu}_{t,T}[X^i_T + X^j_T], \\ \text{Put-ShortPut:} & \mathbb{O}^m_{t,T}[Y^i_T] - \mathbb{O}^m_{t,T}[-Y^j_T] \leqslant \mathbb{O}^{g_\mu}_{t,T}[Y^i_T + Y^j_T], \\ \text{Call-ShortPut:} & \mathbb{O}^m_{t,T}[X^i_T] - \mathbb{O}^m_{t,T}[-Y^j_T] \leqslant \mathbb{O}^{g_\mu}_{t,T}[X^i_T + Y^j_T]. \end{cases} \tag{5.2}$$

In the above inequalities the data of the left hand sides is the market prices of options taken from CME parameter files. In our testing the transaction cost is neglected, i.e., we assume that $\mathbb{O}^m_{t,T}[-X] = -\mathbb{O}^m_{t,T}[X]$. The right hand sides is the corresponding values of $g_\mu$-expectations. We fix $\mu = 25$ uniformly for all tested inequalities. We have calculated all these values on the right hand side by using standard binomial tree algorithm of BSDE. Here an improved version of the algorithms of BSDE proposed Peng and Xu [2005] has been applied to solve the following 1-dimensional BSDE:

$$y_t = X + \int_t^T \mu(|y_s| + |z_s|)ds - \int_t^T z_s dB_s \tag{5.3}$$

with different terminal values $y_T = X^i_T - X^j_T$, $Y^i_T - Y^j_T$, $X^i_T - Y^j_T$, $Y^i_T - X^j_T$, $X^i_T + X^j_T$, $Y^i_T + Y^j_T$, $X^i_T + Y^j_T$, respectively. The closing prices of S&P500 futures options of 69 trading days from year 2000 to 2003 have been put in the test. With the above mentioned combinations, we have tested a total number of 6,200,828 inequalities of (5.1) and (5.2). This means that our BSDE (5.3) have been calculated 6,200,828 times. A very positive result was obtained: among the totally 6,200,828 tested inequalities, only 17 are against the criteria (5.1). Among those 12 cases of violations, 5 are singular situation since they themselves all violate Axiomatic monotonicity condition (A1). 5 cases are all from the same file cme0701s.par, 2003, Put-Put. They are all the following singular cases:

$$\mathbb{O}^m_{t,T}[(S_T - k_i)^-] > \mathbb{O}^m_{t,T}[(S_T - k_j)^-], \quad \text{for } k_i > k_j.$$

The other 12 violations are the cases where the time $T - t$ is too short (less than 2 days).

Since we have not found available data of bid-ask prices (Figure 4) of the above options from CME, we then have tested the bid-ask pricing mechanism of S&P500 index options operated by the system of market makers of CBOE. The data source is from Yahoo's finance quotes of the option prices from 07 December to 08 May 2006 (Figure 5). We have collected the prices of 5,000 time points, i.e., 5,000 different $t$ among 100 trading days. We denote this pricing mechanism by $\mathbb{O}^{mm}_{t,T}[X]$ for the ask price of an option $X$. According to our point of view the bid price of the same $X$ is $-\mathbb{O}^{mm}_{t,T}[-X]$ and thus the bid-ask spread is $\mathbb{O}^{mm}_{t,T}[X] + \mathbb{O}^{mm}_{t,T}[-X]$. We have tested a total number of 589,360 inequalities of (5.1) and (5.2), with $\mathbb{O}^{mm}$ in the place of $\mathbb{O}^m$. Only 1 case of violation appears.

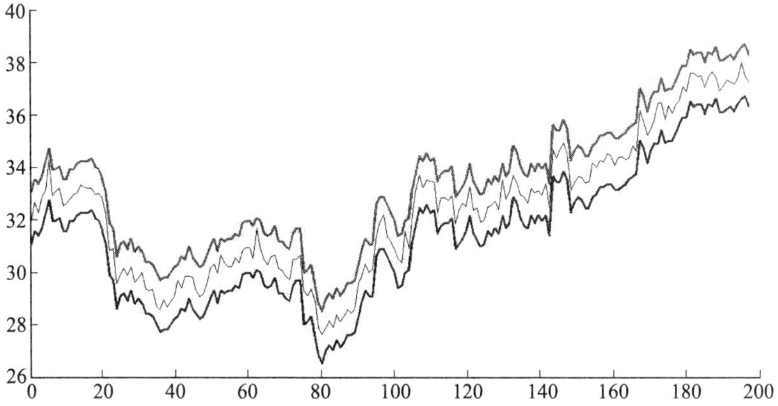


Figure 4 Ask, Bid & Last

We will report these test results in details in our forthcoming paper [3].

| | | | | | | | | | | | | | | |
|---|---|---|---|---|---|---|---|---|---|---|---|---|---|---|
| SZPDQ | 29.20 | +8.70 | 4.00 | 30.10 | 31.60 | 5,351.00 | **1,285.00** | SZPPQ | 3.10 | -1.00 | 1,322.00 | 2.75 | 3.40 | 12,182.00 |
| SZPDR | 28.20 | +4.50 | 3.00 | 25.90 | 27.40 | 13,031.00 | **1,290.00** | SZPPR | 3.80 | -1.30 | 3,675.00 | 3.70 | 3.90 | 28,198.00 |
| SZPDS | 24.10 | +5.10 | 45.00 | 21.90 | 23.40 | 6,420.00 | **1,295.00** | SZPPS | 4.70 | -1.70 | 4,166.00 | 4.70 | 5.10 | 28,184.00 |
| SXYDT | 19.10 | +2.90 | 1,691.00 | 18.50 | 19.50 | 58,185.00 | **1,300.00** | SXYPT | 6.10 | -1.60 | 3,164.00 | 5.90 | 6.20 | 46,912.00 |
| SXYDA | 16.50 | +3.30 | 586.00 | 14.90 | 16.10 | 14,188.00 | **1,305.00** | SXYPA | 7.60 | -1.50 | 4,478.00 | 7.20 | 7.80 | 9,624.00 |
| SXYDB | 12.80 | +2.70 | 1,895.00 | 12.20 | 12.90 | 24,616.00 | **1,310.00** | SXYPB | 9.50 | -2.00 | 982.00 | 9.10 | 9.50 | 6,692.00 |
| | | | | | | Stock Price ▸ | **1,311.56** | [illegible] | | | | | | |
| SXYDC | 10.00 | +2.50 | 3,432.00 | 9.20 | 9.90 | 10,329.00 | **1,315.00** | SXYPC | 11.80 | -2.70 | 104.00 | 11.40 | 12.40 | 898.00 |
| SXYDD | 7.40 | +1.70 | 3,566.00 | 7.10 | 7.50 | 11,403.00 | **1,320.00** | SXYPD | 14.10 | -2.70 | 744.00 | 14.00 | 14.50 | 4,367.00 |
| SXYDE | 5.50 | +1.30 | 14,490.00 | 4.90 | 5.50 | 50,337.00 | **1,325.00** | SXYPE | 16.80 | -3.80 | 528.00 | 16.90 | 18.10 | 3,331.00 |
| SXYDF | 4.20 | +1.30 | 1,092.00 | 3.50 | 4.20 | 11,412.00 | **1,330.00** | SXYPF | 23.50 | -0.20 | 11.00 | 20.30 | 21.80 | 796.00 |

Figure 5 SPX options data

See Figures (6–9) for same test results. We will report these test results in details in our forth coming paper [3].

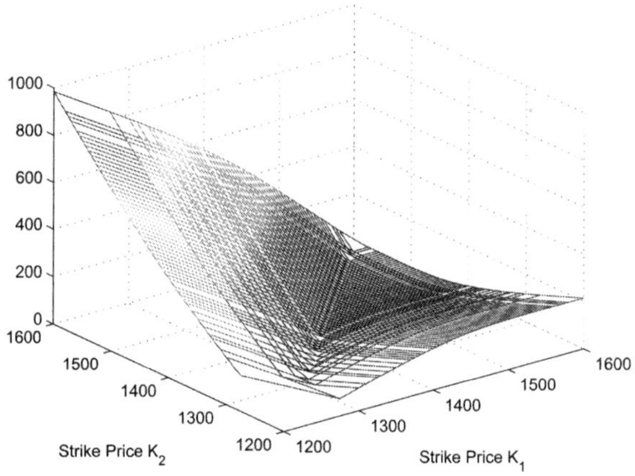


Figure 6 $E^{g_\mu}[(X_T - K_1)^+ - (X_T - K_2)^+] - E^m[(X_T - K_1)^+] - E^m[(X_T - K_2)^+]$

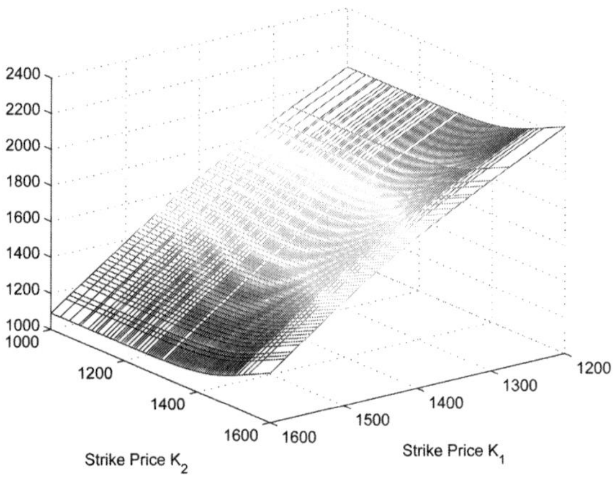


Figure 7 $E^{g_\mu}[(X_T - K_1)^+ - (K_2 - X_T)^+] - E^m[(X_T - K_1)^+] - E^m[(K_2 - X_T)^+]$

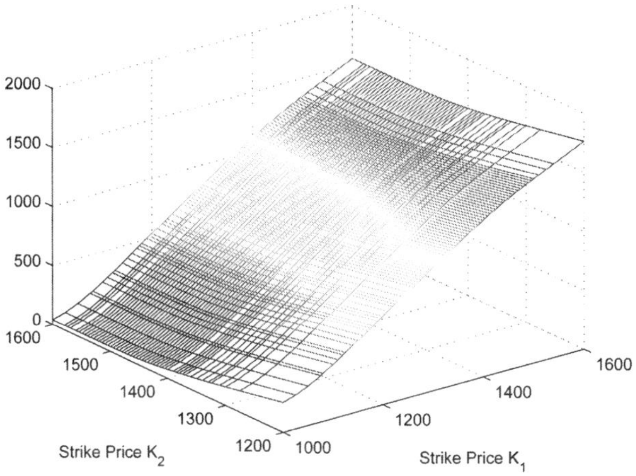


Figure 8 $E^{g_\mu}[(K_1 - X_T)^+ - (X_T - K_2)^+] - E^m[(K_1 - X_T)^+] - E^m[(X_T - K_2)^+]$

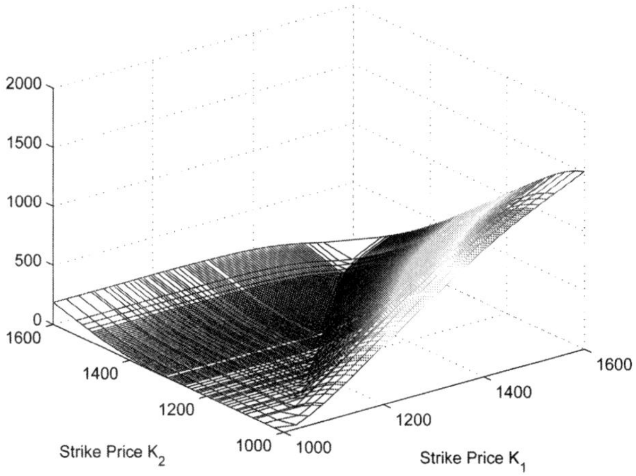


Figure 9 $E^{g_\mu}[(K_1 - X_T)^+ - (K_2 - X_T)^+] - E^m[(K_1 - X_T)^+] - E^m[(K_2 - X_T)^+]$

**Markovian pricing mechanisms.** We limit ourselves to consider, for each fixed maturity $T$, the derivatives $X$ depending only on the price $S_T$, i.e., $X$ is a path independent derivative. $X$ is then in the class of

$$X = \Phi(S_T) \text{ with } \Phi \in L^2(S_T),$$

where $L^2(S_T)$ denotes the collection of all real functions $\Phi$ defined on $R^n$ such that $\Phi(S_T) \in L^2(\mathcal{F}_T)$. A dynamic pricing mechanism $\mathbb{O}$ is called a Markovian pricing mechanism if for each $0 \leqslant t \leqslant T < \infty$ and $\Phi \in L^2(S_T)$ there exists $\phi \in L^2(S_t)$ such that $\mathbb{O}_{t,T}[\Phi(S_T)] = \phi(S_t)$.

In other words, *the price of a path-independent option by a Markovian pricing mechanism is still path-independent.*

**Example 5.1.** *We consider a situation where the underlying price $S$ is a diffusion process:*

$$dS_t = b(S_t)dt + \Lambda(S_t)dB_t, \quad S_0 = s_0 \in R^n,$$

*where $b$ and $\Lambda$ are given Lipschitz functions of $R^n$ valued on $R^n$ and $R^{n\times d}$ respectively. If a generating function $g$ has the following form:*

$$g(t, y, z) = f(S_t, y, z),$$

*where $f$ is a Lipschitz function of $(s, y, z) \in R^n \times R \times R^d$. By the nonlinear Feynman-Kac formula introduced in [19, Peng 1991] and developed in [18, Pardoux-Peng 1992], for each option $X = \Phi(S_T)$ with smooth function $\Phi$ the price of the related $g$-expectation is*

$$\mathbb{O}^g_{t,T}[\Phi(S_T)] = u(t, S_t),$$

*where $u : R_+ \times R^n \longmapsto R$ is the (viscosity) solution of the following PDE defined on $(t, s) \in [0, T] \times R^n$:*

$$\frac{\partial u}{\partial t} + \frac{1}{2}\sum_{i,j=1}^{n}(\Lambda(s)\Lambda^T(s))_{ij}\frac{\partial^2 u}{\partial s_i \partial s_j} + \sum_{i=1}^{n} b_i(s)\frac{\partial u}{\partial s_i} + f(s, u, \Lambda^T(s)\nabla u) = 0$$

*with terminal condition $u(T, s) = \Phi(s)$. If $S_t$ is a 1-dimensional geometric Brownian motion, i.e., $\Lambda(s) = \sigma s$ and $b(s) = \mu s$, then the above PDE becomes*

$$\frac{\partial u}{\partial t} + \frac{1}{2}\sigma^2 s^2 \frac{\partial^2 u}{\partial s^2} + \mu s \frac{\partial u}{\partial s} + f(s, u, \sigma s \frac{\partial u}{\partial s}) = 0.$$

*The Black-Scholes formula corresponds to the case $f = -ry - (\mu - r)\sigma^{-1}z$. We then have*

$$\frac{\partial u}{\partial t} + \frac{1}{2}\sigma^2 s^2 \frac{\partial^2 u}{\partial s^2} - ru + rs\frac{\partial u}{\partial s} = 0, \quad u(T, s) = \Phi(s).$$

# 6 Characterization of $g$-pricing mechanism by its generating function $g$

For a pricing mechanism, it is important to distinguish the selling price and buying price of a same pricing mechanism, corresponding to the ask price and the bid price if the mechanism under investigation is generated through the system of market makers of an option market (cf. [12] Section

6.5 and [16]). If $\mathbb{O}_{t,T}[X]$ is the ask price at the time $t$ of a derivative $X$ with maturity $T$, then the bid price must be $-\mathbb{O}_{t,T}[-X]$ and we have, in general,

$$\mathbb{O}_{t,T}[X] > -\mathbb{O}_{t,T}[-X].$$

Here we stress our point of view that, in fact, the ask price and bid price are produced by a single mechanism, called bid-ask pricing mechanism of market makers. Our result of data analysis to test the criteria (A5) of the domination condition (4.1) strongly supports this point of view. Moreover, this analysis also supports our point of view that, for a well-developed market, there exist a function $g$ satisfying Lipschitz condition (3.2) such that the corresponding ask-bid pricing mechanism is modeled by the $g$-expectation $\mathbb{O}^g[\cdot]$.

A rational dynamic pricing mechanism also possesses some other important properties, such as convexity, sub-additivity. See [9], [22], [30], [13], [14] among many others. We will see that the generating function $g$ perfectly reflects the behavior of $\mathbb{O}^g$. This may play an important role to statistically find $g$ by using the corresponding data of prices. In the following we provide several theoretical results with proofs given in Appendix. This problem was treated also by [30], [13] and [14].

**Proposition 6.1.** *Let $g, \bar{g} : (\omega, t, y, z) \in \Omega \times [0, \infty) \times R \times R^d \to R$ be two generating functions satisfying (3.2). Then the following two conditions are equivalent:*
*(i) $g(\omega, t, y, z) \geqslant \bar{g}(\omega, t, y, z)$, $\forall (y, z) \in R \times R^d$, $dP \times dt$ a.s.,*
*(ii) The corresponding $g$-pricing mechanisms $\mathbb{O}^g[\cdot]$ and $\mathbb{O}^{\bar{g}}[\cdot]$ satisfy*

$$\mathbb{O}^g_{t,T}[X] \geqslant \mathbb{O}^{\bar{g}}_{t,T}[X], \quad \forall 0 \leqslant t \leqslant T < \infty, \quad \forall X \in L^2(\mathcal{F}_T).$$

*In particular, $\mathbb{O}^g[X] \equiv \mathbb{O}^{\bar{g}}[X]$ if and only if $g \equiv \bar{g}$.*

**Corollary 6.2.** *The following two conditions are equivalent:*
*(i) The generating function $g$ satisfies, for each $(y, z) \in R \times R^d$,*

$$g(t, y, z) \geqslant -g(t, -y, -z), \text{ a.e., a.s.,}$$

*(ii) $\mathbb{O}^g_{t,T}[\cdot] : L^2(\mathcal{F}_T) \longmapsto L^2(\mathcal{F}_t)$ satisfies, for each $0 \leqslant t \leqslant T$.*

**Proposition 6.3.** *The following two conditions are equivalent:*
*(i) The generating function $g = g(t, y, z)$ is convex (resp. concave) in $(y, z)$, i.e., for each $(y, z)$ and $(\bar{y}, \bar{z})$ in $R \times R^d$ and for a.e. $t \in [0, T]$,*

$$g(t, \alpha y + (1-\alpha)\bar{y}, \alpha z + (1-\alpha)\bar{z}) \leqslant \alpha g(t, y, z) + (1-\alpha) g(t, \bar{y}, \bar{z}), \text{ a.s.}$$
$$(\text{resp.} \geqslant \alpha g(t, y, z) + (1-\alpha) g(t, \bar{y}, \bar{z}), \text{ a.s.}).$$

*(ii) The corresponding pricing mechanism* $(\mathbb{O}^g_{t,T}[\cdot])_{0\leqslant t\leqslant T<\infty}$ *is convex (resp. concave), i.e., for each fixed* $\alpha\in[0,1]$, *we have*

$$\mathbb{O}^g_{t,T}[\alpha X+(1-\alpha)\bar{X}]\leqslant \alpha\mathbb{O}^g_{t,T}[X]+(1-\alpha)\mathbb{O}^g_{t,T}[\bar{X}],\ a.s. \tag{6.1}$$
$$(resp.\ \geqslant \alpha\mathbb{O}^g_{t,T}[X]+(1-\alpha)\mathbb{O}^g_{t,T}[\bar{X}],\ a.s.)$$
$$for\ each\ t\leqslant T,\ \ and\ X,\bar{X}\in L^2(\mathcal{F}_T).$$

**Proposition 6.4.** *The following two conditions are equivalent:*
*(i) The generating function g is positively homogenous in* $(y,z)\in R\times R^d$, *i.e.,*

$$g(t,\lambda y,\lambda z)=\lambda g(t,y,z),\ a.e.,\ a.s.,$$

*(ii) The corresponding pricing mechanism* $\mathbb{O}^g_{t,T}[\cdot]:L^2(\mathcal{F}_T)\longmapsto L^2(\mathcal{F}_t)$ *is positively homogenous: for each* $0\leqslant t\leqslant T$ *, i.e.,* $\mathbb{O}^g_{t,T}[\lambda X]=\lambda\mathbb{O}^g_{t,T}[X]$, *for each* $\lambda\geqslant 0$ *and* $X\in L^2(\mathcal{F}_T)$.

From the above two propositions we immediately have

**Corollary 6.5.** *The following two conditions are equivalent:*
*(i) The generating function g is sub-additive: for each* $(y,z)$, $(\bar{y},\bar{z})\in R\times R^d$,

$$g(\omega,t,y+\bar{y},z+\bar{z})\leqslant g(\omega,t,y,z)+g(\omega,t,\bar{y},\bar{z}),\quad dt\times dP,\ a.s.,$$

*(ii) The corresponding pricing mechanism* $\mathbb{O}^g_{t,T}[\cdot]:L^2(\mathcal{F}_T)\longmapsto L^2(\mathcal{F}_t)$ *is sub-additive: for each* $0\leqslant t\leqslant T$ *and* $X$, $\bar{X}\in L^2(\mathcal{F}_T)$

$$\mathbb{O}^g_{t,T}[X+\bar{X}]\leqslant \mathbb{O}^g_{t,T}[X]+\mathbb{O}^g_{t,T}[\bar{X}].$$

**Proposition 6.6.** *The generating function g is independent of y if and only if the corresponding g-expectation satisfies the following "cash translatability" property: for each* $t\leqslant T$,

$$\mathbb{O}^g_{t,T}[X+\eta]=\mathbb{O}^g_{t,T}[X]+\eta,\ a.s.,\ for\ each\ X\in L^2(\mathcal{F}_T),\ \eta\in L^2(\mathcal{F}_t).$$

We consider the following self-financing condition:

$$\mathbb{O}^g_{t,T}[0]\equiv 0,\quad \forall 0\leqslant t\leqslant T.$$

**Proposition 6.7.** $\mathbb{O}^g[\cdot]$ *satisfies the self-financing condition if and only if its generating function g satisfies (3.3)-(a).*

*Proof.* The "if" part is obvious. The "only if part": $Y_t:=\mathbb{O}^g_{t,T}[0]\equiv 0$, implies

$$Y_t\equiv 0\equiv 0+\int_t^T g(s,0,Z_s)ds-\int_t^T Z_s dB_s,\quad t\in[0,T].$$

Thus $Z_t\equiv 0$ and then $g(t,0,Z_t)=g(t,0,0)\equiv 0$. □

"Zero-interest rate" condition:

$$\mathbb{O}^g_{t,T}[\eta] = \eta, \quad \forall 0 \leqslant t \leqslant T < \infty, \eta \in L^2(\mathcal{F}_t).$$

**Proposition 6.8.** *$\mathbb{O}^g[\cdot]$ satisfies the zero-interest rate condition if and only if its generating function $g$ satisfies (3.3)-(b).*

**Proposition 6.9.** *The following condition are equivalent:*
*(i) For each $0 \leqslant t \leqslant T$ and $X \in L^2(\mathcal{F}^t_T)$, the g-pricing mechanism $\mathbb{O}^g_{t,T}[X]$ is a deterministic number;*
*(ii) The corresponding pricing generating function $g$ is a deterministic function of $(t, y, z) \in [0, T] \times R \times R^d$.*

The proof is similar as the others. We omit it.

**Example 6.10.** *An interesting problem is: if we know that a pricing mechanism under our investigation is a g-expcctation $\mathbb{O}^g$, how to find the generating function $g$? If we limited ourselves to only take data of prices quoted by markets, this is still an open problem. We now consider a case of "toy model" where $g$ depends only on $z$, i.e., $g = g(z) : \mathbf{R}^d \to \mathbf{R}$. We will find such $g$ by the following testing method. Let $\bar{z} \in R^d$ be given. We denote $Y_s := \mathbb{O}^g_{s,T}[\bar{z} \cdot (B_T - B_t)]$, $s \in [t, T]$, where $t$ is the present time. It is the solution of the following BSDE*

$$Y_s = \bar{z} \cdot (B_T - B_t) + \int_s^T g(Z_u)du - \int_s^T Z_u dB_u, \quad s \in [t, T].$$

*It is seen that the solution is $Y_s = \bar{z} \cdot (B_s - B_t) + \int_s^T g(\bar{z})ds$, $Z_s \equiv \bar{z}$. Thus*

$$\mathbb{O}^g_{t,T}[\bar{z} \cdot (B_T - B_t)] = Y_t = g(\bar{z})(T - t),$$

*or*

$$g(\bar{z}) = (T - t)^{-1} \mathbb{O}^g_{t,T}[\bar{z} \cdot (B_T - B_t)]. \tag{6.2}$$

*Thus the function $g$ can be tested as follows: at the present time $t$: if the valuation $\mathbb{O}^g_{t,T}[\bar{z} \cdot (B_T - B_t)]$ of (a toy model of) derivative $\bar{z} \cdot (B_T - B_t)$ is obtained, then $g(\bar{z})$ is explicitly given by (6.2). We observe that, in the case where $S$ is a geometric Brownian motion, $B_T - B_t$ can be expressed as a function of $S_T/S_t$. But this cannot be applied to a general situation.*

**Remark 6.11.** *The above test is also applied for the case $g : [0, \infty) \times \mathbf{R}^d \to \mathbf{R}$, or for a more general situation $g = \gamma y + g_0(t, z)$.*

An interesting problem is, in general, how to find the generating function $g$ by a testing of the input-output behavior of $\mathbb{O}^g[\cdot]$? Let $b : R^n \longmapsto R^n$, $\bar{\sigma} : R^n \longmapsto R^{n \times d}$ be two Lipschitz functions. For each $(t, x) \in R_+ \times R^n$, we consider the SDE of the form

$$X^{t,x}_s = x + \int_t^s b(X^{t,x}_s)ds + \int_t^s \sigma(X^{t,x}_s)dB_s, \quad s \geqslant t.$$

This SDE is regarded as the equation of the price of the underlying stock. The following result was obtained in Proposition 2.3 of [2].

**Proposition 6.12.** *We assume that the generating function $g$ satisfies (3.2). We also assume that, for each fixed $(y,z)$, $g(\cdot,y,z) \in D^2_{\mathcal{F}}(0,T)$ (the space of all $\mathcal{F}_t$-adapted processes with RCLL paths). Then for each $(t,x,p,y) \in [0,\infty) \times R^n \times R^n \times R$, we have*

$$L^2\text{-}\lim_{\epsilon\to 0} \frac{1}{\epsilon}[\mathbb{O}^g_{t,t+\epsilon}[y + p\cdot(X^{t,x}_{t+\epsilon} - x)] - y] = g(t,y,\sigma^T(x)p) + p\cdot b(x).$$

# 7 Conclusion

Our huge member of tests, using market data and numerical BSDE calculation technique, strongly support that the market ask and bid pricing mechanism of CME options on S&P500 is a $g$-pricing mechanism. We have also proposed different types of generating functions $g$ (convex function, concave sublinear function, superlinear function) to be good candidates of different pricing mechanisms for sellers, buyers, large investors and small investors. But to solve the problem of how to test a specific financial company's pricing mechanism, we need more specific data which is, in general, a high secret.

# References


[1] J.-M. Bismut. Conjugate convex functions in optimal stochastic control. *J. Math. Anal. Appl.* **44**, 384-404, 1973.

[2] P. Briand, F. Coquet, Y. Hu, J. Mémin and S. Peng. A converse comparison theorem for BSDEs and related properties of *g-expectations*, *Electron. Comm. Probab.* **5**, 2000.

[3] L. Chen, S. Peng and P. Sun Testing the domination inequality of dynamic pricing mechanisms, working paper, 2006.

[4] F. Coquet, Y. Hu, J. Memin and S. Peng. Filtration-consistent nonlinear expectations and related $g$-expectations. *Probab. Theory Relat. Fields* **123**, 1-27, 2002.

[5] J. Cvitanic and I. Karatzas. Hegding contingent claims with constrained portfolios *Ann. Appl. Probab.* **2**, 767-818, 1993.

[6] Z. Chen and S. Peng. A Nonlinear Doob-Meyer type Decomposition and its Application. *SUT Journal of Mathematics* (Japan), 34(2), 197-208, 1998.

[7] Z. Chen and S. Peng. Continuous Properties of $g$-martingales. *Chin. Ann. of Math.* **22B: 1**, 115-128, 2001.

[8] Z. Chen and S. Peng. A general down crossing inequality for $g$-martingales. *Statist. Probab. Lett.*, 46(2), 169-175, 2000.

[9] N.El Karoui, N.S. Peng and M.-C. Quenez. Backward stochastic differential equation in finance. *Math. Finance*, 7(1), 1-71, 1997.

[10] N.El Karoui, C. Kapoudjian, E. Pardoux, S. Peng and M.-C. Quenez. Reflected Solutions of Backward SDE and Related Obstacle Problems for PDEs. *Ann. Probab.*, 25(2), 702-737, 1997.

[11] H. He and N.D. Pearson. Consumption and portfolio policies with incomplete markets and short-sale constrains: the infinite dimensional case. *J. Econ. Theory* **54**, 259-304, 1991.

[12] J. Hull. *Options, Futures and other Derivative Securities*, Prentice Hall, 1997.

[13] L. Jiang. Some results on the uniqueness of generators of backward stochastic differential equations. *C. R. Acad. Sci. Paris*, Ser. I 338 575-580, 2004.

[14] L. Jiang. A converse comparison theorem for g-expectations. *Acta Mathematicae Applticae Sinica*, English Series, 20(4), 701-706, 2004.

[15] E. Jouini and H. Kallal. Arbitrage in securities markets with short-sale constraints. *Math. Finance* **5**, 178-197, 1995.

[16] M. Musiela and M. Rutkowski. *Martingale Methods in Financial Modelling*, Springer, 1997.

[17] E. Pardoux and S. Peng. Adapted solution of a backward stochastic differential equation. *Systems and Control Letters*, 14(1), 55-61, 1990.

[18] E. Pardoux and S. Peng. Backward Stochastic Differential equations and Quasilinear Parabolic Partial Differential Equations. *Lecture Notes in CIS* **176**, 200-217, Springer, 1992.

[19] S. Peng. Probabilistic Interpretation for Systems of Quasilinear Parabolic Partial Differential Equations. *Stochastics* **37**, 61-74, 1991.

[20] S. Peng. BSDE and Stochastic Optimizations. *Topics in Stochastic Analysis*, J. Yan, S. Peng, S. Fang and L.M. Wu, Ch.2 (Chinese vers.), Science Publication, 1997.

[21] S. Peng. BSDE and related $g$-expectation. *Pitman Research Notes in Mathematics Series*, 364, "Backward Stochastic Differential Equation", eds. N. El Karoui & L. Mazliak, 141-159, 1997.

[22] S. Peng and F. Yang. Duplicating and Pricing Contingent Claims in Incomplete Markets. *Pacific Economic Review*, 4(3): 237-260, 1999.

[23] S. Peng. Nonlinear expectations and nonlinear Markov chains. *Chin. Ann. Math.*, 26B(2), 159-184, 2005.

[24] S. Peng. Filtration Consistent Nonlinear Expectations and Evaluations of Contingent Claims. *Acta Mathematicae Applicatae Sinica*, English Series, 20(2), 1-24, 2004.

[25] S. Peng. Nonlinear expectation, nonlinear evaluations and risk measurs, K. Back T. R. Bielecki, C. Hipp, S. Peng, W. Schachermayer. *Stochastic Methods in Finance Lectures*, (eds. M. Frittelli and W. Runggaldier) 143-217, LNM 1856, Springer-Verlag, 2004.

[26] S. Peng. Dynamical evaluations. *C. R. Acad. Sci. Paris*, Ser. I 339, 585-589, 2004.

[27] S. Peng. Dynamically Consistent Nonlinear Evaluations and Expectations. arXiv: math.PR/0501415 v1 24 Jan 2005.

[28] S. Peng. Modelling Derivatives Pricing Mechanisms with Their Generating Functions. arXiv: math.PR/0605599 v1 23 May 2006.

[29] The Numerical Algorithms and simulations for BSDEs, Preprint in arXiv:math.PR/0611864 v1 28 Nov 2006.

[30] E. R. Gianin. Some examples of risk measures via g-expectations, preprint. *Insurance: Mathematics and Economics*, 2003.

# Analysis of Nonconforming Rotated $\mathcal{Q}_1$ Element for the Reissner-Mindlin Plate problem*

Jun Hu
*LMAM and School of Mathematical Sciences, Peking University, Beijing 100871, China*

Zhongci Shi
*Institute of Computational Mathematics, Chinese Academy of Sciences, Beijing 100080, China*

**Abstract**

In this paper, we analyze a recently proposed nonconforming finite element method for the Reissner-Mindlin plate problem [10, 13], namely the MS 2 element. It is proved that the rotation is bounded in the $H^1$ norm and converges at a suboptimal rate in the $L^2$ norm, while the displacement converges at the optimal rate with respect to both the $H^1$ and the $L^2$ norms as the plate thickness $t$ is fixed.

# 1 Introduction

Assume that the plate is clamped along the boundary $\partial\Omega$ and denote by $\omega$ and $\boldsymbol{\phi}$ the transverse displacement and rotation, respectively, of the solution of the following boundary value problem, called the Reissner-Mindlin plate problem: Find $(\boldsymbol{\phi}, \omega) \in \boldsymbol{H}_0^1(\Omega) \times H_0^1(\Omega)$ such that

$$a(\boldsymbol{\phi}, \boldsymbol{\psi}) + (\boldsymbol{\gamma}, \nabla v - \boldsymbol{\psi}) = (g, v), \quad \forall (\boldsymbol{\psi}, v) \in \boldsymbol{H}_0^1(\Omega) \times H_0^1(\Omega) \qquad (1.1)$$

with the shear strain $\boldsymbol{\gamma}$ defined as

$$\boldsymbol{\gamma} = \lambda t^{-2}(\nabla \omega - \boldsymbol{\phi}).$$

*2000 *Mathematics Subject Classification.* 65N30. This research was supported by the Special Funds for Major State Basic Research Project. The first author was in part supported by the NSFC Grant 10601003.

In (1.1) $g$ is the scaled transverse loading, $t$ the plate thickness, $\lambda = E\kappa/2(1+\nu)$ the shear modulus with $E$ Young's modulus, $\nu$ the Poisson ratio and $\kappa$ the shear correction factor. The bilinear form $a$ is defined as $a(\boldsymbol{\eta}, \boldsymbol{\psi}) = (\mathcal{C}\mathcal{E}(\boldsymbol{\eta}), \mathcal{E}(\boldsymbol{\psi}))$ with $\mathcal{E}(\boldsymbol{\psi})$ the symmetric part of the gradient $\nabla\boldsymbol{\psi}$, and $\mathcal{C}\tau$ defined for any $2 \times 2$ symmetric matrix $\tau$ as

$$\mathcal{C}\tau := \frac{E}{12(1-\nu^2)}\left[(1-\nu)\tau + \nu\,\mathrm{tr}(\tau)\boldsymbol{I}\right].$$

Recently, a lower order quadrilateral finite element has been proposed in [13] (hereafter MS 2 element for short) for the Reissner-Mindlin plate problem. It is proved [12, 13] that for this new element, the discrete Korn inequality does not hold uniformly in the mesh size $h$ since both components of the rotation are approximated by the nonconforming rotated $\mathcal{Q}_1$ element [15] (hereafter NR $\mathcal{Q}_1$ element) that causes the lack of uniform ellipticity of the discrete problem with respect to $h$ and $t$. Using a $(t, h)$-dependent constant, it is shown the optimal error estimate under the condition $t = \mathcal{O}(h)$ [13]. This result is unsatisfactory since it does not indicate that the convergence is uniform with respect to $h$ and $t$. In this paper we want to improve the above result in the following sense that the finite element solution actually converges as the mesh size $h$ tends to zero while the plate thickness $t$ is fixed. The analysis is based on a superconvergence property of the consistency error for the nonconforming rotated $\mathcal{Q}_1$ element and an improved result for the lowest order Raviart-Thomas element. In particular, we prove that the rotation is bounded in the $H^1$ norm and converges at a suboptimal rate in the $L^2$ norm while the displacement converges at the optimal rate with respect to both the $H^1$ and the $L^2$ norms as $t$ is fixed. Our result can be summarized in the following theorem

**Theorem 1.1.** *Let $(\boldsymbol{\phi}, \omega)$ and $(\boldsymbol{\phi}_h, \omega_h)$ be the solutions of Problem* (1.1) *and Problem* (2.5) *in Section 2 respectively. Then,*

$$\begin{aligned}&\|\mathcal{E}_h(\boldsymbol{\phi}-\boldsymbol{\phi}_h)\|_0 + \|\boldsymbol{\phi}-\boldsymbol{\phi}_h\|_0 + \|\nabla(\omega-\omega_h)\|_0 + t\|\gamma-\gamma_h\|_0\\ &\leqslant Ch(\|\boldsymbol{\phi}\|_3 + \|\gamma\|_1 + \|g\|_0).\end{aligned}$$

The proof will be given in Section 3.

**Remark 1.2.** *Another proof of this theorem can be found in [9] and [11]. However, the proof presented herein is much simpler.*

An outline of the remaining parts of the paper is as follows. Section 2 introduces the finite element spaces as well as the discrete problem, and an improved error estimate for the reduction operator and the superconvergence of the consistency error are presented. The proof of Theorem 1.1 will be given in Section 3.

This paper will use the standard notations for both Sobolev spaces and differential operators [1]. The generic constant $C$ is always assumed to be independent of the plate thickness $t$ and the mesh size $h$.

# 2 Finite element approximation

This section presents the discrete problem, the error estimates of the interpolation operator and the consistency, and the discrete Korn inequality.

## 2.1 Triangulation

We use finite element method to approximate the Problem 1.1 and introduce a rectangular mesh subdivision $\mathcal{T}_h$ of the rectangular domain $\Omega$. The regularity of the mesh $\mathcal{T}_h$ is assumed in the sense of Ciarlet [7] such that $\bigcup_{K\in\mathcal{T}_h} K = \bar{\Omega}$. Two distinct elements $K$ and $K'$ in $\mathcal{T}_h$ are either disjoint, or share the common edge $e$, or a common vertex. Let $F$ denote the set of all edges in $\mathcal{T}_h$ with $F'$ the set of interior edges. Given any edge $e \in F$ we assign an unit normal $\boldsymbol{n}_e$. In relation to $\boldsymbol{n}_e$ one can define the element $K^+ \in \mathcal{T}_h$ and element $K^- \in \mathcal{T}_h$ with $e = K^+ \cap K^-$. Let $\partial K$ denote the boundary of $K$.

For each $K \in \mathcal{T}_h$, we introduce the following affine invertible transformation

$$F_K : \hat{K} \to K, \quad x = \frac{h_{x,K}}{2}\xi + x_{0,K}, \quad y = \frac{h_{y,K}}{2}\eta + y_{0,K}$$

with $(x_{0,K}, y_{0,K})$ the center and $h_{x,K}$ and $h_{y,K}$ the horizontal and vertical edge length of $K$, respectively, and $\hat{K} = [-1,1]^2$ the reference element.

## 2.2 Finite element spaces and lemmas

Given $S \subset R^2$, denote by $Q_{i,j}(S)$, the space of polynomials of degree not greater than $i$ for the first variable and not greater than $j$ for the second variable defined over $S$, and set $Q_k(S) = Q_{k,k}(S)$.

We use the conforming bilinear element space

$$W_h = \{\, v \in H_0^1(\Omega) \mid v_{|K} \in Q_1(K), \quad \forall K \in \mathcal{T}_h \,\}$$

for the approximation of the displacement. The rotation is approximated by the NR $\mathcal{Q}_1$ element space [15] defined as

$$N_h = \{\, v \in L^2(\Omega) \mid v_{|K} \in NR(K) \,\text{and}\, v \,\text{is continuous regarding}\, \Pi_{\mathcal{F}} \,\},$$

where

$$NR(K) = \{\, q \circ F_K^{-1} \mid q \in \mathrm{Span}\langle 1, \xi, \eta, \xi^2 - \eta^2 \rangle \,\},$$

and the nodal functional $\Pi_{\mathcal{F}}, \forall \mathcal{F} \subset \partial K$ is

$$\Pi_{\mathcal{F}} = \frac{1}{|\mathcal{F}|} \int_{\mathcal{F}} v \, ds, \quad \forall v \in H^1(K).$$

The homogeneous NR $\mathcal{Q}_1$ space is defined as

$$N_{0,h} = \{\, v \in N_h \mid \Pi_{\mathcal{F}}(v) = 0, \quad \text{if} \, \mathcal{F} \subset \partial\Omega \,\}.$$

Furthermore, we introduce the following broken norm and seminorm on $N_h$

$$\|v\|_{1,h} = \sum_{K \in \mathcal{T}_h} \|v\|_{1,K}^2, \quad |v|_{1,h} = \sum_{K \in \mathcal{T}_h} |v|_{1,K}^2.$$

It is obvious that $|\cdot|_{1,h}$ is a norm on $N_{0,h}$.

Define

$$\boldsymbol{V}_h = N_{0,h} \times N_{0,h}.$$

Since $\boldsymbol{V}_h$ is nonconforming, when differential operators such as $\mathcal{E}$, div, curl, rot and $\nabla$ may be applied to functions in $\boldsymbol{V}_h$ in a piecewise manner, we shall write $\mathcal{E}_h, \mathrm{div}_h, \mathrm{curl}_h, \mathrm{rot}_h$ and $\nabla_h$ in all these cases.

For any $v \in H^1(\Omega)$, define its interpolation $\Pi_h v \in N_h$ and

$$\Pi_{\mathcal{F}_i}(\Pi_h v \mid_K) = \Pi_{\mathcal{F}_i}(v \mid_K), \quad \mathcal{F}_i \subset \partial K, \quad i = 1, \cdots, 4, \quad \forall K \in \mathcal{T}_h. \tag{2.1}$$

We have

**Lemma 2.1.** [15] *The interpolation operators $\Pi_{\mathcal{F}}$ and $\Pi_h$ admit the following estimates*

$$\|v - \Pi_{\mathcal{F}}(v)\|_{0,\mathcal{F}} \leqslant C h_K^{1/2} |v|_{1,K}, \quad \forall v \in H^1(K) \tag{2.2}$$

*and*

$$\|v - \Pi_h v\|_0 + h\|v - \Pi_h v\|_{1,h} \leqslant C h^2 \|v\|_2, \quad \forall v \in H_0^1(\Omega) \cap H^2(\Omega). \tag{2.3}$$

For the nonconforming rotated $\mathcal{Q}_1$ element, we prove the following superconvergence for the consistency error.

**Lemma 2.2.** *For any $u \in H^2(\Omega)$, there exists a constant $C$ independent of $h$ such that*

$$\Big| \sum_{K \in \mathcal{T}_h} \int_{\partial K} u v \nu_j ds \Big| \leqslant C h^2 |u|_2 |v|_{1,h}, \quad \forall v \in N_{0,h}, \quad j = 1, 2,$$

*with $\nu_j$, $j = 1, 2$, the two components of the outer unit normal vector to $\partial K$.*

*Proof.* By symmetry, we only proves this result for the case $j = 1$. Since $v$ is continuous regarding $\Pi_{\mathcal{F}}$, we have

$$\sum_{K\in\mathcal{T}_h}\int_{\partial K} uv\nu_1 ds = \sum_{K\in\mathcal{T}_h}\sum_{\mathcal{F}\subset\partial K}\int_{\mathcal{F}}(u-\Pi_{\mathcal{F}}u)(v-\Pi_{\mathcal{F}}v)\nu_1 ds$$
$$= \sum_{K\in\mathcal{T}_h}\Big(\int_{\mathcal{F}_3}(u-\Pi_{\mathcal{F}_3}u)(v-\Pi_{\mathcal{F}_3}v)dy - \int_{\mathcal{F}_1}(u-\Pi_{\mathcal{F}_1}u)(v-\Pi_{\mathcal{F}_1}v)dy\Big),$$

with $\mathcal{F}_3$ and $\mathcal{F}_1$ two vertical edges of $K$. To get the superconvergence of the consistency error, we need some cancelation between integrals $\int_{\mathcal{F}_1}(u-\Pi_{\mathcal{F}_1}u)(v-\Pi_{\mathcal{F}_1}v)$ and $\int_{\mathcal{F}_3}(u-\Pi_{\mathcal{F}_3}u)(v-\Pi_{\mathcal{F}_3}v)ds$ on each element. In fact,

$$\int_{\mathcal{F}_3}(u-\Pi_{\mathcal{F}_3}u)(v-\Pi_{\mathcal{F}_3}v)dy - \int_{\mathcal{F}_1}(u-\Pi_{\mathcal{F}_1}u)(v-\Pi_{\mathcal{F}_1}v)dy$$
$$= \int_{y_{0,K}-h_{y,K}/2}^{y_{0,K}+h_{y,K}/2}((u-\Pi_{\mathcal{F}_3}u)|_{\mathcal{F}_3} - (u-\Pi_{\mathcal{F}_1}u)|_{\mathcal{F}_1})(v-\Pi_{\mathcal{F}_1}v)dy,$$

since $(v-\Pi_{\mathcal{F}_1}v)|_{\mathcal{F}_1} = (v-\Pi_{\mathcal{F}_3}v)|_{\mathcal{F}_3}$. Let $a = x_{0,K}-\frac{h_{x,K}}{2}$, $b = x_{0,K}+\frac{h_{x,K}}{2}$, $c = y_{0,K}-h_{y,K}/2$ and $d = y_{0,K}+h_{y,K}/2$, we deduce it as

$$|(u-\Pi_{\mathcal{F}_3}u)|_{\mathcal{F}_3} - (u-\Pi_{\mathcal{F}_1}u)|_{\mathcal{F}_1}|$$
$$= |h_{y,K}^{-1}\int_c^d (u(b,y)-u(b,z))dz - h_{y,K}^{-1}\int_c^d (u(a,y)-u(a,z))dz|$$
$$= |h_{y,K}^{-1}\int_c^d\int_z^y \Big(\frac{\partial u(b,s_2)}{\partial s_2} - \frac{\partial u(a,s_2)}{\partial s_2}\Big)ds_2|$$
$$= |h_{y,K}^{-1}\int_c^d\int_z^y\int_a^b \frac{\partial^2 u(s_1,s_2)}{\partial s_1\partial s_2}ds_1 ds_2| \leqslant Ch|u|_{2,K},$$

which together with Lemma 2.1 yield

$$\Big|\int_{\mathcal{F}_3}(u-\Pi_{\mathcal{F}_3}u)(v-\Pi_{\mathcal{F}_3}v)dy - \int_{\mathcal{F}_1}(u-\Pi_{\mathcal{F}_1}u)(v-\Pi_{\mathcal{F}_1}v)dy\Big|$$
$$\leqslant Ch^2|u|_{2,K}|v|_{1,K}.$$

A summation over all elements shows the desired result. □

**Remark 2.3.** *Another proof of the above result can be found in [12, Lemma 3].*

Finally, we use the lowest-order rotated Raviart-Thomas space (R-T hereafter) for the approximation of the shear force as follows:

$$\mathbf{\Gamma}_h = \{\,\boldsymbol{\chi}\in \boldsymbol{H}_0(\mathrm{rot},\Omega) \mid \boldsymbol{\chi}_{|K}\in Q_{0,1}(K)\times Q_{1,0}(K), \forall K\in\mathcal{T}_h\,\}.$$

On each element we define the local interpolation operator $\boldsymbol{R}_{h|_K} = \boldsymbol{R}_K$

$$\int_{\mathcal{F}} (\boldsymbol{\psi} - \boldsymbol{R}_K \boldsymbol{\psi}) \cdot \boldsymbol{t}\, ds = 0, \quad \forall \mathcal{F} \subset \partial K$$

for any $\boldsymbol{\psi} \in \boldsymbol{H}^1(\Omega) \cap \boldsymbol{H}_0(\mathrm{rot}, \Omega)$.

Since $\int_{\mathcal{F}} \boldsymbol{\psi} \cdot \boldsymbol{t} ds$ is also well-defined for $\boldsymbol{\psi} \in \boldsymbol{V}_h$, then the operator $\boldsymbol{R}_h$ is also well-defined for such $\boldsymbol{\psi}$ and $\boldsymbol{R}_h \boldsymbol{\psi} \in \boldsymbol{\Gamma}_h$ for $\boldsymbol{\psi} \in \boldsymbol{V}_h$. The interpolation estimate for $\boldsymbol{R}_h$ reads as

**Lemma 2.4.** *For any $K \in \mathcal{T}_h$, there holds*

$$\|(\boldsymbol{I} - \boldsymbol{R}_K)\boldsymbol{\sigma}\|_{0,K} \leqslant Ch \|\mathcal{E}(\boldsymbol{\sigma})\|_{0,K}, \quad \forall \boldsymbol{\sigma} \in \boldsymbol{H}^1(K), \tag{2.4}$$

$$\|\mathrm{rot}(\boldsymbol{\sigma} - \boldsymbol{R}_K \boldsymbol{\sigma})\|_{0,K} \leqslant Ch |\mathrm{rot}\, \boldsymbol{\sigma}|_{1,K}, \quad \forall \boldsymbol{\sigma} \in \boldsymbol{H}^1(\mathrm{rot}, K) \tag{2.5}$$

*with $\boldsymbol{\sigma} = (\sigma_1, \sigma_2)$.*

*Proof.* It follows immediately from the definition of the interpolation operator $\boldsymbol{R}_K$ that

$$\begin{aligned} \|(\boldsymbol{I} - \boldsymbol{R}_K)\boldsymbol{\sigma}\|_{0,K}^2 &= \int_K |(\boldsymbol{\sigma} - \boldsymbol{R}_h \boldsymbol{\sigma})|^2 dxdy \\ &\leqslant Ch^2 (\|\frac{\partial \sigma_1}{\partial x}\|_{0,K}^2 + \|\frac{\partial \sigma_2}{\partial y}\|_{0,K}^2) \\ &\leqslant Ch^2 \|\mathcal{E}(\boldsymbol{\sigma})\|_{0,K}^2. \end{aligned} \tag{2.6}$$

The proof of the second estimate can be found, for instance, in [9]. □

## 2.3 The discrete problem and the Korn inequality

The MS 2 element [13] for the R-M plate model can be stated as the solution of the following problem.

**Problem 2.5.** *Find $(\boldsymbol{\phi}_h, \omega_h) \in \boldsymbol{V}_h \times W_h$ such that*

$$a_h(\boldsymbol{\phi}_h, \boldsymbol{\psi}) + \lambda t^{-2} (\nabla \omega_h - \boldsymbol{R}_h \boldsymbol{\phi}_h, \nabla v - \boldsymbol{R}_h \boldsymbol{\psi}) = (g, v), \quad \forall (\boldsymbol{\psi}, v) \in \boldsymbol{V}_h \times W_h, \tag{2.7}$$

*and the shear force is defined locally as*

$$\boldsymbol{\gamma}_h = \lambda t^{-2} (\nabla \omega_h - \boldsymbol{R}_h \boldsymbol{\phi}_h).$$

*The broken bilinear form $a_h(\cdot, \cdot)$ reads*

$$a_h(\boldsymbol{\phi}_h, \boldsymbol{\psi}) = \sum_{K \in \mathcal{T}_h} \int_K \mathcal{C}\mathcal{E}(\boldsymbol{\phi}_h) : \mathcal{E}(\boldsymbol{\psi}) dxdy. \tag{2.8}$$

Since both components of vector-valued functions in $\boldsymbol{V}_h$ are the NR $\mathcal{Q}_1$ element, the discrete Korn inequality does not hold. In fact, using a counter example, it is shown in [13] that there exists $\boldsymbol{\psi} \in \boldsymbol{V}_h$ such that

$$\|\mathcal{E}_h(\boldsymbol{\psi})\|_0 \leqslant Ch^{1/2}|\boldsymbol{\psi}|_{1,h}. \tag{2.9}$$

It is also proved in the same paper that $\|\mathcal{E}_h(\cdot)\|_0$ is a norm over the space $\boldsymbol{V}_h$. By the equivalence of norms over a finite dimension space, we have

$$\|\mathcal{E}_h(\boldsymbol{\psi})\|_0 \geqslant \beta|\boldsymbol{\psi}|_{1,h}. \tag{2.10}$$

Obviously $\beta$ depends on the mesh size $h$. The following lemma gives a lower bound of $\beta$.

**Lemma 2.6.** [12, Theorem 2] *There exists a positive constant $C$ independent of $h$ such that*

$$\|\mathcal{E}_h(\boldsymbol{\psi})\|_0 \geqslant Ch|\boldsymbol{\psi}|_{1,h}, \qquad \forall \boldsymbol{\psi} \in \boldsymbol{V}_h, \tag{2.11}$$

*namely $\beta \geqslant Ch$.*

# 3 Error estimates

In this section, we consider the error estimate of MS 2 element. Let $\pi_h^1$ be the standard bilinear interpolation operator and $\boldsymbol{\Pi}_h = (\Pi_h, \Pi_h)$ be the vector counterpart of the interpolation operator $\Pi_h$ defined in (2.1).

For the operators $\pi_h^1$ and $\boldsymbol{R}_h$, a straightforward argument shows

$$\nabla \pi_h^1 v = \boldsymbol{R}_h \nabla v\,, \text{ for any } v \in H_0^1(\Omega). \tag{3.1}$$

For the analysis, we need the error estimate of MS 1 element proposed in [13] that reads: Find $(\boldsymbol{\phi}_I, \omega_I) \in (W_h \times N_{0,h}) \times W_h$ with

$$\begin{aligned} a_h(\boldsymbol{\phi}_I, \boldsymbol{\psi}_h) + (\boldsymbol{\gamma}_I, \nabla v_h - \boldsymbol{R}_h\boldsymbol{\psi}_h) &= (g, v_h), \\ \forall (v_h, \boldsymbol{\psi}_h) &\in (W_h \times N_{0,h}) \times W_h, \\ \boldsymbol{\gamma}_I &= \lambda t^{-2}(\nabla \omega_I - \boldsymbol{R}_h\boldsymbol{\phi}_I). \end{aligned} \tag{3.2}$$

We have

**Lemma 3.1.** *[13] Let $(\boldsymbol{\phi}, \omega)$ and $(\boldsymbol{\phi}_I, \omega_I)$ be the solutions to Problem 1.1 and Problem ((3.2)), respectively. Then,*

$$\|\omega - \omega_I\|_1 + \|\boldsymbol{\phi} - \boldsymbol{\phi}_I\|_1 + t\|\boldsymbol{\gamma} - \boldsymbol{\gamma}_I\|_0 + \|\boldsymbol{\gamma} - \boldsymbol{\gamma}_I\|_{-1} \leqslant Ch(\|g\|_{-1} + t\|g\|_0). \tag{3.3}$$

Now we turn to the proof of Theorem 1.1.

*Proof.* The proof is divided into two steps.

*Step 1: We estimate the term* $\|\mathcal{E}_h(\boldsymbol{\phi}-\boldsymbol{\phi}_h)\|_0+t\|\boldsymbol{\gamma}-\boldsymbol{\gamma}_h\|_0$. Denote the consistency error by

$$cs=-a_h(\boldsymbol{\phi},\boldsymbol{\psi}_h)-(\boldsymbol{\gamma},\nabla v_h-\boldsymbol{R}_h\boldsymbol{\psi}_h)+(g,v_h),$$

for any $(v_h,\boldsymbol{\psi}_h)\in W_h\times \boldsymbol{V}_h$. Thanks to (2.5), we have for any $\boldsymbol{\psi}_h\in\boldsymbol{V}_h$ and $v_h\in W_h$ that

$$a_h(\boldsymbol{\phi}_h-\boldsymbol{\phi},\boldsymbol{\psi}_h)+(\boldsymbol{\gamma}_h-\boldsymbol{\gamma},\nabla v_h-\boldsymbol{R}_h\boldsymbol{\psi}_h)=cs.$$

Let $\boldsymbol{\phi}_I\in W_h\times N_{0,h}$, $\omega_I\in W_h$, and $\boldsymbol{\gamma}_I$ be defined in (3.2). We use $\boldsymbol{\phi}_h-\boldsymbol{\phi}=\boldsymbol{\phi}_h-\boldsymbol{\Pi}_h\boldsymbol{\phi}_I+\boldsymbol{\Pi}_h\boldsymbol{\phi}_I-\boldsymbol{\phi}$ and $\boldsymbol{\gamma}_h-\boldsymbol{\gamma}=\boldsymbol{\gamma}_h-\boldsymbol{\gamma}_I+\boldsymbol{\gamma}_I-\boldsymbol{\gamma}$ to get

$$\begin{aligned}a_h(\boldsymbol{\phi}_h-\boldsymbol{\Pi}_h\boldsymbol{\phi}_I,\boldsymbol{\psi}_h)+a_h(\boldsymbol{\Pi}_h\boldsymbol{\phi}_I-\boldsymbol{\phi},\boldsymbol{\psi}_h)+(\boldsymbol{\gamma}_h-\boldsymbol{\gamma}_I,\nabla v_h-\boldsymbol{R}_h\boldsymbol{\psi}_h)&\\+(\boldsymbol{\gamma}_I-\boldsymbol{\gamma},\nabla v_h-\boldsymbol{R}_h\boldsymbol{\psi}_h)=cs.&\end{aligned}\tag{3.4}$$

Since $\boldsymbol{\Pi}_h\boldsymbol{\phi}_I\in\boldsymbol{V}_h$, we can take $\boldsymbol{\psi}_h=\boldsymbol{\phi}_h-\boldsymbol{\Pi}_h\boldsymbol{\phi}_I$. A close observation finds

$$\boldsymbol{R}_h\boldsymbol{\Pi}_h\boldsymbol{\phi}_I=\boldsymbol{R}_h\boldsymbol{\phi}_I.\tag{3.5}$$

Taking $v_h=-\omega_I+\omega_h$, this yields

$$\nabla v_h-\boldsymbol{R}_h\boldsymbol{\psi}_h=-\frac{t^2}{\lambda}(\boldsymbol{\gamma}_I-\boldsymbol{\gamma}_h).\tag{3.6}$$

Then, it follows from (3.4) that

$$\begin{aligned}&\|\mathcal{E}_h(\boldsymbol{\phi}_h-\boldsymbol{\Pi}_h\boldsymbol{\phi}_I)\|_0^2+t^2\|\boldsymbol{\gamma}_h-\boldsymbol{\gamma}_I\|_0^2\\&\quad\leqslant C\|\mathcal{E}_h(\boldsymbol{\phi}_h-\boldsymbol{\Pi}_h\boldsymbol{\phi}_I)\|_0\|\mathcal{E}_h(\boldsymbol{\phi}-\boldsymbol{\Pi}_h\boldsymbol{\phi}_I)\|_0\\&\qquad+Ct^2\|\boldsymbol{\gamma}_I-\boldsymbol{\gamma}\|_0\|\boldsymbol{\gamma}_h-\boldsymbol{\gamma}_I\|_0+|cs|.\end{aligned}\tag{3.7}$$

Applying Lemma 3.1 to the terms $\|\mathcal{E}_h(\boldsymbol{\phi}-\boldsymbol{\phi}_I)\|_0$ and $t\|\boldsymbol{\gamma}_I-\boldsymbol{\gamma}\|_0$, Lemma 2.2 and Lemma 2.4 to the consistency error $cs$, Lemma 2.1 to the interpolation operator $\boldsymbol{\Pi}_h$, and using Young's inequality, we obtain

$$\begin{aligned}\|\mathcal{E}_h(\boldsymbol{\phi}_h-\boldsymbol{\Pi}_h\boldsymbol{\phi}_I)\|_0^2+t^2\|\boldsymbol{\gamma}_h-\boldsymbol{\gamma}_I\|_0^2\leqslant Ch^2(\|\boldsymbol{\phi}\|_3+\|\boldsymbol{\gamma}\|_1+\|g\|_0)|\boldsymbol{\psi}_h|_{1,h}&\\+Ch\|\boldsymbol{\gamma}\|_0\|\mathcal{E}_h(\boldsymbol{\psi}_h)\|_0.&\end{aligned}\tag{3.8}$$

Now the Korn inequality (2.6) gives

$$\|\mathcal{E}_h(\boldsymbol{\phi}_h-\boldsymbol{\Pi}_h\boldsymbol{\phi}_I)\|_0+t\|\boldsymbol{\gamma}_h-\boldsymbol{\gamma}_I\|_0\leqslant Ch(\|\boldsymbol{\phi}\|_3+\|\boldsymbol{\gamma}\|_1+\|g\|_0).\tag{3.9}$$

Finally, it follows from the triangle inequality and Lemma 3.1 that

$$\|\mathcal{E}_h(\boldsymbol{\phi}-\boldsymbol{\phi}_h)\|_0+t\|\boldsymbol{\gamma}-\boldsymbol{\gamma}_h\|_0\leqslant Ch(\|\boldsymbol{\phi}\|_3+\|\boldsymbol{\gamma}\|_1+\|g\|_0).\tag{3.10}$$

**Remark 3.2.** *A direct consequence of the above estimate and* (2.6) *is*

$$\|\boldsymbol{\phi}_h\|_{1,h} \leqslant C(\|\boldsymbol{\phi}\|_3 + \|\boldsymbol{\gamma}\|_1 + \|g\|_0). \tag{3.11}$$

*Step 2: We estimate the term* $\|\boldsymbol{\phi} - \boldsymbol{\phi}_h\|_0 + \|\nabla\omega - \nabla\omega_h\|_0$. In order to use the dual argument, we introduce the following auxiliary problem: For given $d \in (L^2(\Omega))^2$, find $(\boldsymbol{\phi}_d, \omega_d) \in \boldsymbol{H}_0^1(\Omega) \times H_0^1(\Omega)$ with

$$\begin{aligned} &a(\boldsymbol{\phi}_d, \boldsymbol{\psi}) + (\boldsymbol{\gamma}_d, \nabla v - \boldsymbol{\psi}) = (d, \boldsymbol{\psi}), \text{ for any } (\boldsymbol{\psi}, v) \in \boldsymbol{H}_0^1(\Omega) \times H_0^1(\Omega), \\ &\boldsymbol{\gamma}_d = \lambda t^{-2}(\nabla\omega_d - \boldsymbol{\phi}_d). \end{aligned} \tag{3.12}$$

The solution of this problem admits the following regularity

$$\|\boldsymbol{\phi}_d\|_2 + \|\omega_d\|_2 + t\|\boldsymbol{\gamma}_d\|_1 + \|\boldsymbol{\gamma}_d\|_0 + \|\operatorname{div}\boldsymbol{\gamma}_d\|_0 \leqslant C\|d\|_0. \tag{3.13}$$

Now we decompose the term

$$\begin{aligned} (d, \boldsymbol{\phi} - \boldsymbol{\phi}_h) &= (d, \boldsymbol{\phi} - \boldsymbol{\phi}_h) - a_h(\boldsymbol{\phi}_d, \boldsymbol{\phi} - \boldsymbol{\phi}_h) - (\boldsymbol{\gamma}_d, \nabla(\omega - \omega_h) \\ &\quad -(\boldsymbol{\phi} - \boldsymbol{\phi}_h)) \quad (I_1) \\ &\quad +a_h(\boldsymbol{\phi}_d, \boldsymbol{\phi} - \boldsymbol{\phi}_h) + (\boldsymbol{\gamma}_d, \nabla(\omega - \omega_h) - (\boldsymbol{\phi} - \boldsymbol{\phi}_h)) \quad (I_2) \\ &= I_1 + I_2. \end{aligned} \tag{3.14}$$

Using Lemma 2.2, the first term $I_1$ can be bounded as

$$|I_1| \leqslant Ch\|\boldsymbol{\phi}_d\|_2 |\boldsymbol{\phi} - \boldsymbol{\phi}_h|_{1,h}. \tag{3.15}$$

The second term $I_2$ can be treated as follows. Noticing $\boldsymbol{R}_h\boldsymbol{\Pi}_h\boldsymbol{\phi}_d = \boldsymbol{R}_h\boldsymbol{\phi}_d$ and (3.1), we have

$$\begin{aligned} I_2 &= (g, \omega_d) - a_h(\boldsymbol{\phi}_d, \boldsymbol{\phi}_h) - (\nabla\omega_d - \boldsymbol{\phi}_d, \boldsymbol{\gamma}_h) + (\boldsymbol{\gamma}_d, \boldsymbol{\phi}_h - \boldsymbol{R}_h\boldsymbol{\phi}_h) \\ &= (g, \omega_d) - a_h(\boldsymbol{\Pi}_h\boldsymbol{\phi}_d, \boldsymbol{\phi}_h) - (\nabla\pi_h^1\omega_d - \boldsymbol{R}_h\boldsymbol{\phi}_d, \boldsymbol{\gamma}_h) \\ &\quad +a_h(\boldsymbol{\Pi}_h\boldsymbol{\phi}_d - \boldsymbol{\phi}_d, \boldsymbol{\phi}_h) + (\nabla(\pi_h^1\omega_d - \omega_d) \\ &\quad -\boldsymbol{R}_h\boldsymbol{\phi}_d + \boldsymbol{\phi}_d, \boldsymbol{\gamma}_h) + (\boldsymbol{\gamma}_d, \boldsymbol{\phi}_h - \boldsymbol{R}_h\boldsymbol{\phi}_h) \\ &= (g, \omega_d) - (g, \pi_h^1\omega_d) + \lambda^{-1}t^2(\boldsymbol{R}_h\boldsymbol{\gamma}_d - \boldsymbol{\gamma}_d, \boldsymbol{\gamma}_h) \\ &\quad +(\boldsymbol{\gamma}_d, \boldsymbol{\phi}_h - \boldsymbol{R}_h\boldsymbol{\phi}_h) + a_h(\boldsymbol{\Pi}_h\boldsymbol{\phi}_d - \boldsymbol{\phi}_d, \boldsymbol{\phi}_h). \end{aligned} \tag{3.16}$$

Due to (3.10) and (3.13), the interpolation properties of $\pi_h^1$, $\boldsymbol{\Pi}_h$ and $\boldsymbol{R}_h$, it implies

$$\begin{aligned} |I_2| &\leqslant Ch^2\|g\|_0\|\omega_d\|_2 + Cht^2\|\boldsymbol{\gamma}_d\|_1\|\boldsymbol{\gamma}_h\|_0 + Ch\|\boldsymbol{\gamma}_d\|_0\|\mathcal{E}_h(\boldsymbol{\phi}_h)\|_0 \\ &\leqslant Ch(\|g\|_0 + \|\boldsymbol{\phi}\|_3 + \|\boldsymbol{\gamma}\|_1)\|d\|_0, \end{aligned} \tag{3.17}$$

which together with (3.15) prove

$$\|\boldsymbol{\phi} - \boldsymbol{\phi}_h\|_0 \leqslant Ch(\|g\|_0 + \|\boldsymbol{\phi}\|_3 + \|\boldsymbol{\gamma}\|_1). \tag{3.18}$$

Taking into account the definitions of $\boldsymbol{\gamma}$ and $\boldsymbol{\gamma}_h$, we have

$$\begin{aligned}\|\nabla(\omega-\omega_h)\|_0 &\leqslant \lambda^{-1}t^2\|\boldsymbol{\gamma}-\boldsymbol{\gamma}_h\|_0+\|\boldsymbol{\phi}-\boldsymbol{R}_h\boldsymbol{\phi}_h\|_0\\ &\leqslant Ch(\|\boldsymbol{\phi}\|_3+\|\boldsymbol{\gamma}\|_1+\|g\|_0),\end{aligned} \tag{3.19}$$

which ends the proof. □

**Remark 3.3.** *A dual argument can show*

$$t\|\omega-\omega_h\|_0 \leqslant Ch^2(\|\boldsymbol{\phi}\|_3+\|\boldsymbol{\gamma}\|_1+\|g\|_0). \tag{3.20}$$

*For brevity, we omit the proof, see [9] for details.*

# References


[1] R.A. Adams. *Sobolev Spaces*, Academic Press, 1975.

[2] D.N. Arnold and R.S. Falk. Analysis of a linear-linear-element for the Reissner-Mindlin model. *Math. Models Methods. Appl. Sci.* **7**, 486-514, 1997.

[3] K.J. Bathe, F. Brezzi and M. Fortin. Mixed-interpolated elements for Reissner-Mindlin plates, Internat. *J. Numer. Methods Engrg.* **28**, 1787-1801, 1989.

[4] F. Brezzi and M. Fortin. Numerical approximation of Reissner-Mindlin plates. *Math. Comp.* **47**, 151-158, 1986.

[5] F. Brezzi and M. Fortin. *Mixed and hybrid finite element methods*, Springer-Verlag, 1991.

[6] F. Brezzi, M. Fortin and R. Stenberg. Error analysis of mixed-interpolated elements for Reissner-Mindlin plate. *Math. Models Methods Appl. Sci.* **1**, 125-151, 1991.

[7] P.G. Ciarlet. *The finite element method for elliptic problem*, North-Holland, Amserdam, 1978.

[8] R. Duran, E. Hernández, L Hervella-Nieto, E.Liberman and R. Rodríguez. Error estimates for lower-order isoparametric quadrilateral finite elements for plates. *SIAM. J. Numer. Anal.*, 4(1), 1771-1772, 2003.

[9] J. Hu. *Quadrilateral locking free elements in elasticity.* Doctorate Dissertation, Institute of Computaional Mathematics, CAS, 2004.

[10] J. Hu, P.B. Ming and Z.C. Shi. Nonconforming quadrilateral rotated $Q_1$ element for Reissner-Mindlin plate. *J. Comp. Math.* **21**, 25-32, 2003.

[11] J. Hu and Z.C. Shi. Analysis of nonconforming-nonconform-ing quadrilateral rotated $Q_1$ element for Reissner-Mindlin plate. *Technique Report of Institute of Computational Mathematics*, CAS, Report No.ICM-03-010.

[12] P. Knobloch and L. Tobiska. On Korn's first inequality for quadrilateral nonconforming finite elements of first order approximation properties, Internat. *J. Numer. Anal. Model* **2**, 439-458, 2005.

[13] P.B. Ming and Z.C. Shi. Nonconforming rotated $Q_1$ finite element for Reissner-Mindlin plate. *Math. Models Methods Appl. Sci.* **11**, 1311-1342, 2001.

[14] E. Oñte, F. Zarate and F. Flores. A simple triangular element for thick and thin plate and shell analysis, Internat. *J. Numer. Methods. Engrg.* **37**, 2569-2582, 1994.

[15] R. Rannacher and S. Turek. A simple nonconforming quadrilateral Stokes element. *Numer. Methods Partial Differential Eq.* **8**, 97-111, 1992.

[16] M. Süri, I. Babuška and C. Schwab. Locking effects in the finite element approximation of plate models. *Math. Comp.* **64**, 461-482, 1995.

[17] J.M. Thomas. *Sur l'analyse numérique des méthods d'éléments finis hybrides et mixtes.* Thése d'Etat, Université Pierre et Marie Curie, Paris, 1977.

[18] Z.M. Zhang and S.Y. Zhang. Wilson element for the Reissner-Mindlin plate. *Comput. Methods Appl. Mech. Engrg.* **113**, 55-65, 1994.

# Monitoring the Corrosion of the Blast Furnace by Perturbation Method

Yongji Tan
*School of Mathematical Sciences, Fudan University*
*Shanghai 200433, China*

**Abstract**

This paper discusses an inverse boundary problem for the axisymmetric steady-state heat equation, which arises in monitoring the boundary corrosion for the blast-furnace. Measure temperature at some locations are used to identify the shape of the corrosion boundary. The numerical inversion is complicated and consuming since the wear-line varies during the process, and the boundary in the heat problem is not fixed. This paper suggests a perturbation method by which both the direct and inverse problems can be solved with fixed boundary, and a lot of computing time will be saved. The numerical results are in good agreement with test model data as well as industrial data, even in severe corrosion case.

## 1 Introduction

In steel industry there are many interesting mathematical problems. In recent years many Chinese applied mathematicians have solved many such problems, such as the temperature control of the blast furnace [1], the cooling control of continuously casting [2] and hot rolling [3], and monitoring the corrosion of the blast furnace and the steel ladle. Here we only use the problem of monitoring the corrosion of the blast furnace as a case to show that in industry there are many problems are expected to solve by mathematics, some times mathematics play a key role and new models, new mathematical methods and new algorithms can save the expense for experiments and reduce the computing time.

A blast furnace is a huge steel container many meters high and lined with heat-resistant material (see Figure 1). The solid raw materials (iron ore, coke and limestone) are added from the top, and hot air is blasted in from the bottom. The blast furnace is hottest at the bottom where the coke burns. It is coolest at the top where the iron forms and trickles down to the bottom, from where it is tapped off.

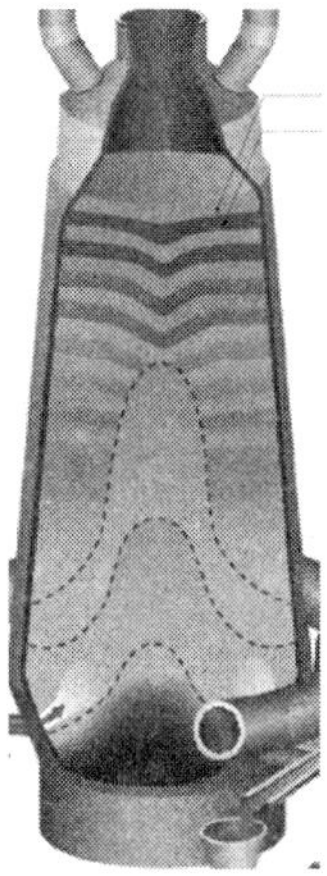

Figure 1 The blast furnace

The walls of the furnace are subject to both physical and chemical wear. Thus it is important to monitor the wearing to avoid molten metal from breaking out through the furnace wall and causing damage. So the shape of the inner wall surface must be determined.

Since the high temperature, it is impossible to investigate the corrosion inside the furnace directly. People preinstall some thermal couples in the lining of the furnace and try to use the measuring data of temperature to identify the wearing boundary in the inner wall of the bottom part of the furnace.

# 2 Mathematical model of heat conduction

The blast furnace is almost rotation symmetric. The sidewall and the bottom of it are cooled by water and by air respectively. We use the axi-symmetric configuration to describe the shape of the blast furnace, and the half cross section through the symmetric axis is shown as Figure 2.

If the production process is in a stable state, the temperature distribution in the wall is not varying with time. Therefore the governing equation for the heat conduction in the wall is a steady state heat equation. In axi-symmetric configuration, it is

$$\frac{1}{r}\cdot\frac{\partial}{\partial r}\left(rk\frac{\partial u}{\partial r}\right)+\frac{\partial}{\partial z}\left(k\frac{\partial u}{\partial z}\right)=0 \quad \text{in } \Omega, \tag{2.1}$$

where $u(r,z)$ is the temperature at point $(r,z)\in\Omega$ , where $r$ and $z$ are the radial and axial coordinates, respectively. The thermal conductivity

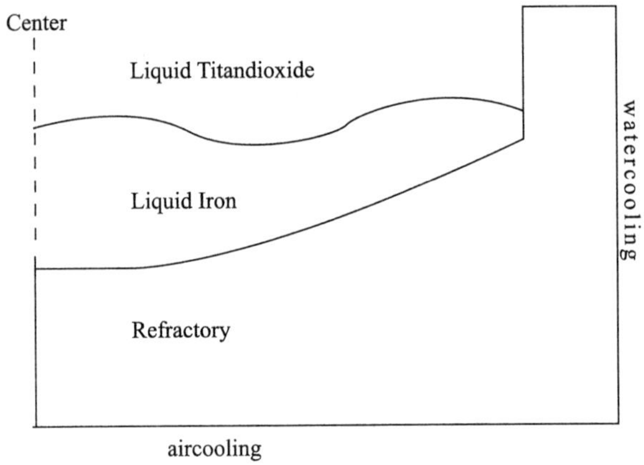


Figure 2 Cross section of the blast furnace

of the material, i.e. magnesia bricks, $k$, is a piecewise constant since the lining is constructed by different bricks in different region.

The boundary of $\Omega, \Gamma$ is split in 5 segments, as shown in Figure 3,

$$\Gamma = \Gamma_1 \cup \Gamma_2 \cup \Gamma_3 \cup \Gamma_4 \cup \Gamma_5. \tag{2.2}$$

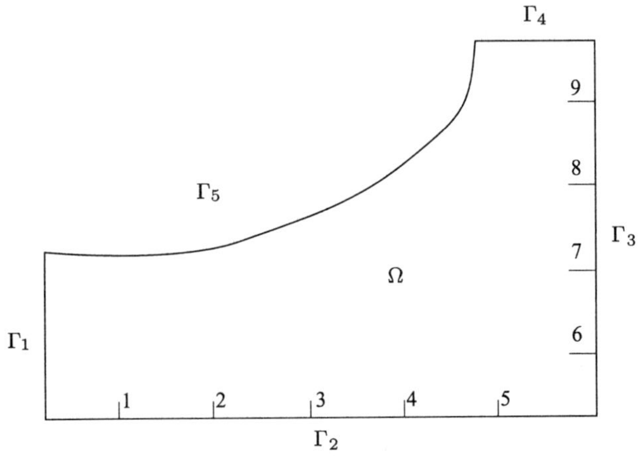


Figure 3 Domain and measurement locations

The boundary conditions for the problem are as follows. On $\Gamma_1$, the heat flux is zero, since the model is rotational symmetric,

$$\frac{\partial u}{\partial r} = 0 \quad \text{on } \Gamma_1. \tag{2.3}$$

The bottom segment $\Gamma_2$ is air-cooled, so we have a mixed condition. $\alpha_2$ is the heat transfer coefficient between the bottom of the furnace and air, and $u_2$ is the ambient air temperature,

$$-k\frac{\partial u}{\partial z} + \alpha_2 u = \alpha_2 u_2 \quad \text{on } \Gamma_2. \tag{2.4}$$

The sidewall $\Gamma_3$ is water-cooled, so we have a mixed condition as well. $\alpha_3$ is the heat transfer coefficient between the sidewall and water, and $u_3$ is the ambient water temperature,

$$k\frac{\partial u}{\partial n} + \alpha_3 u = \alpha_3 u_3 \quad \text{on } \Gamma_3. \tag{2.5}$$

The upper boundary $\Gamma_4$ is also assumed to be insulated,

$$\frac{\partial u}{\partial z} = 0 \quad \text{on } \Gamma_4. \tag{2.6}$$

At the wear-line boundary $\Gamma_5$, we have a Dirichlet condition. In practice, it is the isotherm with temperature 1450°C, which is the melting temperature of iron,

$$u = f(r, z) \quad \text{on } \Gamma_5. \tag{2.7}$$

If the wearing boundary $\Gamma_5$ is known, we are able to solve the above boundary value problem by the finite element method, finite difference method or other numerical method and obtain the approximation of the temperature distribution $u$. This is called direct or forward problem.

# 3 Inverse problem: least square method

Our purpose is to monitor the corrosion of the furnace, i.e. to determine the boundary $\Gamma_5$ by the temperature at the thermal couples. Therefore, the problem is to determine a part of the boundary of the domain by knowing the solution of a boundary value problem at some locations. This kind problem is called boundary inverse problem.

There are many works have been done on this problem [4]–[8], the main idea of them are as follows.

### 3.1 Mathematical model for boundary inverse problem

Suppose that there are $L$ thermal couples located at position $(r_l, z_l), l = 1, \cdots, L$, and the measurement is $T_l$ at $(r_l, z_l)$. Since the solution of boundary value problem (2.1),(2.3)–(2.7) depends the wearing boundary $\Gamma_5$, we denote it by $u(r, z, \Gamma_5)$. Our problem can be formulated to find $\Gamma_5$ such that $u(r_l, z_l, \Gamma_5) = T_l, l = 1, \cdots, L$ satisfied for the solution $u(r, z, \Gamma_5)$ of boundary value problem (2.1), (2.3)–(2.7).

By considering the measuring error and computational error, we can replace the rigorously equality condition by the condition of minimizing the summation of the error for calculating and measuring temperature at the measuring points. By introducing

$$E(\Gamma_5) = \sum_{l=1}^{L} (u(r_l, z_l, \Gamma_5) - T_l)^2, \tag{3.1}$$

the inverse problem model is to find $(u(r, z, \Gamma_5^0), \Gamma_5^0)$ such that

$$E(\Gamma_5^0) = \min_{\Gamma_5} E(\Gamma_5) \tag{3.2}$$

and $u(r, z, \Gamma_5^0)$ is the solution of the boundary value problem (2.1), (2.3)–(2.7) with $\Gamma_5 = \Gamma_5^0$.

### 3.2 Discretization

Since $\Gamma_5$ is a curve in $(r, z)$ plane, the minimization of $E(\Gamma_5)$ is a complicated variational problem and is hard to solve. To simplify the problem, we discrete $\Gamma_5$ by points $P_i(r_i, z_i), i = 1, \cdots, m$, and think that $\Gamma_5$ is the cubic spline interpolation of them. We define a multi-variable function $E(r_1, z_1, ..., r_m, z_m)$ by

$$E(r_1, z_1, ..., r_m, z_m) = \sum_{l=1}^{L} (u(r, z; r_1, z_1, ..., r_m, z_m) - T_l)^2, \tag{3.3}$$

where $u(r, z; r_1, z_1, ..., r_m, z_m)$ is the solution of boundary value problem (2.1), (2.3)–(2.7) with $\Gamma_5$ as the cubic spline interpolation of $\{P_i\}_{i=1,...,m}$.

Now the mathematical model for the boundary inverse problem becomes to find $(u(r, z; r_1^0, z_1^0, ..., r_m^0, z_m^0), \{r_1^0, z_1^0, ..., r_m^0, z_m^0\})$ such that

$$E(r_1^0, z_1^0, ..., r_m^0, z_m^0) = \min_{\{r_1, z_1, ..., r_m, z_m\}} E(r_1, z_1, ..., r_m, z_m), \tag{3.4}$$

and with $u(r, z; r_1^0, z_1^0, ..., r_m^0, z_m^0)$ being the solution of boundary value problem (2.1), (2.3)–(2.7) with $\Gamma_5$ as the cubic spline interpolation of $\{P_i^0 = (r_i^0, z_i^0)\}_{i=1,...,m}$. That problem is a minimization problem of a multi-variable function.

## 3.3 Inversion

As soon as $\{r_1, z_1, ..., r_m, z_m\}$ is given, we obtain a curve $\Gamma_5$ by cubic spline interpolation. Solving the boundary value problem by finite element method after meshing the domain $\Omega$ with sub-boundary $\Gamma_5$ determined above, we obtain an approximate of $u(r, z; r_1, z_1, \cdots, r_m, z_m)$ and it's values at points $(r_l, z_l), l = 1, \cdots, L$, then use formula (3.1) to determine $E(r_1, z_1, ..., r_m, z_m)$. In this way the multi-variable function $E(r_1, z_1, ..., r_m, z_m)$ is well defined. By optimization technique such as quasi-Newton technique or genetic algorithm, we obtain the solution for the least square inverse problem.

## 3.4 Shortcoming of the method

The procedure of optimization is usually an iteration procedure. In each step of iteration, it is necessary to calculate the value of the function $E(r_1, z_1, ..., r_m, z_m)$ (even its derivatives) several times and many steps is needed to realized the optimization. To calculate the value of the function $E(r_1, z_1, ..., r_m, z_m)$ we have to find the solution $u(r, z; r_1, z_1, ..., r_m, z_m)$ of boundary value problem (2.1), (2.3)–(2.7) with different part of boundary $\Gamma_5$. To fit the change of the domain, we need to do re-meshing for finite method and it is time consuming. One choice to avoid re-meshing is that to do some transform according the new boundary in the initial mesh without change the topological structure. However, this method usually causes serious aberrance of the mesh (see Figure 4) and reduces the accuracy.

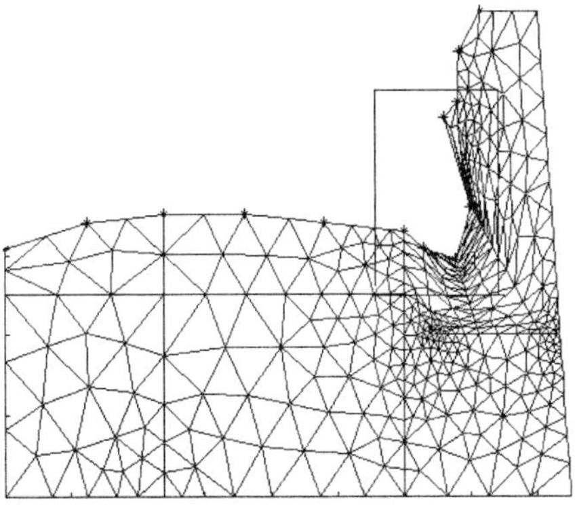

Figure 4 Aberrance of the mesh

After re-meshing we have to solve the liner system of finite element method again. It also costs much computing time. Therefore, the algorithm above is a very much time consuming computation tusk.

# 4 Perturbation method

To monitor the corrosion of the inner wall of the blast furnace, people usually do computation once several days and try to know how new corrosion happened in the inner wall of the furnace during this time interval. Because the corrosion process of the inner wall is very slow, we are able to think that the new wear-line as the old wear-line with a small perturbation. Therefore we can use perturbation technique and reduce the problem to a problem with fixed boundary.

## 4.1 Asymptotic solution

Suppose that the old wear-line is $\widetilde{\Gamma}_5$ which can be represented by $z = s(r)$, and the new wear-line is $\Gamma_5$ which can be represented as $z = s(r) + \varepsilon g(r)$ where $\varepsilon$ is a small parameter. The boundary value problem for the temperature function $u(r, z)$ with new wear-line is

$$\begin{cases} \dfrac{1}{r} \cdot \dfrac{\partial}{\partial r}\left(rk\dfrac{\partial u}{\partial r}\right) + \dfrac{\partial}{\partial z}\left(k\dfrac{\partial u}{\partial z}\right) = 0 \quad \text{in } \Omega, \\ \dfrac{\partial u}{\partial r} = 0 \quad \text{on } \Gamma_1, \\ -k\dfrac{\partial u}{\partial z} + \alpha_2 u = \alpha_2 u_2 \quad \text{on } \Gamma_2, \\ k\dfrac{\partial u}{\partial n} + \alpha_3 u = \alpha_3 u_3 \quad \text{on } \Gamma_3, \\ \dfrac{\partial u}{\partial z} = 0 \quad \text{on } \Gamma_4, \\ u = f(r, z) \quad \text{on } \Gamma_5. \end{cases} \tag{4.1}$$

We do formally asymptotic expansion([9],[10],[11]) for $u$ about small parameter $\varepsilon$, we have

$$u = u_0 + \varepsilon u_1 + O(\varepsilon^2). \tag{4.2}$$

By substituting (4.2) into the partial differential equation in (4.1) we have

$$\frac{1}{r} \cdot \frac{\partial}{\partial r}\left(rk\frac{\partial u_0}{\partial r}\right) + \frac{\partial}{\partial z}\left(k\frac{\partial u_0}{\partial z}\right) + \\ \varepsilon\left\{\frac{1}{r} \cdot \frac{\partial}{\partial r}\left(rk\frac{\partial u_0}{\partial r}\right) + \frac{\partial}{\partial z}\left(k\frac{\partial u_1}{\partial z}\right)\right\} + O(\varepsilon^2) = 0.$$

By comparing the coefficients of $\varepsilon^0$ and $\varepsilon$ at the both sides of the above equality respectively we obtain

$$\frac{1}{r} \cdot \frac{\partial}{\partial r}\left(rk\frac{\partial u_0}{\partial r}\right) + \frac{\partial}{\partial z}\left(k\frac{\partial u_0}{\partial z}\right) = 0,$$

and

$$\frac{1}{r} \cdot \frac{\partial}{\partial r}\left(rk\frac{\partial u_1}{\partial r}\right) + \frac{\partial}{\partial z}\left(k\frac{\partial u_1}{\partial z}\right) = 0.$$

By substituting (4.2) into the boundary condition on $\Gamma_1$ of (4.1) and comparing the coefficients of $\varepsilon^0$ and $\varepsilon$, we have

$$\frac{\partial u_0}{\partial r} = 0, \quad \frac{\partial u_1}{\partial r} = 0 \quad \text{on } \Gamma_1.$$

In same way we obtain the boundary conditions for $u_0$, and $u_1$ on $\Gamma_4$ respectively

$$\frac{\partial u_0}{\partial z} = 0, \quad \frac{\partial u_1}{\partial z} = 0 \quad \text{on } \Gamma_4.$$

By substituting (4.2) into the boundary condition on $\Gamma_2$ of (4.1) we have

$$-k\frac{\partial u_0}{\partial z} + \alpha_2 u_0 + \varepsilon\left(-k\frac{\partial u_0}{\partial z} + \alpha_2 u_0\right) + O(\varepsilon^2) = \alpha_2 u_2.$$

By comparing the coefficients of $\varepsilon^0$ and $\varepsilon$, we obtain

$$-k\frac{\partial u_0}{\partial z} + \alpha_2 u_0 = \alpha_2 u_2, \quad -k\frac{\partial u_1}{\partial z} + \alpha_2 u_1 = 0 \quad \text{on } \Gamma_2.$$

Similarly, the boundary conditions satisfied by $u_0$ and $u_1$ on $\Gamma_3$ are

$$k\frac{\partial u_0}{\partial n} + \alpha_3 u_0 = \alpha_3 u_3, \quad k\frac{\partial u_1}{\partial n} + \alpha_3 u_1 = 0 \quad \text{on } \Gamma_3.$$

On boundary $\Gamma_5$, the boundary condition $u = f$ can be written as

$$u(r, s(r) + \varepsilon g(r)) = f.$$

By asymptotic expansion about $\varepsilon$, it becomes

$$u_0(r, s(r) + \varepsilon g(r)) + \varepsilon u_1(r, s(r) + \varepsilon g(r)) + O(\varepsilon^2) = f.$$

Doing Taylor's expansion for $u_0$ and $u_1$ near $(r, s(r))$ in above equality, we get

$$u_0(r, s(r)) + \varepsilon\left(\frac{\partial u_0}{\partial z}(r,, s(r))g(r) + u_1(r, s(r))\right) + O(\varepsilon^2) = f.$$

By comparing the coefficients of $\varepsilon^0$ and $\varepsilon$, we finally obtain the boundary conditions on $\widetilde{\Gamma_5}$ satisfied by $u_0$ and $u_1$ are

$$u_0 = f, \quad u_1 = -\frac{\partial u_0}{\partial z} g(r).$$

Now we obtain two boundary value problems for $u_0$ and $u_1$, in domain $\widetilde{\Omega}$ which is enclosed by boundary $\Gamma = \Gamma_1 \cup \Gamma_2 \cup \Gamma_3 \cup \Gamma_4 \cup \widetilde{\Gamma}_5$,

$$\begin{cases} \frac{1}{r} \cdot \frac{\partial}{\partial r}\left(rk\frac{\partial u_0}{\partial r}\right) + \frac{\partial}{\partial z}\left(k\frac{\partial u_0}{\partial z}\right) = 0 \quad \text{in } \Omega, \\ \frac{\partial u_0}{\partial r} = 0 \quad \text{on } \Gamma_1, \\ -k\frac{\partial u_0}{\partial z} + \alpha_2 u_0 = \alpha_2 u_2 \quad \text{on } \Gamma_2, \\ k\frac{\partial u_0}{\partial n} + \alpha_3 u_0 = \alpha_3 u_3 \quad \text{on } \Gamma_3, \\ \frac{\partial u_0}{\partial z} = 0 \quad \text{on } \Gamma_4, \\ u_0 = f(r,z) \quad \text{on } \Gamma_5, \end{cases} \tag{4.3}$$

and

$$\begin{cases} \frac{1}{r} \cdot \frac{\partial}{\partial r}\left(rk\frac{\partial u_1}{\partial r}\right) + \frac{\partial}{\partial z}\left(k\frac{\partial u_1}{\partial z}\right) = 0 \quad \text{in } \Omega, \\ \frac{\partial u_1}{\partial r} = 0 \quad \text{on } \Gamma_1, \\ -k\frac{\partial u_1}{\partial z} + \alpha_2 u_1 = 0 \quad \text{on } \Gamma_2, \\ k\frac{\partial u_1}{\partial n} + \alpha_3 u_1 = 0 \quad \text{on } \Gamma_3, \\ \frac{\partial u_1}{\partial z} = 0 \quad \text{on } \Gamma_4, \\ u_1 = -\frac{\partial u_0}{\partial z} g(r) \quad \text{on } \Gamma_5. \end{cases} \tag{4.4}$$

We will use

$$\widetilde{u} = u_0 + \varepsilon u_1$$

to approximate the solution $u$ of (2.1), (2.3)–(2.7). It is easy to see that both $u_0$ and $u_1$ are solved in a domain $\widetilde{\Omega}$ with fixed boundary.

## 4.2 The error of asymptotic solution

Even if the derivation above is in formal and it seems that $\tilde{u}$ and $u$ are defined in different domains $\tilde{\Omega}$ and $\Omega$, we can rigorously prove that the error between $\tilde{u}$ and $u$ is of degree $O(\varepsilon^2)$ in Domain $\Omega$.

In fact, let $v = u - \bar{u}$ , then we have

$$\left.\frac{\partial v}{\partial n}\right|_{\Gamma_1} = \left.\frac{\partial v}{\partial n}\right|_{\Gamma_4} = \left.\left(k\frac{\partial v}{\partial n} + \alpha_2 v\right)\right|_{\Gamma_2} = \left.\left(k\frac{\partial v}{\partial n} + \alpha_3 v\right)\right|_{\Gamma_3} = 0 \tag{4.5}$$

and

$$
\begin{aligned}
v|_{\Gamma_5} &= u|_{\Gamma_5} - (u_0|_{\Gamma_5} + \varepsilon u_1|_{\Gamma_5}) \\
&= \bar{u_0} - (u_0|_{\bar{\Gamma}_5} + \left(\frac{\partial u_0}{\partial z}\varepsilon g(r)\right)\Big|_{\bar{\Gamma}_5} + O(\varepsilon^2) + \varepsilon(u_1|_{\bar{\Gamma}_5} \\
&\quad + \left(\frac{\partial u_1}{\partial z}\varepsilon g(r)\right)\Big|_{\bar{\Gamma}_5} + O(\varepsilon^2))) \\
&= \bar{u_0} - (\bar{u_0} + \varepsilon\left(\frac{\partial u_0}{\partial z} g(r)\right)\Big|_{\bar{\Gamma}_5} + O(\varepsilon^2)) - \varepsilon(-\left(\frac{\partial u_0}{\partial z} g(r)\right)\Big|_{\bar{\Gamma}_5} \\
&\quad + \varepsilon\left(\frac{\partial u_1}{\partial z} g(r)\right)\Big|_{\bar{\Gamma}_5} + O(\varepsilon^2)) \\
&= O(\varepsilon^2) - \varepsilon^2\left(\frac{\partial u_1}{\partial z} g(r)\right)\Big|_{\bar{\Gamma}_5} + O(\varepsilon^3) \\
&= O(\varepsilon^2).
\end{aligned}
$$

According to the extremum principle [12], we can get from (4.5) that the extremum value of $v$ can not be yielded on boundary $\Gamma_1, \Gamma_4$. If it is yielded on $\Gamma_5$, we can get that

$$|v| = O(\varepsilon^2).$$

Suppose that the maximal value of $v$ is obtained on $\Gamma_2$ (or $\Gamma_3$), we can get from the extremum principle that

$$\frac{\partial v}{\partial n} > 0, \quad \text{on } \Gamma_2 \quad (\text{or } \Gamma_3).$$

So we have $v_{\max} < 0$ from (4.5). Therefore,

$$
\begin{aligned}
0 > v_{\max} \geqslant v|_{\Gamma_5} &= O(\varepsilon^2), \\
|v| &\leqslant O(\varepsilon^2).
\end{aligned}
$$

In similiar manner, suppose the minimal value of $v$ is obtained on $\Gamma_2$ (or $\Gamma_3$), we can get from the extremum principle that

$$\frac{\partial v}{\partial n} < 0, \quad \text{on } \Gamma_2 \quad (\text{or } \Gamma_3).$$

So we have $v_{\min} > 0$ from (4.5). Therefore,

$$
\begin{aligned}
0 < v_{\min} \leqslant v|_{\Gamma_5} &= O(\varepsilon^2), \\
|v| &\leqslant O(\varepsilon^2).
\end{aligned}
$$

Thus we get $|u - \bar{u}| = O(\varepsilon^2)$, the error between the true value and the approximation value is about magnitude of $\varepsilon^2$.

## 4.3 Solution of direct problem

If $g(r)$ is known, we are able to solve boundary value problem (4.3) and obtain $u_0$ then solve boundary value problem (4.4) by substituting $u_0$ into its last boundary condition and obtain $u_1$. $\tilde{u} = u_0 + \varepsilon u_1$ is the asymptotic solution.

To stress the dependence of $\tilde{u}$ on $g(r)$, we use $\tilde{u}(r, z, g)$ to replace $\tilde{u}(r, z)$. Correspondingly, we use $u_0(r, z, g), u_1(r, z, g)$ instead of $u_0(r, z)$, $u_1(r, z)$. It should be noticed that both the domains for (4.3) and (4.4) are all $\tilde{\Omega}$ which is a fixed domain, and the partial differential equations in these two problems are same, and all the left hand sides of the boundary conditions in two problems are all same. In this case, when we use finite element method to solve these two problems, we only need mesh the domain $\tilde{\Omega}$ once, and compute the stiffness matrix once. It saves much computation time.

## 4.4 Inverse problem and solution

By least square technique, the inverse problem of determining the corrosion boundary can be formulated as follows.

To find $(\tilde{u}(r, z, g_0), g_0(r))$ such that

$$E(g_0) = \min_{g(r)} E(g), \tag{4.6}$$

where

$$E(g) = \sum_{l=1}^{L} (\tilde{u}(r_l, z_l, g) - T_l)^2, \tag{4.7}$$

with $\tilde{u}(r, z, g) = u_0(r, z, g) + \varepsilon u_1(r, z, g)$ where $u_0(r, z, g)$ and $u_1(r, z, g)$ are the solutions of (4.3) and (4.4) respectively. We discrete $g(r)$ by some points $(r_i, z_i), i = 1, \cdots, m$ with which we reconstruct the approximation of $g(r)$ by cubic spline interpolation. We simplify the problem by fixing the radial coordinates $r_1, \cdots, r_m$ of above points. Then as soon as the axial coordinates $z_1, \cdots, z_m$ are given $g(r)$ is reconstructed by cubic spline interpolation. By denoting $\vec{Z} = (z_1, \cdots, z_m)^T$, we are able to get $g(r)$ from it and then by solving (4.3) and (4.4) to obtain $u_0, u_1$ and $\tilde{u} = u_0 + \varepsilon u_1$. So we use $\tilde{u}(r, z, \vec{Z})$ instead of $\tilde{u}(r, z, g)$.

Therefore the problem is reduced to minimize a multi-variable function as follows.

To find $(\tilde{u}(r, z, \vec{Z}_0), \vec{Z}_0)$ such that

$$E(\vec{Z}_0) = \min_{\vec{Z}} E(\vec{Z}), \tag{4.8}$$

where

$$E(\vec{Z}) = \sum_{l=1}^{L} (\tilde{u}(r_i, z_l, \vec{Z}) - T_l)^2, \tag{4.9}$$

with

$$\tilde{u}(r, z, \vec{Z}) = u_0 + \varepsilon u_1,$$

where $u_0, u_1$ are the solutions of (4.3), (4.4) with $g(r)$ interpolating by $\vec{Z}$. Since the inverse problem is usually instable, we use Tickhonoff regularization ([13]) to make the problem stable by adding a regularization term to the cost function (4.9). Usually the regularization term is some kind of discretization of

$$\beta \int_0^R (g^2(r) + g'^2(r))dr, \tag{4.10}$$

where $(0, R)$ is the domain of $g(r)$, $\beta$ is the regularization parameter.

Usually we use Quasi-Newton method or genetic algorithm to solve the optimization problem. During the iteration process it is necessary to compute the value of function $E(\vec{Z})$ and solve the boundary value problems (4.3), (4.4) many times. However, in this case we only need to revise the right hand side of the linear equation system of the finite element method. The coefficient matrix is not a changed a little bit. If we solve this equation system by decomposition method, e.g. Crout decomposition or $LL^T$ decomposition, only decomposition once is necessary. For each iteration, what we should to do is just revision of the right hand side and doing back solution. In finite element method, the computing time for meshing and forming the coefficient matrix and decomposition of the matrix almost cost 90 percent of the total computation time. Now these works only need to carry out once. Therefore the perturbation method save more than 90 percent of the computation cost.

# 5 Numerical results

Some test cases were created to test the algorithm expressed above for solving the inverse problem.

## 5.1 Test 1

For simplicity, the side wall $\Gamma_3$ is set to be vertical to the ground. Suppose that the shape of $\Gamma_5$ and $\tilde{\Gamma}_5$ are given, that is, the expressions of $s(r)$, and $g(r)$ are known.

Giving value of $k, \alpha_2, \alpha_3, u_2, u_3$, solving direct problem (2.1), (2.3)–(2.7), we obtain calculating temperature at the thermocouple locations which is denoted by $\vec{V} = (v_1, \cdots, v_L)^T$. Then we add a random error vector to $\vec{V}$ and use it as the measured temperature $\vec{T} = (T_1, \cdots, T_L)^T$. The range of the random error is chosen to be 1%. By use of this $\vec{T}$ as the measurement, we solve the inverse problem by iteration and obtain the curve parameter vector $\vec{Z}_0$, then we can calculate the curve by interpolation and compare it with the originally given curve $s(r)$ and $g(r)$. Here

$$s(r) = 2.5 + e^{\frac{r}{6.7} ln(3.5)}, \quad g(r) = \begin{cases} 0.5832(r-5)^2 - 14.58, & 0 \leqslant r \leqslant 5, \\ 2(r-5)^2 - 14.58, & 5 \leqslant r \leqslant 6.7, \end{cases}$$

$$k = 10, \quad \alpha_2 = 30, \quad \alpha_3 = 70, \quad u_2 = 35, \quad u_3 = 33,$$
$$\varepsilon = 0.02, \quad \beta = 0.005.$$

The maximal relative error of the wear-line is about $9.6 \times 10^{-3}$, as shown in Figure 5.

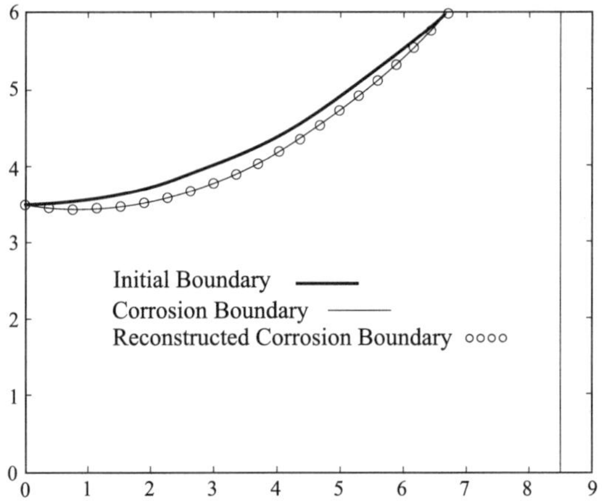


Figure 5 Result of the Test1

## 5.2 Real blast furnace, artificial data

For a real blast furnace, suppose that the shape of the bottom and the wear-line are know, i.e. the expressions of $s(r)$, and $g(r)$ are known.

Giving value of $k, \alpha_2, \alpha_3, u_2, u_3$, solving direct problem (2.1), (2.3)–(2.7), we obtain calculating temperature at the thermocouple locations which is denoted by $\vec{V} = (v_1, \cdots, v_L)^T$. Then we add a random error vector to $\vec{V}$ and use it as the measured temperature $\vec{T} = (T_1, \cdots, T_L)^T$. The range of the random error is chosen to be 1%.

By use of this $\vec{T}$ as the measurement, we solve the inverse problem by iteration and obtain the curve parameter vector $\vec{Z}_0$, then we can calculate the curve by interpolation and compare it with the originally given curve $s(r)$ and $g(r)$.

The computation parameters we used here are $k = 10$, $\alpha_2 = 30$, $\alpha_3 = 70$, $u_2 = 35$, $u_3 = 33$, and the temperature of the wear-line is 1450°C.

The original temperature data $\vec{V}$ and the computation data $\vec{T}$ (with 1% random error) and the temperature value at the thermal couple locations $\vec{\Theta}$, computed by use of the reconstructed wear-line are shown in the following table.

| Locations | $\vec{V}$ | $\vec{T}$ | $\vec{\Theta}$ |
|---|---|---|---|
| 1(0,0) | 117.491 | 117.967 | 117.758 |
| 2(1.75,0) | 136.361 | 136.488 | 136.656 |
| 3(3.51,0) | 133.718 | 133.571 | 133.505 |
| 4(5.25,0) | 126.999 | 127.493 | 127.490 |
| 5(7,0) | 99.099 | 99.339 | 99.379 |
| 6(8.888,0) | 31.494 | 31.68 | 31.472 |
| 7(8.829,0.6) | 36.732 | 37.068 | 36.654 |
| 8(8.77,1.2) | 44.957 | 44.977 | 44.774 |
| 9(8.7109,1.8) | 59.433 | 59.885 | 59.271 |
| 10(8.6617,2.3) | 71.259 | 70.793 | 71.571 |
| 11(8.6126,2.8) | 86.629 | 87.46 | 87.080 |
| 12(8.5634,3.3) | 101.14 | 100.678 | 100.587 |
| 13(8.5142,3.8) | 108.463 | 107.926 | 108.139 |
| 14(8.465,4.3) | 112.707 | 113.554 | 113.326 |
| 15(8.4158,4.8) | 125.395 | 125.99 | 126.076 |

The comparing of the computational wear-line and the real wear-line is shown in Figure 6.

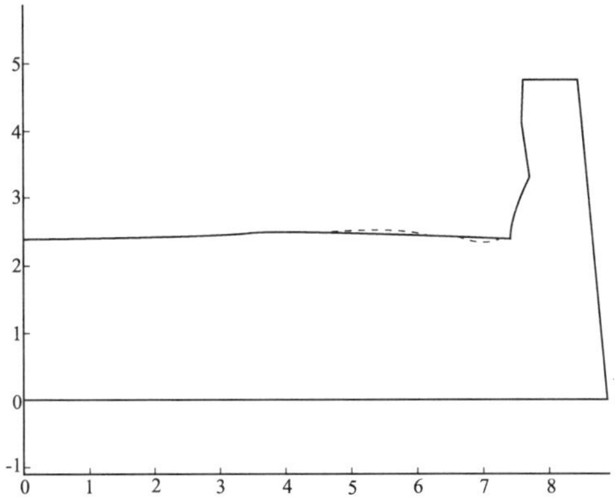


Figure 6 Comparing of the computational wear-line and the real wear-line

## 5.3 Real measurement

Finally, we present simulations of the wear-line using actual temperature measurements from a blast furnace. Unlike the above two cases, the actual wear-line $\Gamma_5$ can only be expressed by a parametric curve, as

$$\begin{cases} r = r(t) + \varepsilon h(t), \\ z = z(t) + \varepsilon l(t). \end{cases} \tag{5.1}$$

The boundary condition $u|_{\Gamma_5} = f$ converts to

$$u_0|_{\tilde{\Gamma}_5} = f$$

and

$$u_1|_{\tilde{\Gamma}_5} = -\left(\frac{\partial u_0}{\partial r} \cdot h(t) + \frac{\partial u_0}{\partial z} \cdot l(t)\right)\Big|_{\tilde{\Gamma}_5}.$$

Because of the difference of material, the conductivity varies. In actual furnace, the domain $\Omega$ is divided into two parts, and $k_1 = 10$, $k_2 = 3.3$, see Figure 7. The other parameters are $\alpha_2 = 30$, $\alpha_3 = 70$, $u_2 = 35$, $u_3 = 33$, and the temperature of the wear-line is 1450°C.

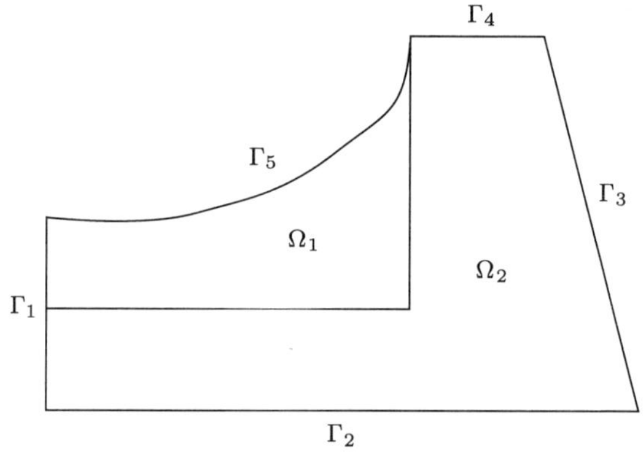


Figure 7 Domain of the blast furnace

Solving inverse problem (2.1), (2.3)–(2.7), we get the calculated shape of the wear-line. As shown in Figure 8, the difference between the calculated curve and the wear-line shape computed by FEM without perturbation, is quite acceptable.

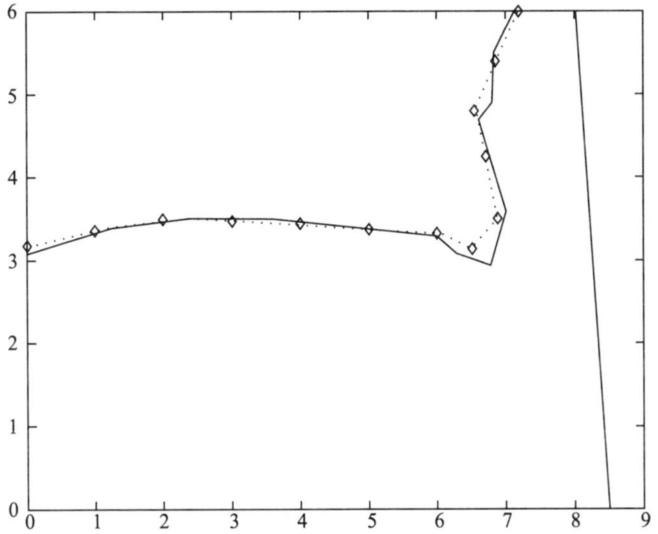


Figure 8 Comparing of the results

We then calculate the temperature at the measuring points. Comparing with the measured temperature, the difference is partly cause by measurement error, so it is also acceptable (see Figure 9).

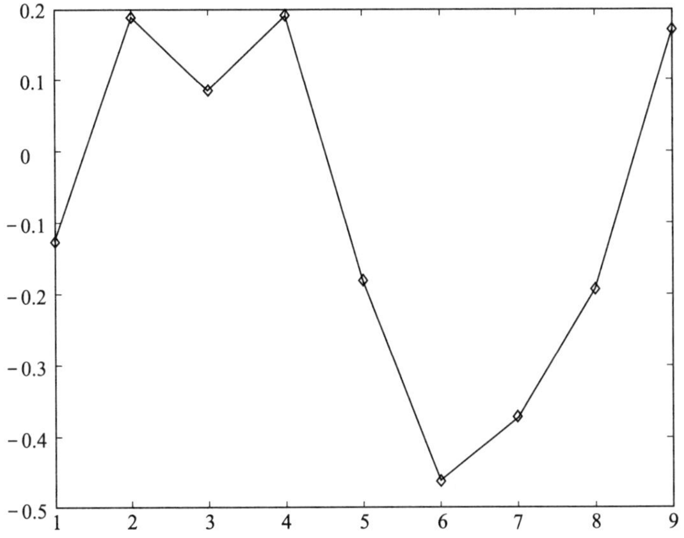


Figure 9 Relative error of measuring and calculating temperature

# References


[1] X.G. Liu, F. Liu. *Optimization and intelligent control system of BF ironmaking process*, 31-82, Beijing Metallurgical Industry, Beijing (in Chinese), 2003.

[2] Y. Tan, H. Xing, H. Wu, B. Fan, J. Yan, B. Zhao, J. Chen. FEM and parameter optimization for continuous casting process. *Mathematica Applicata* **13**, 44-50, 2000.

[3] H. Huang. Temperature control by Laminar flows. itshape Workshop on industrial applications Report, City university of Hong Kong, 39-44, Hong Kong, 2002.

[4] H. Yoshikawa et al.. Estimation of erosion line of refractory and solidification layer in blast furnace hearth. *Proc. of 4th conf. on simulation technology*, Japan Society of Simulation Technology, 75-78, 1984.

[5] K. Sorli, I.M. Skaar. Monitoring the wear-line of a melting furnace. *Inverse problems in engineering: theory and practice*, June 13-18, Port Ludlow, WA, USA, 1999.

[6] F. Berntsson. Boundary identification for an elliptic equation. *Inverse problems* **18**, 1579-1592, 2002.

[7] I. Szczygiel, A. Fic. Inverse Convection-Diffusion Problem of Estimating Boundary Veloctiy Based on Internal Temperature Measurements. *Inverse Problems in Engineering*, 10(3), 271-291, 2002.

[8] D.P. Baker, G.S. Dulikravich, B.H. Dennis, T.J. Martin. Inverse determination of eroded smelter wall thickness variation using an elastic membrane concept. *Proceeding of NHTC'03 ASME summer heat transfer conference*, July 21-23, Las Vegas, Nevada, 2003.

[9] Y.J. Tan, X.X. Chen. Identifying corrosion boundary by perturbation method. *Differential Equations & Asymptotic Theory in Mathematical Physics*, eds. Chen Hua & Roderick Wong, World Scientific, Singapore, 2004.

[10] J. Hinch. *Perturbation methods.* Cambridge University Press, Cambridge, 1991.

[11] J.A. Murdock. Perturbations: Theory and Methods. *SIAM, Classics In Applied Mathematics*, 1991.

[12] J. Ockendon, S. Howison, A. Lacey, A. Movchan. *Applied Partial Differential Equations* (revised edition). Oxford University Press, Oxford, 2003.

[13] H.W. Engl, M. Hanke, A. Neubauer. *Regularization of Inverse Problems*, Kluwer Academic Publishers, 1996.

# Numerical Study of Magnetic Properties of Nanowire Arrays

Yong Xiao
*Department of Mathematics and Natural Science*
*Harbin Institute of Technology Shenzhen Graduate School*
*Shenzhen 518055, China*

Sufen Zhao
*Department of Physics*
*Beijing University of Aeronautics & Astronautics*
*Beijing 100083, China*

Xiaoping Wang
*Department of Mathematics*
*The Hong Kong University of Science and Technology*
*Clear Water Bay, Kowloon, Hong Kong, China*
*E-mail: mawang@ust.hk*

**Abstract**

The Gauss-Seidel projection method is used to study the magnetic properties of nanowire arrays. We make a number of improvements of the numerical scheme so that the method is more efficient for the nanowire array configurations. A 3-D parallel code was developed and implemented. We study the correlation between the microstructure and the magnetic properties of the nanowire array. It is found that coercivity decreases when wire number increases. Our results also show that coercivity for an nanowire array is significantly smaller than that for a single wire. Decreasing the interwire distance would reduce the switch field and at the same time increase the recording density. Another way to reduce the switching field is to reduce the wire length. Dynamics of the magnetic switching for nanowire array is also investigated. These results have important implications in the high density magnetic recording applications.

# 1 Introduction

Highly ordered arrays of magnetic nanowires together with their intrinsic nature, give rise to outstanding cooperative properties different from bulk and even from film systems. Magnetic nanowire array is very important both for basic research and for its potential applications such as high-density recording media and magnetic sensors. Recently, micromagnetic simulation based on Landau-Lifshitz-Gilbert model has become an important tool for quantitative studies of the magnetic behavior of various types of magnetic materials. In the simulation of the magnetization reversal process, it is important to be able to resolve the different small length scales involved, in particular magnetic domain walls, and magnetic vortices, since they play important roles in the switching process. Such a simulation demands high accuracy and efficiency of the method being used. One of the main difficulties in micromagnetics simulations is the severe time step constraint introduced by the exchange field. Using standard explicit integrators leads to a physical time step of sub-picoseconds, which is often two orders of magnitude smaller than the fastest physical time scales. Direct implicit integrators require solving complicated, coupled systems. In [5] and [3], we introduced an implicit method whose complexity is comparable to solving the scalar heat equation implicitly. This method is based on a combination of a Gauss-Seidel implementation of a fractional-step implicit solver for the gyromagnetic term, and the projection method for the heat flow of harmonic maps. This method speeds up the numerical simulations by several order of magnitude and allows us to carry out fully resolved calculations for the 3-D domain structures and reversal process.

In this paper, the Gauss-Seidel projection scheme is used to study the magnetic properties of nanowire arrays. We make a number of improvements so that the scheme is more efficient for the nanowire array configurations. A 3-D parallel code was developed and implemented. Together, the algorithms improved the computational efficiency by a factor of 20. The method is then used to systematically investigate the magnetic properties of the nanowire array.

The paper is organized as the following: In Section 2, we present the Landau-Lifshitz-Gilbert model and the Gauss-Seidel projection method. An efficient way to compute the stray field for the nanowire array is explained and the parallel procedure is outlined. In Section 3, we give the results and analysis of our simulations. Section 4 is the conclusion.

# 2 Simulation model and the numerical methods

## 2.1 The Landau-Lifshitz-Gilbert (LLG) equation

The dynamics of magnetization distribution is described by the Landau-Lifshitz-Gilbert equation

$$\frac{\partial \boldsymbol{M}}{\partial t} = -\gamma \boldsymbol{M} \times \mathscr{H} - \frac{\gamma\alpha}{M_s} \boldsymbol{M} \times (\boldsymbol{M} \times \mathscr{H}), \tag{1}$$

where $|\boldsymbol{M}| = M_s$ is the saturation magnetization, $\gamma$ is the gyromagnetic ratio and $\alpha$ is the damping constant. $\mathscr{H}$ is the effective magnetic field, computed from the Landau-Lifshitz free energy functional:

$$\begin{aligned} F[\boldsymbol{M}] &= \frac{1}{2} \int_\Omega \left\{ K_u \Phi\left(\frac{\boldsymbol{M}}{M_s}\right) + \frac{A}{M_s^2} |\nabla \boldsymbol{M}|^2 - 2\mu_0 \boldsymbol{H}_e \cdot \boldsymbol{M} \right\} dx \\ &\quad + \frac{\mu_0}{2} \int_\Omega \nabla U \cdot \boldsymbol{M}\, dx, \qquad (2) \\ \mathscr{H} &= -\frac{\delta F}{\delta \boldsymbol{M}}. \qquad (3) \end{aligned}$$

In the above, $A$ is the exchange constant, $\mu_0$ is the permeability of vacuum, $K_u$ is the anisotropy constant, $\boldsymbol{H}_e$ is the external magnetic field and $\Omega$ is the volume occupied by the material, and the demagnetization field potential $U$ can be computed by solving

$$\Delta U = \begin{cases} \nabla \cdot \boldsymbol{M}, & \text{in } \Omega, \\ 0, & \text{outside } \Omega \end{cases} \tag{4}$$

together with the jump conditions

$$[U]_{\partial\Omega} = 0, \tag{5}$$

$$\left[\frac{\partial U}{\partial \nu}\right]_{\partial\Omega} = -\boldsymbol{M} \cdot \nu. \tag{6}$$

The solution to (4), with boundary conditions is

$$\nabla U = \nabla \int_\Omega \nabla N(\boldsymbol{r} - \boldsymbol{r}') \cdot \boldsymbol{M}(\boldsymbol{r}')\, d\boldsymbol{r}', \tag{7}$$

where $\boldsymbol{r} = (x, y, z)$ and $N(\boldsymbol{r}) = -\frac{1}{4\pi|\boldsymbol{r}|}$ is the Newtonian potential in $\mathbb{R}^3$.

## 2.2 Gauss-Seidel projection method

The full Landau-Lifshitz equation (1) can be rewritten in dimensionless form. Let $\mathscr{H} = M_s \boldsymbol{h}$, $\boldsymbol{H}_s = M_s \boldsymbol{h}_s$, $\boldsymbol{H}_e = M_s \boldsymbol{h}_e$, $\boldsymbol{M} = M_s \boldsymbol{m}$, $t \to$

$(\mu_0 \gamma M_s)^{-1} t$ and $x \to Lx$; we can write (1) as

$$\boldsymbol{m}_t = -\boldsymbol{m} \times \boldsymbol{h} - \alpha \boldsymbol{m} \times (\boldsymbol{m} \times \boldsymbol{h}), \tag{8}$$

where

$$\boldsymbol{h} = \varepsilon \Delta \boldsymbol{m} + \boldsymbol{h}_s + \boldsymbol{h}_e + \boldsymbol{h}_k. \tag{9}$$

Here, $\epsilon = A/(\mu_0 M_s^2 L^2)$. $\boldsymbol{h}_s$, $\boldsymbol{h}_e$ and $\boldsymbol{h}_k$ are effective field from stray field, external field and the uniaxial anisotropy. We define the vector field

$$\boldsymbol{f} = \boldsymbol{h}_s + \boldsymbol{h}_e + \boldsymbol{h}_k. \tag{10}$$

We solve the equation

$$\boldsymbol{m}_t = -\boldsymbol{m} \times (\varepsilon \Delta \boldsymbol{m} + \boldsymbol{f}) - \alpha \boldsymbol{m} \times \boldsymbol{m} \times (\varepsilon \Delta \boldsymbol{m} + \boldsymbol{f}) \tag{11}$$

by the Gauss-Seidel projection method developed in [3–5] which includes three steps in the following:

**Step 1** Implicit Gauss-Seidel:

$$g_i^n = (I - \varepsilon \Delta t \Delta_h)^{-1} (m_i^n + \Delta t f_i^n), \quad i = 1, 2, 3, \tag{12}$$

$$g_i^* = (I - \varepsilon \Delta t \Delta_h)^{-1} (m_i^* + \Delta t f_i^*), \quad i = 1, 2, 3, \tag{13}$$

$$\begin{pmatrix} m_1^* \\ m_2^* \\ m_3^* \end{pmatrix} = \begin{pmatrix} m_1^n + (g_2^n m_3^n - g_3^n m_2^n) \\ m_2^n + (g_3^n m_1^* - g_1^* m_3^n) \\ m_3^n + (g_1^* m_2^* - g_2^* m_1^*) \end{pmatrix}. \tag{14}$$

**Step 2** Heat flow without constrains:

$$\boldsymbol{f}^* = Q(m_1^* \alpha_1 + m_2^* \alpha_2 + m_3^*) \boldsymbol{\alpha} + \boldsymbol{h}_s^* + \boldsymbol{h}_e, \tag{15}$$

$$\begin{pmatrix} m_1^{**} \\ m_2^{**} \\ m_3^{**} \end{pmatrix} = \begin{pmatrix} m_1^* + \alpha \Delta t (\varepsilon \Delta_h m_1^{**} + f_1^*) \\ m_2^* + \alpha \Delta t (\varepsilon \Delta_h m_2^{**} + f_2^*) \\ m_3^* + \alpha \Delta t (\varepsilon \Delta_h m_3^{**} + f_3^*) \end{pmatrix}. \tag{16}$$

**Step 3** Projection onto $S^2$:

$$\begin{pmatrix} m_1^{n+1} \\ m_2^{n+1} \\ m_3^{n+1} \end{pmatrix} = \frac{1}{|\boldsymbol{m}^{**}|} \begin{pmatrix} m_1^{**} \\ m_2^{**} \\ m_3^{**} \end{pmatrix}. \tag{17}$$

## 2.3 An efficient FFT in computing stray field

In each time step, we need to compute the stray field $\boldsymbol{H}_s$ from (7) which is the most time consuming part. The most efficient way to compute (7) is by FFT. From (7), the stray field $\boldsymbol{H}_s = -\nabla U$ can be expressed by

$$\boldsymbol{H}_s = \frac{M_s}{4\pi} \nabla \left\{ \int_\Omega \frac{\nabla \cdot \boldsymbol{m}(\boldsymbol{r}')}{|\boldsymbol{r} - \boldsymbol{r}'|} \, d\boldsymbol{r}' - \int_{\partial\Omega} \frac{\boldsymbol{m}(\boldsymbol{r}') \cdot \boldsymbol{n}}{|\boldsymbol{r} - \boldsymbol{r}'|} \, dS(\boldsymbol{r}') \right\}, \tag{18}$$

where $\boldsymbol{n}$ is the outward normal direction of the material surface.

Now we consider ferromagnetic cuboid. By dividing the material $V$ into cells $V_{ijk}$ such that

$$V_{ijk} = [x_{i-1/2}, x_{i+1/2}] \times [y_{j-1/2}, y_{j+1/2}] \times [z_{k-1/2}, z_{k+1/2}]$$

with $x_i = idx$ and $x_{i\pm 1/2} = (i \pm 1/2)dx$, etc., one has

$$V = \sum_{i,j,k} V_{ijk}.$$

In each cell $V_{ijk}$, the computation point is located at the cell center $(x_i, y_j, z_k)$ and $\boldsymbol{m}$ is regarded as a constant. Therefore the integral of divergence vanishes. For one observing point $\boldsymbol{r}_{ijk} = \boldsymbol{r}(x_i, y_j, z_k)$, the stray field can be approximated by

$$\mathbf{H}_s(\boldsymbol{r}_{ijk}) = \frac{M_s}{4\pi} \sum_{p,q,r} \int_{\partial V_{pqr}} \frac{\boldsymbol{r}_{ijk} - \boldsymbol{r}'}{|\boldsymbol{r}_{ijk} - \boldsymbol{r}'|^3} \boldsymbol{m}(\boldsymbol{r}') \cdot \mathbf{n} \, dS(\boldsymbol{r}'). \tag{19}$$

Or in componentwise, one has

$$\begin{pmatrix} H_x(x_i, y_j, z_k) \\ H_y(x_i, y_j, z_k) \\ H_z(x_i, y_j, z_k) \end{pmatrix} = \sum_{p,q,r} \begin{pmatrix} K_{xx} & K_{xy} & K_{xz} \\ K_{yx} & K_{yy} & K_{yz} \\ K_{zx} & K_{zy} & K_{zz} \end{pmatrix}_{x_{i-p}, y_{j-q}, z_{k-r}} \begin{pmatrix} m_x(x_p, y_q, z_r) \\ m_y(x_p, y_q, z_r) \\ m_z(x_p, y_q, z_r) \end{pmatrix}, \tag{20}$$

Note that each element in the demagnetization tensor $K$ can be calculated analytically and only depends on $(x_{i-p}, y_{j-q}, z_{k-r})$, which enables us to implement FFT (with zero padding) to compute the stray field.

We consider a nanowire array configuration showing in Figure 1. For convenience, we use wires with square (instead of circular) cross section in our simulations. To compute the stray field efficiently in the nanowire array configuration, we note that $\boldsymbol{m}$ is zero in the nonmagnetic spacer. The amount of computation can be significantly reduced with a modified FFT procedure. To illustrate the procedure, consider a convolution in 1-D with 2 parts. In discrete form, the convolution $H = K * m$ is

$$H_j = \sum_{i=0}^{N-1} K(x_{j-i}) m(x_i), \quad j = 0, \cdots, N-1, \tag{21}$$

where $N = 3n$, $K(x)$ is a globe function and $m$ is magnetization of nanowire arrays, with values

$$m(x_i) = m(x_0), \cdots, m(x_{n-1}), 0, \cdots, 0, m(x_{2n}), \cdots, m(x_{3n-1}),$$

where $i = 0, \cdots, N-1$. The middle portion with $i = n, ..., 2n-1$ is the nonmagnetic spacer and $m$ is zero. We only need to compute the

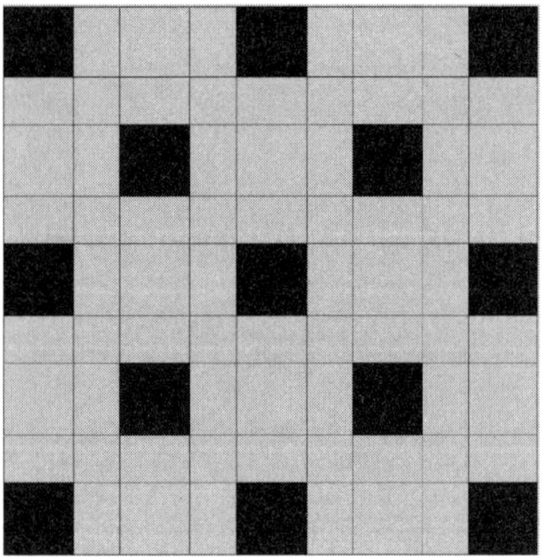

Figure 1 Top view of nanowire model for numerical simulation

convolution in the region of $m(x_i) \neq 0$, i.e., $H_j$ for $j = 0, \cdots, n-1$ and $j = 2n, \cdots, 3n-1$. So (21) can be rewritten as

$$H_j = \sum_{i=0}^{n-1} K(x_{j-i})m(x_i) + \sum_{i=2n}^{3n-1} K(x_{j-i})m(x_i). \tag{22}$$

The second summation is a convolution after a shift of index number. Therefore $H_j$ for $j = 1, ..., n$ is a sum of two convolutions of the same size which can be efficiently computed by FFT

$$\begin{aligned} \mathrm{FFT}(\{H_j\}) = \mathrm{FFT}\left(\{\sum_{i=0}^{n-1} K(x_{j-i})m(x_i))\}\right) \\ +\mathrm{FFT}\left(\{\sum_{i=2n}^{3n-1} K(x_{j-i})m(x_i))\}\right). \end{aligned} \tag{23}$$

With inverse FFT, we then obtain the stray field of the first part. Zero padding is needed to satisfy the periodic boundary condition. The convolution of the second part can be computed similarly with the index $j = 2n, \cdots, 3n-1$. The procedure can then be generalized to three dimensions.

## 2.4 Parallel computing

To improve the computational speed and capacity further, we employ parallel computing for the nanowire array configuration. The numerical tests have shown that our parallel scheme improves the speed of calculation by a factor of five on an eight-node cluster. With all the improvement on the computation efficiency, we were able to carry out a simulation of hysteresis loop for a 221-wire array on a 32-node cluster in 25 hours. The maximum number of wires simulated in publication is 36 by Rahmana in 2005 [2].

# 3 Results and Discussions

Three-dimensional micromagnetic simulations of the nanowire arrays (Figure 1) were carried out by numercially solving the Landau-Lifshitz equation (1). To study the magnetic reversal properties, an external field is applied along the longitudinal direction. The parameters used in these simulations are shown in Table 1.

Table 1 Set of parameters for the nanowires used in micromagnetic simulations

| | |
|---|---|
| number of magnetic nanowires | 5–221 |
| saturation magnetization ($M_s$) | $1.353 \times 10^6$ A/m |
| exchange parameter ($A$) | $1.55 \times 10^{-11}$ J/m |
| bulk anisotropy ($K_u$, along wire axe) | $4.0 \times 10^5$ J/m$^3$ |
| Damping constant ($\alpha$) | 0.1 |

We study the correlation between the microstructure and the magnetic properties of the nanowire array. In particular, we will look at the dependence of the magnetic properties of nanowire array on the geometric parameters of the array, i.e., wire number, wire length, and wire size.

## 3.1 The effects of wire number and length

To systematically investigate the effect of the magnetostatic interaction between the wires, simulations of the hysteresis loops are performed for different wire number and wire length, with a maximum number of 221 interacting nanowires and maximum wire length of 500 nm. The external field is applied in the longitudinal direction. Closed hysteresis loops are simulated for values of external field between $+10^6$ A/m and $-10^6$ A/m. We take the typical size of each wire with diameter 30 nm, interwire distance 20 nm, and length from 20 to 500 nm, with the geometric configuration according to Figure 1. The nanowires switch between two states along wire axis. This behavior results in a hysteresis loop of nearly rectangular shape which are similar to that of a single nanowire. Hysteresis loops of interacting nanowire arrays with different wire number and length (with interwire distance 20 nm and wire size 30 nm) are shown in Figure 2.

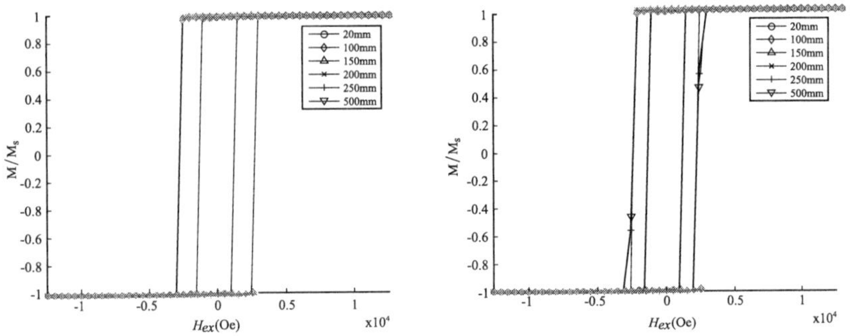


5 Wires 41 Wires

Figure 2 Hysteresis loops for different wire length

The coercivity $H_c$ for different wire lengths are plotted against increasing wire numbers in Figure 3. It is shown that magnetostatic coupling between the wires strongly influences the coercive field $H_c$. For small wire number, a reduction of coercivity $H_c$ is clearly visible as more wires are added. However, $H_c$ reaches a limit as more and more wires are added. This limiting value seems to be same for wire length over 100 nm. For fixed wire number, the $H_c$ increases with the length of the wire. However, there seems to be a sudden jump of $H_c$ from 1200 Oe to 2200 Oe as wire length increases from 20 nm to 50 nm indicating a qualitative change in switching behavior. This is due to the long range

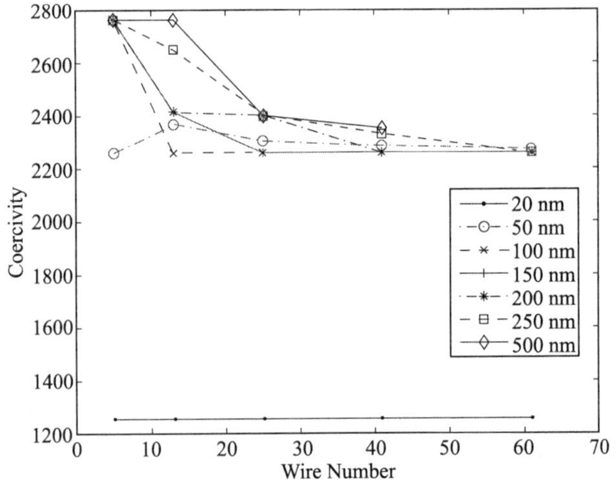


Figure 3 The coercive field decreases with increasing number of interacting nanowires for different wire length

effect of the magnetostatic coupling. The contribution of the magnetostatic energy is much smaller in short wire array than that for long wire array. Therefore short wire array would behave like a single wire leading to small coercive field $H_c$.

## 3.2 The effect of wire diameter, length, and interwire distance

Now we study how the magnetic property of the nanowire array depends on the wire size (diameter) and interwire distance. We vary both the wire size and the interwire distance from 5nm to 50nm. Their coercivity are shown in Tables 2 and 3. It is observed that the coercivity decreases with wire size but increases with interwire distance. Furthermore, as we fix the wire size and increase the interwire distance, there is a critical interwire distance above which the coercivity remains almost constant. In practice, one would want to decrease the interwire distance in order to increase the magnetic recording density. The above results suggest that decreasing interwire distance induces strong magnetostatic coupling between the neighboring wires and therefore decreasing the switching field. On the other hand, increasing wire length would enhance the effect the anisotropy thereby increasing the switching field.

Table 2 Coercivity (Oe) of wire length 20 nm

| size \ distance | 5 nm | 10 nm | 15 nm | 20 nm | 30 nm | 40 nm |
|---|---|---|---|---|---|---|
| 5 nm | 6.7848e3 | 7.7908e3 | 8.3142e3 | 8.2938e3 | | |
| 10 nm | 4.3178e3 | 5.7756e3 | 6.1701e3 | 6.3034e3 | | |
| 20 nm | 2.4175e3 | 3.2863e3 | 3.2855e3 | 3.2658e3 | 3.2659e3 | 3.2659e3 |
| 30 nm | 1.1896e3 | 1.2552e3 | 1.2552e3 | 1.2552e3 | 1.2553e3 | 1.2553e3 |
| 40 nm | 176.8227 | 134.1576 | 139.2495 | 250.7712 | 250.7880 | 250.7975 |
| 50 nm | 54.6632 | 26.9171 | 66.7516 | 65.1177 | 65.0679 | 64.4937 |

Table 3 Coercivity (Oe) of wire length 50 nm

| size \ distance | 5 nm | 10 nm | 15 nm | 20 nm | 30 nm | 40 nm |
|---|---|---|---|---|---|---|
| 5 nm | 5.6846e3 | 5.8013e3 | 6.4378e3 | 6.7614e3 | | |
| 10 nm | 4.2912e3 | 4.6483e3 | 4.7958e3 | 4.7960e3 | | |
| 20 nm | 2.7808e3 | 2.7838e3 | 3.2858e3 | 3.2869e3 | 3.3415e3 | 3.6740e3 |
| 30 nm | 1.7545e3 | 2.2560e3 | 2.2584e3 | 2.3693e3 | 2.3703e3 | 2.6033e3 |
| 40 nm | 1.5945e3 | 1.6591e3 | 1.5967e3 | 1.5971e3 | 1.4097e3 | 1.5975e3 |
| 50 nm | 1.1188e3 | 1.0903e3 | 931.8527 | 1.2227e3 | 1.2447e3 | 1.2478e3 |

## 3.3 Switching dynamics

In this section, we look at the details of the magnetic reversal process. Figure 4 shows a magnetic reversal process for an array (wire length × wire diameter × interwire distance=$100 \times 20 \times 30$ nm). It shows the cross section of the array in which the color represents the out of plane component. The four intermediate states correspond to the $a, b, c, d$ points on the hysteresis loop (Figure 5) which represent the start, intermediate and the end of the switching. It is clear that the switching starts at the boundary wires of the array. The middle wire is the last to switch. The reason is that the magnetostatic field is long range so that the demagnetization field is weaker near the boundary wires which make them easier to switch.

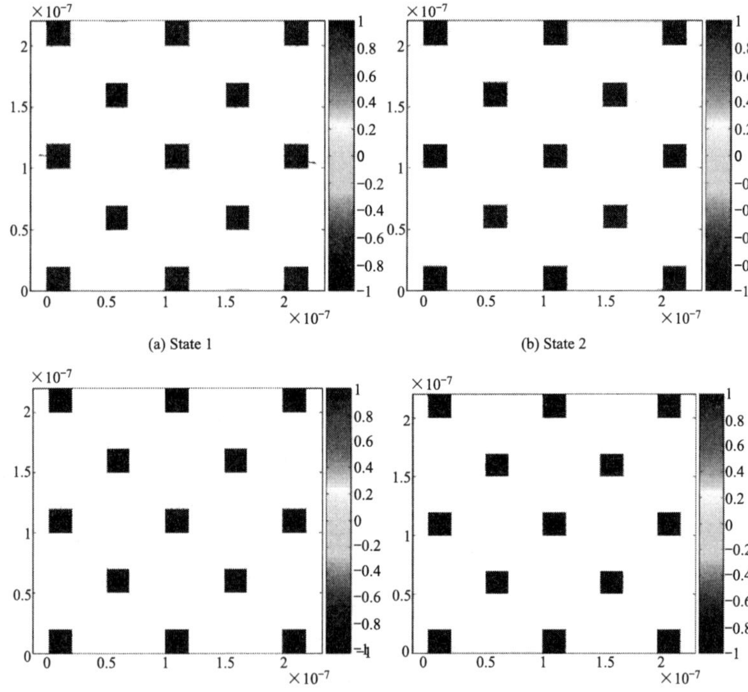


(a) State 1 (b) State 2

(c) State 3 (d) State 4

Figure 4 Switching dynamics of nanowire arrays along wire axis

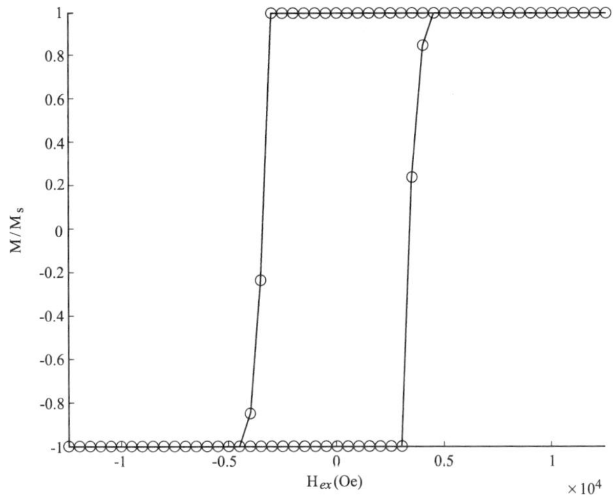


Figure 5 Switch positions on Hysteresis loop

# 4 Conclusion

Micromagnetic simulations based on the finite different method with parallel computing allow for detailed and large scale predictions of the micromagnetic properties of magnetic nanowire arrays. Numerical simulations have played a major role in improving and understanding the micromagnetic model and experimental results. This had led to increased confidence in both the computations and the experiments.

The newly developed algorithms improved both the computational efficiency and capacity. A 3-D parallel code was developed and implemented. The algorithms improved the computational efficiency at least by a factor of 20 times.

Magnetic properties of nanowire arrays, especially the effects of the microstructure of the nanowire array were studied. It is found that coercivity decreases when wire number increases and wire size has a dominant effect on the properties of nanowires. Our results also show that coercivity for an nanowire array is significantly smaller than that for a single wire (about 25% smaller in our example). Decreasing the interwire distance would reduce the switch field and at the same time increase the recording density. Another way to reduce the switching field (by about 15% in our example) is to reduce the length of the wires in the array. Our results also show that in the nanowire array, the switching field for the wires in the middle of the array is higher than that for the wires at the boundary of the array. These results have important implications in the high density magnetic recording applications.

## Acknowledgements

This work is supported in part by Hong Kong RGC-CERG grant HKUST 603503, HKUST 604105 and RGC-DAG 04/05.sc17.

## References

[1] S.H. Charap, P.-L. Lu, Y. He. *IEEE Trans. Magn.* **33**, 978, 1997.

[2] I.Z. Rahman, A. Boboca, K.M. Razeeba, M.A. Rahman. *J. Magn. Magn. Mater.* **290-291**, 246, 2005.

[3] X.P. Wang, J. Carlos. Garcia-Cervera and Weinan E. A Gauss-Seidel Projection Method for Micromagnetics Simulations. *J. Comp. Phys.* **171**, 357, 2001.

[4] J. Carlos García-Cervera, E. Weinan. *IEEE Trans. Mag.* **39**, 1766, 2003.

[5] E. Weinan, X.P. Wang. *SIAM J. Numer. Anal.* **38**, 1647, 2000.

[6] S. Shingubara, K. Morimoto, M. Nagayanagi, T. Shimizu, O. Yaegashi, G.R. Wu, H. Sakaue, T. Takahagi, K. Takase. *J. Magn. Magn. Mater.* **272**, 1598, 2004.

# Generalized B-spline*

Zongmin Wu
*School of Mathematics, Fudan University*
*Shanghai 200433, China*
*zmwu@fudan.edu.cn*

**Abstract**

We try to establish a parallel theory to the B-spline for the generalized spline (one kind of Tchebycheffian spline), which is constructed by piecewise functions of the solutions of prescribed ordinary differential equation. Such function spaces contain the piecewise functions of the algebra of the polynomial, the trigonometric polynomial and the hyperbolic polynomial, therefore contain the curves, which are drawn by the rule, the compass and some other common instruments in the engineering. The purpose of the approach is trying to find a more efficient function space, which can be easily used for curve representation, reproduction, approximation and furthermore to the application of pattern recognition, pattern classification etc.

## 1 Introduction

The most common function class, which the people prefer using in the most of applications, is the polynomial. We can use interpolation or the least square approximation to simulate a prescribed curve (parametric or non-parametric type). However the polynomial interpolation possesses Lunge's phenomena and requires solving a large-scale linear system of equation. On the other hand, the least square approximation (as well as the Bernstein-Bezier's approximation) possesses only lower order of the approximation. Therefore the function space of polynomials is not very efficient for the curve simulation.

The spline, which is constructed by piecewise polynomials (more generally the piecewise rational polynomials—the NURBS), is now the most favorite basis of the function space both for Mathematicians and Engineers. The spline is also included in most of standard computer software for curves and surfaces design.

*Supported by National Basic Research Program of China 973-2006CB303102.

The simplest case of the spline is the piecewise linear function, which is also used to construct the quadrature formula for the numerical integration. The approach is used in the theory of finite elements for numerical solution of partial differential equation too.

To the spline theory, the B-spline, the basic (best) spline basis is very important. There are a lot of advantages of using B-spline: such as the approximation capacity, the local control property (compactly supported), the shape preserving property, easy to evaluation (recursive evaluation scheme), multi resolution property (refinable, subdivision algorithm, wavelets) etc. However the spline possesses a disadvantage that it does not include the arc of circle, which is the most common curve in engineering (drawn by rule and compass). One can use rational spline to represent conic function, but it is difficult to develop an algorithm to keep the conic reproducing property, because the weight parameters of the rational polynomial are not linearly dependent on the data.

On the other hand, the type of the curve is usually already known (e.g. Line, Arc, Helix, Cycloid etc.) in the application, and then we have no reason to use piecewise polynomial but not the prescribed type of function space directly. Then we are faced to two key feature of curve representation. One is, which function space should be chosen. The second is, which basis in this function space should be used.

# 2 Definition of the generalized B-spline

[4,15,19] discussed the function space which is constructed by piecewise function of the linear combination of

$$\{1, t, \sin t, \cos t\},$$

and generalized the theory of B-spline to cover the curve containing lines and arcs of circle. They have proved that such function space possesses subdivision algorithm too (but adaptively or non-stationary). [1] discussed the hyperbolic spline, which is constructed by piecewise polynomial and the hyperbolic polynomial. In the references [1, 4, 6-9, 11-18], a lot of people discussed the problem in a lot of special cases and for special topics. The story can be come back to [17], where even a more generalized theory for the piecewise function of Tchebycheffian system (the Tchebycheffian spline) is discussed. The approaches above are all the special cases of the Tchebycheffian spline. In this paper we will discuss one class of the Tchebycheffian spline but in more details and generalize discussion in [1, 4, 6-9, 11-18]. These function spaces are more feasible in the application to compare the general discussion for Tchebycheffian spline. We have observed that the classical cubic spline satisfying $D^4 s(x) = 0$ piecewisely (spline of order $k$ satisfying $D^k s(x) = 0$

piecewisely, the piecewise linear combination of $\{1, t, \sin t, \cos t\}$ satisfying $((D^4 + D^2)s(x) = 0)$ piecewisely and the hyperbolic spline satisfying $(D^4 - D^2)s(x) = 0$ piecewisely. You can give more examples for your purpose of the application.

We will discuss the theory of the generalized spline, which are the piecewise functions in a subset of the algebra of polynomial, the trigonometric polynomial and the hyperbolic polynomial (containing Lines and Arcs of circle, the Helix, Cycloid, etc.).

Back to the history, the spline is at first defined only for odd degree that minimizes the bend energy $\int |D^k s(x)|^2 dx$ ($D$ denote the differential operator) subjected to the interpolatory conditions. The people find that the solutions are piecewise polynomials of degree $2k - 1$ (order $2k$). Later the spline is generalized to the piecewise polynomials of any order, or satisfying $D^n s(x) = 0$ piecewisely. Assume that $P(D) = D^n + p_1 D^{n-1} + p_2 D^{n-2} + \cdots + p_n$ is a real polynomial of degree $n$, where $D$ denote the differential operator that $D^k f = f^{(k)}$. In a lot of cases we assume $P(D) = Q(D) * Q(-D)$ for some special purposes (parallel to the classical spline of the odd degree, that minimizing the energy $\int |Q(D)s(x)|^2 dx$ subjected to the interpolatory condition). Moreover we generalize the approach that the generalized spline of order $n$ is defined to be the function satisfying $P(D)s(x) = 0$ piecewisely and some prescribed continuous condition. We will find at least that the generalized spline minimize the energy $\int |Q(D)s(x)|^2 dx$, if $P(D) = Q(D) * Q(-D)$.

# 3 Differential equation, basic properties

The $n$ dimensional linear function space $S$ is composed of the solutions of $P(D)B(x) = 0$. The function space $S$ is **shift invariant** that $B(x - c) \in S$, if $B(x) \in S$. The function space $S$ is however not always scale invariant, that $B(cx)$ must not necessary to be in the space $S$. This is the key difference to the polynomial space, where the space $S$, defined by $P(D) = D^n$, is both shift and scale invariant.

For the ordinary differential equation $D^n B(x) = 0$, $\{1, x, \cdots, x^{n-1}\}$ is a basis of $S$. Here $x^{n-1}$ play a special role, that the linear combination of the shifts $\{(x - x_j)^{n-1}\}$ generate the whole space of $S$, and the linear combination of the derivatives $\{x^{n-1(k)} = \frac{n!}{(n-k-1)!} x^{n-k-1}\}$ generate the whole space of $S$ too. We call such function to be a **generator**.

A problem is raised, whether we can find such a function (**generator**) for the solution space $S$ of the general ordinary differential equation $P(D)B(x) = 0$?

It can be easily observed that in the complex space,

**Theorem 1.** *If the roots $\{\lambda_j\}$ of the characteristic polynomial $P(\lambda) = 0$*

*are pairwise distinct, then*

$$B(x) = \sum a_j e^{\lambda_j x}, \quad \prod a_j \neq 0$$

*is a generator, i.e.*

*1.* $\{B(x - x_j)\}_{j=1}^{n}$ *generate the whole solution space* $S$, *if* $|\mathrm{Im}\lambda_j(x_j - x_k)| < \pi$ *and* $x_j$ *pairwise distinct.*

*2.* $\{B^{(k)}(x)\}_{k=0}^{n-1}$ *generate the whole solution space* $S$ *too.*

*Proof.*

$$\begin{pmatrix} \ddots & \ddots & \cdots \\ \ddots & e^{-\lambda_j x_k} & \ddots \\ \ddots & \ddots & \cdots \end{pmatrix} \begin{pmatrix} \ddots & & \\ & a_j & \\ & & \ddots \end{pmatrix} \begin{pmatrix} \vdots \\ e^{\lambda_j x} \\ \vdots \end{pmatrix} = \begin{pmatrix} B(x - x_1) \\ \vdots \\ B(x - x_n) \end{pmatrix},$$

$$\begin{pmatrix} \ddots & \ddots & \cdots \\ \ddots & \lambda_j^{k-1} & \ddots \\ \ddots & \ddots & \cdots \end{pmatrix} \begin{pmatrix} \ddots & & \\ & a_j & \\ & & \ddots \end{pmatrix} \begin{pmatrix} \vdots \\ e^{\lambda_j x} \\ \vdots \end{pmatrix} = \begin{pmatrix} B(x) \\ \vdots \\ B^{(n-1)}(x) \end{pmatrix}.$$

The determinant of the matrix of$(\lambda_j^{k-1})$ is the Vandermond's determinant and then the coefficients matrix is non-singular. By using the induction with $n$, the matrix $(e^{-\lambda_j x_k})$ is invertible too (see appendix for details). The result is also hold for $P(\lambda) = 0$ with multiple roots when we use the equivalent basis of the divided difference

$$\{e^{\lambda_1 x}, \frac{e^{\lambda_2 x} - e^{\lambda_1 x}}{\lambda_2 - \lambda_1}, \cdots, \text{high order divided differenece}\}$$

to instead of $\{e^{\lambda_1 x}, e^{\lambda_2 x}, \cdots\}$, and use derivatives to instead of the divided difference, as the roots $\lambda_j$ moved together. □

**Example 1.** *If* $\lambda_j s$ *are pairwise distinct, then we can take* $\{e^{\lambda_1 x}, e^{\lambda_1 x} + e^{\lambda_2 x}, \cdots, e^{\lambda_1 x} + e^{\lambda_2 x} + \cdots + e^{\lambda_n x}\}$ *as a series of generators to construct a subspace series* $U_1 \subset U_2 \subset \cdots \subset U_n = S$ *with shifts or the derivatives of the generators.*

**Example 2.** *Here we will point out a very important generator for our study. Assume* $[\lambda_j, \cdots, \lambda_{j+k}]e^{\lambda x}$ *is the divided difference of the function* $e^{\lambda x}$ *of order* $k$ *on* $\{\lambda_j, \cdots, \lambda_{j+k}\}$*. By using the property that the divided difference is the coefficient of the first term of the interpolatory polynomial (see [3]), that for pairwise distinct* $\lambda_j$,

$$[\lambda_1, \cdots, \lambda_n]e^{\lambda x} = \sum \frac{e^{\lambda_j x}}{\prod_{k \neq j}(\lambda_j - \lambda_k)} \tag{3.1}$$

*is a generator. Assume* $P(\lambda) = \prod_{j=1}^{l}(\lambda - \lambda_j)^{\gamma_j}$, *where* $\sum_{j=1}^{l}\gamma_j = n$ . *The divided difference is composed of linear combination of* $x^l e^{\lambda_j x}$ *with* $l \leqslant \gamma_j - 1$ *and the coefficient of the term* $x^{\gamma_j - 1}e^{\lambda_j x}$ *will not be vanished. Then the divided difference can be served as a generator for the multiple roots of* $\lambda_j$ *too. More clearly, if* $\lambda_j$ *are pairwise distinct and appeared coupled with complex conjugate. The* $P(\lambda)$ *is a real polynomial, which possesses the roots* $\lambda_j$ *with the multiplicity of* $\gamma_j$, *then*

$$\sum_j \frac{x^{\gamma_j - 1}e^{\lambda_j x}}{(\gamma_j - 1)!\prod_{k\neq j}(\lambda_j - \lambda_k)^{\gamma_k}} \tag{3.2}$$

*is a generator.*

**Theorem 2.** *If* $P(\lambda) = \prod_{j=1}^{n}(\lambda - \lambda_j)$ *(or* $P(\lambda) = \prod_{j=1}^{n}(\lambda - \lambda_j)^{\gamma_j}$ *for the root* $\lambda_j$ *with the multiplicity of* $\gamma_j$*), and the* $\lambda_j s$ *are pairwise distinct. Denote*

$$B(x) = [\lambda_1, \cdots, \lambda_n]e^{\lambda x} = \sum_j \frac{e^{\lambda_j x}}{\prod_{k\neq j}(\lambda_j - \lambda_k)}$$

$$\left(= \sum_j \frac{x^{\gamma_j - 1}e^{\lambda_j x}}{(\gamma_j - 1)!\prod_{k\neq j}(\lambda_j - \lambda_k)^{\gamma_k}} \text{ for } \lambda_j \text{with multiplicity of } \gamma_j\right),$$

*then* $P(D)B(x) = 0$, $B^{(k)}(0) = 0$ *for* $k \leqslant n-2$ *and* $B^{(n-1)}(0) = 1$.

*Proof.* We have already got in Example 2 that $B(x)$ is a generator. To prove the derivatives we expand the $e^{\lambda x}$ in Taylor's series

$$e^{\lambda x} = \sum_{j=0}^{n-2}\frac{(\lambda x)^l}{l!} + \frac{(\lambda x)^{n-1}}{(n-1)!} + \sum_{j=n}^{\infty}\frac{(\lambda x)^l}{l!}$$

and use the properties that the divided difference is a linear operator and the coefficient of the first term of the interpolatory polynomial respects to the variant $\lambda$. The first term is a lower order polynomial of $\lambda$, then the divided difference of order $n$ is zero. The interpolatory polynomial of the second term is $(\lambda x)^{n-1}/(n-1)!$ itself, and then the divide difference is $x^{n-1}/(n-1)!$ and the derivatives at origin possess our desired conditions. For the third term all the derivatives of the order not exceeded $n-1$ are zero at origin, since the third term possesses a factor $x^n$. □

# 4 Generalized spline, generalized B-spline

Assume $\{B_j(x)\}_{j=1}^{n}$ are the basis of the solution space $S$, satisfying $B_j^{(k-1)}(0) = \delta_{jk}$ ($B_n(x)$ is already got explicitly in Theorem 2). We are especially interested in the function

$$B_+(x) := \begin{cases} 0, & x < 0, \\ B_n(x), & x \geqslant 0, \end{cases} \quad B_-(x) := \begin{cases} -B_n(x), & x < 0, \\ 0, & x \geqslant 0, \end{cases}$$

which will play the same role as the truncated polynomial in the study of classical polynomial spline theory. We use the function

$$B_\pm(x) = (B_+(x) + B_-(x))/2$$

to be a symmetric kernel function too.

**Remark 1.** $P(D)B_+(x) = 0$ *for* $x \neq 0$. *Since all the lower order derivatives are zero and the* $(n-1)$*th derivative possesses a jump of 1 at the origin, then we have*

$$P(D)B_+(x) = \delta(x)$$

*symbolically, where* $\delta$ *is the Dirac* $\delta$*-function satisfying*

$$\int f(x)\delta(x)dx = f(0)$$

*for any continuous function* $f(x)$. *Analogously*

$$P(D)B_-(x) = \delta(x), \qquad P(D)B(x) = \delta(x).$$

*Then the generalized Fourier transformation of* $B_+$ *(as well as* $B_-, B_\pm$*) is* $1/P(-iw)$.

Parallel to the construction of the polynomial B-spline, we will construct the generalized B-spline basis and discuss the related properties. The B-spline basis $N_j(x)$ is defined to be a linear combination of

$$N_j(x) := \sum_{k=0}^{n} a_{j,k} B_+(x - x_{j+k}), \tag{4.1}$$

which satisfies $N_j(x) = 0$ on $x > x_{j+n}$.

The existence of the generalized B-spline is trivial (see [17] for Tchebycheffian spline too), since the $n+1$ functions $B_+(x - x_{j+k})$ are all in an $n$-dimensional space of $S$ on $x > x_{j+n}$, then linearly dependent. To construct $N_j(x)$, we require only to set the $n$ zero condition on $x_{j+n}$ for (4.1) to solve the coefficients $a_{j,k}$ up to a constant factor. Then the generalized Fourier transformation of $N_j(x)$ can be written to be

$$\hat{N}_j(w) = \frac{\sum_{k=0}^{n} a_{j,k} e^{ix_{j+k}w}}{P(-iw)}. \tag{4.2}$$

Since the function $N_j(x)$ now is a compact supported continuous function, then $N_j(x)$ possesses a classical Fourier transformation that the coefficients $a_{j,k}$ keep the function $\frac{\sum_{k=0}^{n} a_{j,k}e^{ix_{j+k}w}}{P(-iw)}$ bounded. We normalize the $N_j(x)$ with $\int N_j(x) = 1$ that $\hat{N}_j(0) = 1$. For the pairwise distinct $\{\lambda_l\}$, we can easily verify that $a_{j,k}$ are proportional to the $(-1)^k \det(e^{-\lambda_l x_{j+m}})_{m\neq k}$ up to a normalized factor, which depend only on the roots $\lambda_j$ and the knot $\{x_{j+k}\}_{k=0}^{n}$. Since $P(-iw)$ possesses $n$ zero points, there are $n$ linear independent condition to keep the function

$$\frac{\sum_{k=0}^{n} a_{j,k}e^{ix_{j+k}w}}{P(-iw)}$$

bounded. Or equivalently the zero points of $P(-iw)$ should be the zero points of $\sum_{k=0}^{n} a_{j,k}e^{ix_{j+k}w}$ too. The coefficient matrix of the homogeneous equation is of full rank, if $|\bar{w}(x_{j+n} - x_j)| < \pi$, where $\bar{w}$ are the roots of $P(-iw) = 0$. Then there exists a unique solution of $a_{j,k}$ satisfies the bounded condition and the normalized condition $\hat{N}_j(0) = 1$.

From the construction of the generalized B-spline, we get $N_j(x) \equiv 0$ on $(-\infty, x_j)$ and $(x_{j+n}, \infty)$ or equivalently $N_j(x)$ is compact supported on $(x_j, x_{j+n})$ (we will prove that $N_j(x) >\neq 0$ on $(x_j, x_{j+n})$ by generalized Schoenberg-Whitnay's Theorem in next section). Then we have

**Theorem 3.** *Except a constant factor, there exists one and only one non-trivial solution of the generalized spline which is supported on* $(x_j, x_{j+n})$.

We have furthermore

**Theorem 4.** (Minimal support Theorem) *Except the trivial solution, there exists no generalized spline function, whose support is smaller than* $(x_j, x_{j+n})$.

The proof of Theorem 3 and Theorem 4 can be got analogously by following the proof for the polynomial B-spline in [5], we have, the generalized B-spline $\{N_j^n(x)\}_{j=-n+1}^{L-1}$ is linear independent on $(x_0, x_L)$. Therefore the generalized B-spline $\{N_j^n(x)\}_{j=-n+1}^{L-1}$ can be served as a basis on $(x_0, x_L)$ as in the study of polynomial spline.

Count the freedom piece by piece just as in the study of polynomial spline [5], we get

$$\dim(S_{[x_0,\cdots,x_L]}) = n + L - 1.$$

This is equal to the number of the basis of $N_{-n+1}(x), \cdots, N_{L-1}(x)$ and we get the same result as for the polynomial spline.

# 5 Interpolation, knot insertion, recursive computation

**Theorem 5.** (The first Generalized Schoenberg-Whitnay's Theorem) *Assume that $B_+^{(n-1)}(x) > 0$, for $0 < x < H$, and $x_{j+n} - x_j < H$ for any $j$. Then*

(1) *The generalized B-spline $N_j(x) >\neq 0$ on $(x_j, x_{j+n})$.*

(2) *For any $y_{j-1} < y_j \in (x_j, x_{j+n})$, the matrix $(N_j(y_k))$ is non-singular and furthermore for any given data $\{f_j\}$, there exists a unique generalized interpolatory spline*

$$s(x) = \sum \lambda_j N_j(x) \quad satisfying \quad s(y_j) = f_j.$$

Define

$$M_j^{n-1}(x) = \begin{cases} A_j^{n-1} B_+^{(n-1)}(x - x_j), & x_j < x < x_{j+1}, \\ 0, & \text{otherwise} \end{cases} \tag{5.1}$$

and

$$M_j^k(x) = A_j^k \int_{-\infty}^{x} (M_j^{k+1}(t) - M_{j+1}^{k+1}(t))dt, \tag{5.2}$$

where $A_j^k$ is a positive factor that normalizes the integral $\int_{-\infty}^{\infty} M_j^k(x)dx = 1$. Then $M_j^k(x)$ are $C^{n-k-2}$ functions composed of linear combination of $B_+^{(k)}(x - x_j), \cdots, B_+^{(k)}(x - x_{j+n-k})$ and supported only on $(x_j, x_{j+n-k})$. Especially $N_j(x) = M_j^0(x)$. To prove Theorem 5 we prove furthermore

**Theorem 6.** (The second Generalized Schoenberg-Whitnay's Theorem)

(1) *The function $M_j^k(x) >\neq 0$ on $(x_j, x_{j+n-k})$.*

(2) *For any $y_j \in (x_j, x_{j+n-k})$, the matrix $(M_j^k(y_l))$ is non-singular and furthermore for any given data $\{y_j, f_j\}$, $y_j \in (x_j, x_{j+n-k})$ there exists a unique generalized interpolatory spline function $s^k(x) = \sum \lambda_j M_j^k(x)$ satisfying $s^k(y_j) = f_j$.*

*Proof.* We prove the theorem with the induction. The result is hold for $M_j^{n-1}(x)$ trivially. Assume that the results are hold for $M_j^l(x)$, $l \geqslant k+1$. To prove (1), assume that $M_j^k(x^*) = 0$ for some $x^* \in (x_j, x_{j+n-k})$ then $M_j^k(x)' = A_j^k(M_j^{k+1}(x) - M_{j+1}^{k+1}(x))$ possesses two zero points, one is in the interval $(x_j, x^*)$ and the other is in $(x^*, x_{j+n-k})$, if one of the zero points is in the interval $(x_j, x_{j+1})$ (or $(x_{j+n-k-1}, x_{j+n-k})$) then the zero point is just the zero point of $M_j^{k+1}(x)$ (or $M_{j+1}^{k+1}(x)$) and the result is contradicted to the inductions assumption of (1), otherwise, the result is contradictory to the inductions assumption of (2). To prove (2), assume that $\sum \lambda_j M_j^k(y_l) = 0$, $y_j \in (x_j, x_{j+n-k})$, then there are $y_l^* \in (y_l, y_{l+1})$ $(y_{-1}^* \in (x_0, y_0),\ y_{L+1}^* \in (y_{L+1}, x_{L+n-k}))$, that $\sum \lambda_j A_j^k (M_j^{k+1}(y_l^*) - M_{j+1}^{k+1}(y_l^*)) = 0$. If all the $y_l^* \in (x_l, x_{l+n-k-1})$ then the result is contradicted to the inductions assumption of (2), if some of $y_l^* \in (x_{l+n-k-1}, x_{l+n-k})$ then we can reduce the $L$ to a smaller problem and is also contradictory to the induction assumption of (2) or (1) (vise versa for $y_l^* \in (x_{l-1}, x_l)$). □

**Remark 2.** *From the discussion above we have*

$$M_j^k(x) = \sum_{l=0}^{n-k} b_{j,l}^k B_+^{(k)}(x - x_{j+l}).$$

*Then we have a recursive formula for the coefficients* $b_{j,l}^k$ *that*

$$\begin{gathered} b_{j,0}^{n-1} = \int_{x_j}^{x_{j+1}} B_+^{(n-1)}(x - x_j)dx = B_+^{(n-2)}(x_{j+1} - x_j), \quad b_{j,1}^{n-1} = 0, \\ b_{j,l}^k = A_j^k (b_{j,l}^{k+1} - b_{j+1,l-1}^{k+1}), \end{gathered} \tag{5.3}$$

*where the* $A_j^k$ *is the normalized factor that, the inverse of the* $A_j^k$ *is equal to*

$$\sum_{l=0}^{n-k-1} [b_{j,l}^{k+1} B_+^{(k-1)}(x_{j+n-k-1} - x_{j+l}) - b_{j+1,l}^{k+1} B_+^{(k-1)}(x_{j+n-k} - x_{j+1+l})].$$

*Furthermore we have* $a_{j,k} = b_{j,k}^0$, *if we define* $B_+^{(-1)}(x) = \int_0^x B_+(x)dx$.

**Remark 3.** *By using the derivatives as the limit case of the divided difference when some of the knots are moved together, we can especially choose* $y_{-l} = \cdots = y_0 = x_0$, $y_j = x_j, j = 1, \cdots, L-1$, $x_L = y_L = \cdots = y_{L+k}$ *with* $l + k = n - 2$ *to get a unique generalized spline that interpolate the data* $s^{(j)}(x_0) = f^{(j)}(x_0), j = 1, \cdots, l$; $s(x_j) = f(x_j), j = 0, \cdots, L$ *and* $s^{(j)}(x_L) = f^{(j)}(x_L), j = 1, \cdots, k$ *as in the study of polynomial spline too.*

Analogously to the discussion of the polynomial spline, we discuss now the knot insertion formula, then we can get a subdivisions algorithm in the dual sense. If $\bar{x}$ is inserted to the knots sequence $\{x_j\}$ to be a new knots sequence $\{\bar{x}_j\}$, $x_{j^*} < \bar{x} < x_{j^*+n}$. $N_j^n(x)$ and $\bar{N}_j^n$ are defined as above respected to the knots sequence $\{x_j\}$ and $\{\bar{x}_j\}$, then the Fourier transformation of the generalized B-spline are

$$\hat{N}_j^n = \frac{\sum_{k=0}^n a_{j,k} e^{ix_{j+k}w}}{P(-iw)},$$

$$\hat{\bar{N}}_j^n = \frac{\sum_{k=0}^{n-1} \bar{a}_{j,k} e^{ix_{j+k}w} + \bar{b} e^{i\bar{x}w}}{P(-iw)}$$

and

$$\hat{\bar{N}}_{j+1}^{n} = \frac{\sum_{k=0}^{n-1} \bar{a}_{j+1,k} e^{ix_{j+k}w} + \bar{c}e^{i\bar{x}w}}{P(-iw)}.$$

From the definition and the uniqueness of the generalized B-spline, we get

$$N_j^n(x) = \frac{a_{j,0}}{\bar{a}_{j,0}} \bar{N}_j^n(x) + \frac{a_{j,n}}{\bar{a}_{j+1,n-1}} \bar{N}_{j+1}^n(x), \tag{5.4}$$

otherwise the function

$$N_j^n(x) - (\frac{a_{j,0}}{\bar{a}_{j,0}} \bar{N}_j^n(x) + \frac{a_{j,n}}{\bar{a}_{j+1,n-1}} \bar{N}_{j+1}^n(x))$$

will be a generalized spline and supported only on $[x_{j+1}, x_{j+n-1}]$, which is contradicted to the minimal support theorem. Furthermore we have derived that

$$0 = \frac{a_{j,0}}{\bar{a}_{j,0}} \bar{b} + \frac{a_{j,n}}{\bar{a}_{j+1,n-1}} \bar{c}.$$

From the normalized condition we have $\frac{a_{j,0}}{\bar{a}_{j,0}} + \frac{a_{j,n}}{\bar{a}_{j+1,n-1}} = 1$, then

$$N_j^n(x) = \frac{\bar{c}}{\bar{c} - \bar{b}} \bar{N}_j^n(x) - \frac{\bar{b}}{\bar{c} - \bar{b}} \bar{N}_{j+1}^n(x) \text{ too.}$$

# 6 Polynomial reproducing, convergence order

An important property of the polynomial spline is the partition of unity, that all the B-spline basis are summed to unity. This property will not be always satisfied for the generalized B-spline if $P(0) \neq 0$, since then the constant is not the element of the generalized spline function. On the other hand, if $D$ is a factor of $P(D)$, then the constant function is in the space generated by generalized B-spline basis and can be uniquely represented as a linear combination of generalized B-spline $1 = \sum c_j N_j^n(x)$. We can use the property to normalize the generalized B-spline basis in another way and get the property of the partition of unity.

The result can be generalized to that $P(D)$ possesses a factor of $D^k$, then the interpolation with such generalized spline possesses the property of polynomial reproducing property of order $k$ by the uniqueness of the interpolation of generalized Schoenberg Whitnay's Theorem and the uniqueness of representation of the generalized spline. In this case, from the dual point of view, we can get some linear functional $L_j$ that the quasi-interpolation $\sum L_j(f) N_j^n(x)$ is polynomial reproducing of order $k$.

Polynomial reproducing is a key feature of the classical spline theory. Based on the property of the polynomial reproducing, we can use local polynomial approximation (the Taylor's expansion) and then the property of polynomial reproducing to prove that the spline of order $n$ possesses an approximations order of $n$ [3, 17].

Here we failed to have polynomial reproducing property, if $P(D)$ possesses no or only lower order of the factor of $D$. This seems to be a disadvantage of the generalized spline, however for some cases of the application we do not require polynomial reproducing but we do require the arc-reproducing, the helix-reproducing, the cycloid-reproducing etc. The generalized spline possesses such ability to do such a business, since the arc, the helix, the cycloid belongs to the space $S$ and the interpolation of the function in the space $S$ by linear combination of generalized B-spline is unique (then reproducing).

Moreover the high degree polynomial reproducing is a sufficient condition of the high order approximation, but is not the necessary condition. The key feature of the high order approximation is in fact that the reproducing of the function space $S$. The $S$ function reproducing can be served to get the approximations order too. Usually the generalized spline possesses no polynomial reproducing property, however it do possesses the $S$ reproducing property as well as the high order approximation. In fact the order of the approximation depends on the Taylor-like expansion with the function in $\ker(P(D))$ (see [17]). Some of the fundamental theories are discussed below.

Assume that the prescribed function $f(x) \in C^n(a,b)$, we extend the function that $f(x) \in C^n(-\infty,\infty)$ and $f(x) = 0$ for $x \notin (A,B) \supset (a,b)$. Assume that the knots $\{x_j\} \in [A,B]$, $A = x_0$, $x_{j+1} - x_j < h$ and $x_L = B$. Then for any function $g(x)$,

$$\begin{aligned}\int \sum \lambda_j N_j^n(x) g(x) dx &= \sum \lambda_j \int N_j^n(x) g(x) dx \\ &= \sum \lambda_j (g(x_j^*) + \mathcal{O}(nh\|g'(x)\|_\infty)),\end{aligned}$$

where $x_j^*$ can be any point in $(x_j, x_{j+n})$ to form an ordered sequence. Let $y_j = (x_{j-1}^* + x_j^*)/2$, if we take $\lambda_j = f(x_j^*)(y_{j+1} - y_j)$, then

$$\int \sum \lambda_j N_j^n(x) g(x) dx \sim \sum f(x_j^*) g(x_j^*)(y_{j+1} - y_j),$$

which is a Riemanian summation of the $\int f(x)g(x)dx$. Then we have for example the Schumaker's quasi interpolation as for the polynomial B-spline (e.g. see [5]).

**Theorem 7.** *If we take* $x_j^* = (x_{j+1} + x_{j+2} + \cdots + x_{j+n-1})/(n-1)$, $y_j = (x_{j-1}^* + x_j^*)/2$ *and* $\lambda_j = f(x_j^*)(y_{j+1} - y_j) = f(x_j^*)(x_{j+n} - x_j)/(n-1)$, *then for any* $g \in C^{k+1}$

$$\begin{aligned}&|\int D^k(\sum \lambda_j N_j^n(x) - f(x)) \cdot g(x) dx| \\ &= |\int (\sum \lambda_j N_j^n(x) - f(x)) \cdot D^k g(x) dx| \\ &< n\|f\|_\infty (\|g^{(k)}\|_\infty + \|g^{(k+1)}\|_\infty) h = \mathcal{O}(h).\end{aligned} \tag{6.1}$$

*This means, if* $h \to 0$, *then the generalized B-spline quasi-interpolation* $\sum f(x_j^*)(y_{j+1} - y_j)N_j^n(x)$ *and its derivatives will convergent to the function* $f(x)$ *and its derivatives respectively in the meaning of* (6.1), *if all* $\|g^{(k)}\|$ *for* $k \leqslant n$ *are bounded. Especially, up to the order* $n-1$, *we get moreover the point-wise convergence, since all the function above are continuous function defined on closed interval.*

**Generalized Taylor's expansion and Generalized Rolle's Lemma**

To discuss the convergence property more in details, we show at first some lemmas

**Lemma 1.** (Generalized Taylor's expansion) *For any function* $f(x) \in C^n$, *there exists a Taylor like expanse that*

$$f(x) = \sum_{k=0}^{n-1} f^{(k)}(x_0)B_k(x - x_0) + \mathcal{O}(x - x_0)^n,$$

*where* $B_j^{(k-1)}(0) = \delta_{jk}$.

*Proof.* Since the derivatives for $k \leqslant n-1$ at $x_0$ of the both sides are identically, we get the result by comparing the Taylor's expansion of the both sides. □

**Lemma 2.** (Generalized Roll's Lemma) *If* $f(a) = f(b) = 0$ *and* $f \in C^1$, *then there is a point* $\xi \in (a,b)$ *that* $(D-d)f(\xi) = 0$, *where* $D$ *is the differential operator and* $d$ *is any given real number. If* $f(a) = f(c) = f(b) = 0$, $a < c < b$ *and* $f \in C^2$, *then there is point* $\xi \in (a,b)$ *that* $(D-d)(D-\bar{d})f(\xi) = (D^2 - 2A^*D + A^{*2} + B^{*2})f(\xi) = 0$, *where* $d = A^* + iB^*$ *is any given complex number, and* $|(b-a)B^*| < \pi$.

*Proof.* For the first part, we consider the function $g(x) = e^{-dx}f(x)$. Then $g(a) = g(b) = 0$ and by Rolle's Lemma, there is a point $\xi \in (a,b)$ that $Dg(\xi) = 0 = e^{-dx}(D-d)f(\xi) = 0$. We have then the desired result $(D-d)f(\xi) = 0$, since $e^{-dx} \neq 0$. For the second part, we have already proved for $B^* = 0$ in first part. If $B^* \neq 0$, denote $F(t) = f(t/B^* + (a+b)/2)$ and $A = A^*/B^*$, then $F(-(b-a)B^*/2) = F(B^*(c - (a+b)/2)) = F(-(b-a)B^*/2) = 0$ and we require to find $\eta \in (-(b-a)B^*/2, (b-a)B^*/2)$, that $(D^2 - 2AD + A^2 + 1)F(\eta) = 0$. If $g(t) = e^{A^*t}/\cos(t)$ and $h(t) = 1/\cos^2(t)$, then

$$\begin{aligned} D[h(t)D(g(t)F(t))] &= hgF'' + (2hg' + h'g)F' + (h'g' + hg'')F \\ &= hg[D^2 + 2AD + (1 + A^2)]F. \end{aligned}$$

If $|(b-a)B^*| < \pi$, $h$ and $g$ will not be vanished in the interval $(-(b-a)B^*/2, (b-a)B^*/2)$. Moreover since $F$ possesses three zero points in the interval, by using Rolle's Lemma $D(g(t)F(t))$ as well as $h(t)D(g(t)F(t))$ possesses at least two zero points and $D[h(t)D(g(t)F(t))] = hg[D^2 + 2AD + (1 + A^2)]F$ possesses at least one zero point in the interval. □

Based on the lemma we can get the following lemma by induction.

**Lemma 3.** *If $f \in C^n$, $f(x_j) = 0$, $x_0 < \cdots < x_n$, then there is point $\xi \in (x_0, x_n)$ that $P(D)f(\xi) = 0$.*

**Remark 4.** *The condition $|(b-a)B^*| < \pi$ is necessary. Otherwise we can give an example that $f(x) = \sin^2(x)$ possesses a lot of zeros, however $(D^2 - 2D + 5)f(x) = \sin^2(x) + (\sin(x) - \cos(x))^2 > 0$ does not possess any zero point.*

**Lemma 4.** *If $f(x_0) = 0$ and $(D - d)f(x) = f_1(x)$ with real $d$, then $|f(x)| \leqslant C|x - x_0|\|f_1\|_\infty$. If $f(x_0) = f(x_1) = 0$ and $(D - d)(D - \bar{d})f(x) = (D^2 - 2AD + A^2 + B^2)f(x) = f_2(x)$ with complex number $d = A + iB$, then $|f(x)| \leqslant C\max(|x - x_j|^2)\|f_2\|_\infty$. With the induction we have $|f(x)| \leqslant C\max(|x - x_j|^n)\|f_n\|_\infty$, if $f(x_1) = \cdots = f(x_n) = 0$ and $P(D)f(x) = f_n(x)$.*

*Proof.* We need only to prove the lemma for the order one and two. We know the result already in the discussion of classical spline for $P(D) = D^n$ with the integration to $f^{(n)}(x)$ again and again. Now from $e^{dx}D(e^{-dx} f(x)) = (D - d)f(x) = g(x)$ and $(hg)^{-1}D[hD(gf)]$ $f(x) - f(x_0) = \int_{x_0}^{x} e^{d(x-t)g(t)dt}$, where $hg$ possesses no zero point and bounded, we get the assertion. □

We can have furthermore that the Hermitian interpolation with the function in $\ker(P(D))$ possesses an approximation order of $n$ too, more clearly

**Theorem 8.** *If $P(D)s(x) = 0$, $s^{(k)}(x_0) = f^{(k)}(x_0)$, for $k < l$ and $s^{(k)}(x_1) = f^{(k)}(x_1)$, for $k < m$, where $l + m = n$, then $|f(x) - s(x)| < \mathcal{O}|x_1 - x_0|^n$.*

*Proof.* This is a direct consequence of the generalized Rolle's Lemma and the lemma above. □

We would like to remark that the result is hold for any $l < n - 1$ and $m = n - l$.

We have showed a recursive formula of the derivatives of the generalized B-spline in (5.2) and (5.3), now we will give another recursive formula of the derivatives. If $P_n(D) = (D + d)P_{n-1}(D)$ with real number $d$, $N_j^n(x)$ and $N_j^{n-1}(x)$ are the generalized B-spline respected to the $P_n(D)$ and $P_{n-1}(D)$, moreover

$$\hat{N}_j^n(w) = \frac{\sum_{k=0}^{n} a_{j,k} e^{ix_{j+k}w}}{P_n(-iw)},$$

$$\hat{N}_j^{n-1}(w) = \frac{\sum_{k=0}^{n-1} b_{j,k} e^{ix_{j+k}w}}{P_{n-1}(-iw)}$$

and

$$\hat{N}_{j+1}^{n-1}(w) = \frac{\sum_{k=0}^{n-1} b_{j+1,k} e^{ix_{j+k}w}}{P_{n-1}(-iw)},$$

then the derivative of the generalized B-spline can be got via the following equations

$$(D+d)N_j^n(x) = \frac{a_{j,0}}{b_{j,0}} N_j^{n-1}(x) + \frac{a_{j,n}}{b_{j+1,n-1}} N_{j+1}^{n-1}(x),$$

$$DN_j^n(x) = \frac{a_{j,0}}{b_{j,0}} N_j^{n-1}(x) + \frac{a_{j,n}}{b_{j+1,n-1}} N_{j+1}^{n-1}(x) - dN_j^n(x).$$

Usually the derivative of the generalized B-spline is not always a lower order generalized B-spline as in the study of the polynomial B-spline.

If $d$ is a complex number, and if $P(\cdot) = 0$ possesses a coupled roots of $d$ and $\bar{d}$, then we have symbolically

$$(2D + d + \bar{d})N_j^n(x) =$$
$$\tfrac{a_{j,0}}{b_{j,0}} N_j^{n-1}(x) + \tfrac{a_{j,n}}{b_{j+1,n-1}} N_{j+1}^{n-1}(x) + \overline{\tfrac{a_{j,0}}{b_{j,0}} N_j^{n-1}(x) + \tfrac{a_{j,n}}{b_{j+1,n-1}} N_{j+1}^{n-1}(x)}.$$

The right side term is real, and we keep to discuss the problem in the real space.

For the polynomial B-spline, the degree reduction property is very important that derives the recursive computation formula of the B-spline basis. For the generalized B-spline the degree reduction formula is not always well defined. The key feature of the degree reduction of B-spline is that the polynomial multiply a polynomial is a higher order polynomial. For the generalized spline, an example can be showed that if $P(D) = D(D+1)(D+10)$,

$$\frac{B_+^n(x) B_+^{n-1}(x-y)}{B_+^{n-1}(x)} \notin \ker(P(D))_{0<y<x}.$$

Then we cannot get a similar formula of the degree reduction.

However some of the special cases can be considered.
**Case 1:** If $P(D) = (D-d)^n$ with real $d$, then the solutions of the ordinary differential equation are $\{e^{dx}, xe^{dx}, \cdots, x^{n-1}e^{dx}\}$. The generalized B-spline for the differential operator $P(D)$ is just the classical B-spline multiply by $e^{dx}$, and the degree reduction's scheme for classical B-spline can be generalized to the generalized B-spline. This result can be generalized to the case that the solution space is generated by

$$\{\sin(cx)e^{dx}, x\sin(cx)e^{dx}, \cdots, x^{n-1}\sin(cx)e^{dx}\}$$

and

$$\{\cos(cx)e^{dx}, x\cos(cx)e^{dx}, \cdots, x^{n-1}\cos(cx)e^{dx}\}$$

too.

**Case 2:** If $P(D) = D(D-d)\cdots(D-(n-1)d)$, then the solutions of the ordinary differential equation are $\{1, e^{dx}, \cdots, e^{(n-1)dx}\}$. Let $u = e^{dx}$, $u_j = e^{dx_j}$, then we can use the degree reduction formula for polynomial spline $N_j^n(u)$ to derive the degree reduction formula for composite B- spline $N_j^n(e^{dx})$. This approach can be generalized to the case, that the solutions space of the ordinary differential equation is generated by $\{1, u(x), \cdots, u^{n-1}(x)\}$ and shift invariant, where $u(x)$ is an element of the algebra of the polynomial, trigonometric polynomial and the hyperbolic polynomial, and if the function $u(x)$ is monotone.

**Remark 5.** *By using the definitions of* (5.1) *and* (5.2), *we can still get a recursive computation formula to compute the generalized B-spline through an integration or by using* (5.3) *to calculate the coefficients of the B-spline basis recursively.*

The generalized B-spline basis $\{N_j^n(x)\}$ is constructed by the linear combination of

$$\{B_+(x - x_{j+k})\}_{k=0,\cdots,n}.$$

This function space is equivalent to the function space generated by

$$\{[x_j]B_+(x-\cdot), [x_j, x_{j+1}]B_+(x-\cdot), \cdots, [x_j, \cdots, x_{j+n}]B_+(x-\cdot)\}.$$

Then by using the derivative to instead of divided difference, we can discuss the case when the knots $x_{j+k}$ possesses a multiplicity (moved together). More clearly if $x_j$ possesses a multiplicity of $r_j$, then

$$\{\cdots, B_+(x-x_j), \cdots, B_+^{(r_j-1)}(x-x_j), \cdots\}$$

can be used as a basis to generate the generalized spline. Such generalized spline function is still piecewise function in space $\ker(P(D)) = S$, however is $C^{n-r_j}$ at $x_j$.

Analogously to the discussion of polynomial spline, we can get a Bernstein- Bezier's form for generalized spline too with the knot insertion of (5.4) until the knot $x$ with the multiplicity of $n-1$. The result can be get analogously as to the discussion for classical piecewise polynomial. Especially the Bernstein- Bezier's basis for generalized spline on $[0,1]$ are

$$\begin{aligned} B_j(x) &= \sum_{k=0}^{n-j} a_k^j B_\pm^{(k)}(x) + \sum_{k=0}^{j} b_k^j B_\pm^{(k)}(x-1) \\ &= \frac{1}{2\pi}\int \frac{\sum_{k=0}^{n-j} a_k^j (iw)^k + \sum_{k=0}^{j} b_k^j (iw)^k e^{iw}}{P(-iw)} e^{-ixw}\, dw, \end{aligned}$$

where $a_k^j$ $b_k^j$ are the unique solution that keep the integrand bounded, and equal to $1/n$ at the origin. By using knots insertion formula (5.4), we can get the de Casteljau algorithm analogously as the discussion in [5] too.

Another important property of the spline is the minimization of the bent energy. For the generalized spline we can prove the related optimal property too. If $P(D) = Q(D)Q(-D)$, the order of $P(D)$ is then $2n$. We have the following theorem:

**Theorem 9.** *If $\{L_j\}$ are some linear independent functional in the dual space of $(C^{n-2})$ (e.g. $L_jf = f(x_j)$, $L_jf = f'(x_j) + 2f''(x_j)$ etc.), then for any given data $L_jf$ there is a unique interpolatory solution $s(x)$ satisfying $L_js = L_jf$ that minimizes the functional (bend energy) $\int \langle Q(D)s(x), Q(D)s(x)\rangle dx$, if $\ker(Q(D)) \cap \ker(L) = 0$.*

*Proof.* Assume that $\{b_1(x), \cdots b_n(x)\}$ is a basis of $\ker(Q(D))$ and $B_\pm(x)$ is the symmetric kernel as defined above respected to the $P(D)$, then for any $b(x) \in \ker(Q(D))$, $Q(D)b(x) = 0$ or equivalently $Q(-iw)\hat{b}(w) = 0$. This means $\hat{b}(w) = 0$ if $Q(-iw) \neq 0$. If for any $b(x) \in \ker(Q(D))$, $\sum \lambda_j L_j b(x) = 0$, then $\sum \lambda_j L_j e^{-ixw} = 0$, if $Q(-iw) = 0$. Therefore the following integral is well defined and

$$\int |\frac{\sum \lambda_j L_j e^{-iwx}}{Q(-iw)}|^2 dw >\neq 0.$$

Furthermore the matrix

$$\begin{pmatrix} L_{jx}L_{ky}^* B_\pm(x-y) & L_j b_k(x) \\ L_k^* b_j(y) & 0 \end{pmatrix}$$

is non-singular, or equivalently, there exists a unique solution

$$s(x) = \sum \lambda_j L_{jy}^* B_\pm(x-y) + \sum \mu_k b_k(x)$$

satisfying

$$L_j s(x) = L_j f \quad (\text{all } j) \quad \text{and} \quad \sum \lambda_j L_{jy}^* b_k(y) = 0 \quad (\text{all } k).$$

In order to prove the optimal property, we are required only to prove

$$\int \langle Q(D)s(x), Q(D)h(x)\rangle dx = 0,$$

for $h(x) \in \ker(L)$. Because $L_jh(x) = 0$ then $\int \sum \lambda_j L_{jx} e^{-ixw}\hat{h}(w)dw = 0$. The Fourier transformation of the function $Q(D)h(x)$ is $Q(-iw)\hat{h}(w)$ and the Fourier transformation of the function

$$Q(D)s(x) = Q(D)(\sum \lambda_j L_{jy}^* B_\pm(x-y))$$

is $Q(-iw)(\sum \lambda_j L_{jx} e^{-ixw}/|Q(-iw)|^2)$, then

$$\begin{aligned} &\int \langle Q(D)s(x), Q(D)h(x)\rangle dx \\ &= \int \langle Q(-iw)\tfrac{\sum \lambda_j L_{jx} e^{-ixw}}{|Q(-iw)|^2}, Q(-iw)\hat{h}(w)\rangle dw \\ &= \int \sum \lambda_j L_{jx} e^{-ixw} \hat{h}(w) dw = 0. \end{aligned}$$

□

**Remark 6.** *For $Q(D) = D^n$ the classical spline of order $2n$, $k(x,t) = \frac{|x-t|^{2n-1}}{(2n-1)!}$ is called a reproducing kernel that for any function $u(x) \in C^n$, $u(x) - \int \langle D_t^n k(x,t), D_t^n u(t)\rangle dt$ is in the $\ker(Q(D))$. In our case the function $B_\pm(x)$ can be served as a reproducing kernel too. Applying the operator $Q(D_x)$ to the $u(x) - \int \langle Q(D_t)B_\pm(x-t), Q(D_t)u(t)\rangle dt$, derives that $u(x) - \int \langle Q(D_t)B(x-t), Q(D_t)u(t)\rangle dt \in \ker(Q(D_x))$, since $Q(D_x)Q(D_t)B(x-t) = \delta(x-t)$.*

# 7 Uniform distributed knots, wavelets, cardinal spline

Now we will take a close view to the uniform distributed knots that $x_j = j\beta$, then the spline space is shift invariant that $N_j(x) = N_0(x - j\beta)$. Pay attention that the polynomial B-spline on uniform distributed knots play an important role in the study of the wavelet theory.

At first we will point out the shape preserving property for uniform distributed knots, that the $A_j^k$ in recursive formula of (5.2) for $M_j^k(x)$ will be a positive constant independent on $j$. Then if $s(x) = \sum f_j N_j(x)$,

$$s^{(k)}(x) = \sum f_j M_j^0(x) = A^0 \cdots A^{k-1} \sum \nabla^k f_j M_j^k(x),$$

where $\nabla$ is the difference operator. This derives $s^{(k)}(x) \geqslant 0$ $(k < K)$, if $\nabla^k f_j > 0$ for $k < K$, therefore we get the shape preserving property.

Another meaningful result for uniform distributed knots is the convolution property. If $N_{j,1}$ and $N_{k,2}$ is the generalized B-spline respected to the operator $P_1(D)$ and $P_2(D)$ then the convolution $N_{j+k} = N_{k,2} * N_{j,1}$ is the generalized B-spline respected to the operator $P(D) = P_1(D)P_2(D)$. By using the representation of (4.2) the result can be easily verified by the properties of the Fourier transformation.

A direct consequence of the convolution of the generalized B-spline is the degree reduction of the uniform generalized B-spline. If $P(D)$ can be factorized to $P(D) = \prod_{j=1}^n P_j(D)$, $N_k(x)$ and $M_k(x)$ are the generalized B-spline basis respected to the operator $\prod_{j=1}^{k-1} P_j(D)$ and $P_k(D)$, then $N_{k+1}(x) = N_k(x) * M_k(x)$, which derives a degree reduction formula in the convolution sense.

Now we will discuss the subdivisions algorithm and the refinable property of the generalized spline on uniform distributed knots. This problem is discussed in [10] in another way, however we will discuss the problem in more details. Let

$$V_k = \{\cdots, N_{-1}^k, N_0^k, N_1^k, \cdots\},$$

where $N_j^k$ are the generalized B-spline basis on the partition of $\beta_k = \beta/2^k$. Obviously the generalized B-spline on dyadic knots formed a nested linear space

$$\cdots \subset V_k \subset V_{k+1} \subset \cdots .$$

Then any $\sum b_j^k N_j^k(x) \in V_k$ possesses a representation of $\sum b_j^{k+1} N_j^{k+1}(x)$ in $V_{k+1}$.

To find the subdivisions relation of $\{b_j^k\}$ and $\{b_j^{k+1}\}$, it will be enough to find the masks $\{a_j^k\}$ of the refinable relation from $V_k$ to $V_{k+1}$.

$$N_0^k(x) = \sum_{j=0}^{n} a_j^k N_j^{k+1}(x) = \sum_{j=0}^{n} a_j^k N_0^{k+1}(x - j\beta_k/2). \tag{7.1}$$

It is very interesting and useful to get the refinable function and the masks of subdivision algorithm for the generalized B-spline, since the subdivision algorithm is a powerful algorithm for the numerical evaluation of the generalized spline in CAGD. Here we will point out a difference between the generalized B-spline wavelets and the polynomial B-spline wavelets:

**Remark 7.** *Usually*

$$N_0^j(2x) \neq N_0^{j+1}(x),$$

*since the solution of the $P(D)B(x) = 0$ is not scale invariant. $N_0^j(2x) = N_0^{j+1}(x)$ satisfied only for homogenous differential operator $P(D) = D^n$. The masks of the subdivision here are level to level different.*

To find the masks of the subdivision algorithm, we define

$$m_k(w) := P(-iw)\hat{N}_0^k(w) = \sum_{j=0}^{n} e_j^k e^{ij\beta_k w}.$$

The function $m_k(w)$ then possesses the same zero points as $P(-iw)$ and we normalized it to be

$$\int N_0^k(x)dx = \beta_k.$$

Then the coefficients $e_j^k$ can be got explicitly by solving linear system of equation. Apply the Fourier transform to both sides of the equation (7.1), we get

$$\hat{N}_0^k(w) = (\sum a_j^k e^{ij\beta_{k+1}w})\hat{N}_0^{k+1}(w) = A_k(w)\hat{N}_0^{k+1}(w). \tag{7.2}$$

From section 4 we can get the function

$$P(-iw)\hat{N}_0^k(w) = m_k(w),$$

$$P(-iw)\hat{N}_0^{k+1}(w) = m_{k+1}(w)$$

explicitly. These derive

$$A_k(w) = m_k(w)/m_{k+1}(w).$$

Since the spline space is nested, the function $A_k(w)$ will not be a rational trigonometric function but possesses the form $A_k(w) = \sum_{j=0}^n a_j^k e^{ij\beta_{k+1}w}$. Here we get that the length of the mask is at most $n+1$. Moreover

$$\hat{N}_0^k(w) = \frac{m_k(w)}{m_{k+1}(w)}\hat{N}_0^{k+1}(w) = A_k(w)\hat{N}_0^{k+1}(w).$$

We have observed that the refinable function from $V_k$ to $V_{k+1}$ is not same as by the discussion of the classical wavelet that $A_k(w) = A_0(w/2^k)$. Now the masks $A_k(w)$ depend on the subdivision level of $k$. The subdivision algorithm can then be written to be

$$N_0^k(x) = \sum a_j^k N_0^{k+1}(x - j\beta_{k+1}),$$

where the masks $\{a_j^k\}$ depend on the $k$, however can be given from the $\lambda_j$s and the $\beta/2^k$. The condition

$$\sum a_j^k = A_k(0) \equiv 2$$

will be hold as usually, since $\hat{N}_0^k(0) = \beta_k$ and $\hat{N}_0^{k+1}(0) = \beta_{k+1} = \beta_k/2$.

Summarizing the discussion we have: If $s(x) \in V_k \subset V_{k+1}$, then

$$s(x) = \sum b_j^k N_k(x - j\beta_k) = \sum b_j^{k+1} N_{k+1}(x - j\beta_{k+1}),$$

the subdivision algorithm can be written to be

$$b_l^{k+1} = \sum a_{l-2j}^k b_j^k$$

analogously to the classical subdivision algorithm but adaptively (nonstationary) level by level. The masks $a_j^k$ are the coefficients of the trigonometric polynomial of $A_k(w) = m_k(w)/m_{k+1}(w)$.

The cardinal spline plays an important role in the study of spline theory of uniform knots too, that $C_\beta(x)$ is a generalized spline function satisfying

$$C_\beta(x) = \begin{cases} 1, & x = 0, \\ 0, & x = j\beta, j \neq 0. \end{cases}$$

As by the discussion above, there are coefficients $e_j$ that

$$C_\beta(x) = \sum e_j B_\pm(x - j\beta).$$

Symbolically we have

$$P(-iw)\hat{C}_\beta(w) = \sum_{-\infty}^{\infty} e_j e^{ij\beta w}.$$

By using the Fourier transform we have

$$\frac{1}{2\pi}\int \hat{C}_\beta(w) e^{-ij\beta w} dw = \begin{cases} 1, & j = 0, \\ 0, & j \neq 0. \end{cases}$$

If $P(-iw) = Q(iw)Q(-iw) \geqslant 0$ is a real function and $P(-iw) \geqslant \mathcal{O}(|w|^2)$ as $w \to \infty$, we have

$$\begin{aligned} &\frac{1}{2\pi}\int \hat{C}_\beta(w) e^{-ik\beta w} dw \\ &= \frac{1}{2\pi}\int \frac{\sum e_j e^{ij\beta w}}{P(-iw)} e^{-ik\beta w} dw \\ &= \frac{1}{2\pi}\int_0^{2\pi/\beta} \left(\sum_{j=-\infty}^{\infty} \frac{1}{P(-iw + 2\pi ij/\beta)}\right)\left(\sum_{j=-\infty}^{\infty} e_j e^{ij\beta w}\right) e^{-ik\beta w} dw. \end{aligned} \tag{7.3}$$

Then from the theory of the Fourier series we have

$$\left(\sum_{j=-\infty}^{\infty} \frac{1}{P(-iw + 2\pi ij/\beta)}\right)\left(\sum_{j=-\infty}^{\infty} e_j e^{ij\beta w}\right) = \beta$$

or

$$\beta \Big/ \left(\sum_{j=-\infty}^{\infty} \frac{1}{P(-iw + 2\pi ij/\beta)}\right) = \left(\sum_{j=-\infty}^{\infty} e_j e^{ij\beta w}\right).$$

Therefore

$$e_k = \frac{\beta}{2\pi}\int_0^{2\pi/\beta} \beta e^{-ik\beta w} \Big/ \left(\sum_{j=-\infty}^{\infty} \frac{1}{P(-iw + 2\pi ij/\beta)}\right) dw$$

and

$$C_\beta(x) = \sum e_j B_\pm(x - j\beta). \tag{7.4}$$

On the another side we have

$$\hat{C}_\beta(w) = \frac{\beta}{P(-iw)\left(\sum_{j=-\infty}^{\infty} \frac{1}{P(-iw + 2\pi ij/\beta)}\right)}$$

and

$$C_\beta(x) = \frac{1}{2\pi}\int_{-\infty}^{\infty} \frac{\beta}{P(-iw)\left(\sum_{j=-\infty}^{\infty} \frac{1}{P(-iw + 2\pi ij/\beta)}\right)} e^{-ixw} dw. \tag{7.5}$$

Furthermore we have

$$\sum e_j^2 = \frac{\beta}{2\pi}\int_0^{2\pi/\beta} \left| \frac{\beta}{\sum_{j=-\infty}^{\infty} \frac{1}{P(-iw + 2\pi ij/\beta)}} \right|^2 dw.$$

Summarizing the discussion above, we have

**Theorem 10.** *If $P(-iw) \geqslant \mathcal{O}(w^2)$, then there exists a generalized cardinal spline uniquely as showed by (7.4) or (7.5). In this case any generalized spline function can be uniquely represented to be*

$$s(x) = \sum_{j=-\infty}^{\infty} s(j\beta) C(x - j\beta).$$

We can get a subdivision of the cardinal spline too, that

$$C_k(x) = \sum_{j=-\infty}^{\infty} c_j^k C_{k+1}(x - j\beta_{k+1}) = \sum_{j=-\infty}^{\infty} C_k(j\beta_{k+1}) C_{k+1}(x - j\beta_{k+1}).$$

Then we have

$$\hat{C}_k(w) = \left(\sum_{j=-\infty}^{\infty} c_j^k e^{ij\beta_{k+1}w}\right) \hat{C}_{k+1}(w)$$

and then

$$\sum_{j=-\infty}^{\infty} c_j^k e^{ij\beta_{k+1}w} = \frac{2\sum_{j=-\infty}^{\infty} \frac{1}{P(-iw + 2\pi ij/\beta_{k+1})}}{\sum_{j=-\infty}^{\infty} \frac{1}{P(-iw + 2\pi ij/\beta_k)}}. \tag{7.6}$$

Therefore

$$c_j^k = \frac{\beta_{k+1}}{\pi}\int_0^{2\pi/\beta_{k+1}} e^{-ij\beta_{k+1}w}\frac{\sum_{j=-\infty}^{\infty}\overline{\frac{1}{P(-iw+2\pi ij/\beta_{k+1})}}}{\sum_{j=-\infty}^{\infty}\overline{\frac{1}{P(-iw+2\pi ij/\beta_k)}}}. \tag{7.7}$$

Then we get an interpolatory subdivision algorithm:

If $p_j^0 = s(j\beta)$, then the interpolatory subdivision algorithm can be given as to be

$$\begin{aligned} p_{2l}^{k+1} &= p_l^k, \\ p_{2l+1}^{k+1} &= \sum c_{2l-2j+1}^k p_j^k \end{aligned} \tag{7.8}$$

satisfies $p_j^k = s(j\beta_k)$, where $c_j^k$ can be got from (7.7).

The dual basis or the bi-orthogonal basis of the generalized B-spline is very important, that we can expand the generalized spline function to the linear combination of the generalized B-spline basis via its dual basis

$$s(x) = \sum < s(x), \quad N^*(x-j\beta) > N_0(x-j\beta).$$

We can get a projection in the generalized B-spline space that

$$f(x) \sim \sum < f(x), \quad N^*(x-j\beta) > N_0(x-j\beta).$$

The discussion of dual basis of the B-spline basis can be got analogously as for the classical wavelet theory. We want find the generalized spline function $N^*(x)$ that

$$\int N_0(x-j\beta)N^*(x)dx = \delta_{0j},$$

$$\frac{1}{2\pi}\int \hat{N}_0(w)\hat{N}^*(w)e^{ij\beta w}dw = \delta_{0j},$$

$$\frac{1}{2\pi}\int_0^{2\pi/\beta}\sum_{j=-\infty}^{\infty}\hat{N}_0(w+2\pi j/\beta)\hat{N}^*(w+2\pi j/\beta)e^{ij\beta w}dw = \delta_{0j},$$

$$\sum_{j=-\infty}^{\infty}\hat{N}_0(w+2\pi j/\beta)\hat{N}^*(w+2\pi j/\beta) \equiv \beta.$$

If the function $N^*(x)$ is a linear combination of the shifts of the spline $N_0(x-j\beta))$, we have

$$\hat{N}^*(w) = \left(\sum_{j=-\infty}^{\infty}\lambda_j e^{ij\beta w}\right)\hat{N}_0(w),$$

then

$$\left(\sum_{j=-\infty}^{\infty} \lambda_j e^{ij\beta w}\right)\left(\sum_{j=-\infty}^{\infty} \hat{N_0}^2(w + 2\pi j/\beta)\right) \equiv \beta,$$

and

$$\hat{N}^*(w) = \frac{\beta \hat{N}_0(w)}{\sum_{j=-\infty}^{\infty} \hat{N_0}^2(w + 2\pi j/\beta)}$$

is the Fourier transformation of the dual basis of $\{N_0(x - j\beta)\}$.

The more interesting problem is to construct the orthogonal wavelets with the generalized B-spline basis. The wavelet function

$$\psi_k(x) = \sum_{j=-\infty}^{\infty} d_j N_{k+1}(x - j\beta_{k+1}) \in V_{k+1}$$

satisfies

$$\langle N_k(x - j\beta_k), \psi_k(x)\rangle = 0$$

and

$$\langle \psi_k(x - l\beta_k), \psi_k(x - m\beta_k)\rangle = \delta_{lm}.$$

By writing the formula in Fourier transform we have

$$\int e^{ij\beta_k w} A_k(w) D_k(w) \hat{N}_{k+1}^2(w) dw = 0,$$

where $D_k(w) = \sum d_j^k e^{ij\beta_{k+1}}$, and

$$\int e^{ij\beta_k w} D_k^2(w) \hat{N}_{k+1}^2(w) dw = \delta_{0j}.$$

Since the mask $A_k(w)$ and the $D_k(w)$ possess a period of $2\pi/\beta_{k+1}$, we define

$$L_k(w) = \sum \hat{N}_{k+1}^2(w + \frac{2\pi j}{\beta_{k+1}}).$$

Then

$$A_k(w)D_k(w)L_k(w) + A_k(w + \frac{2\pi}{\beta_k})D_k(w + \frac{2\pi}{\beta_k})L_k(w + \frac{2\pi}{\beta_k}) = 0,$$

$$D_k^2(w)L_k(w) + D_k^2(w + \frac{2\pi}{\beta_k})L_k(w + \frac{2\pi}{\beta_k}) = \frac{\beta_k}{2\pi}.$$

Solve the $D_k(w + \frac{2\pi}{\beta_k})$ from the first equation and set it into the second equation we get

$$D_k^2(w) = \frac{\beta_k}{2\pi L_k(w)} \cdot \frac{L_k(w + \frac{2\pi}{\beta_k})A_k^2(w + \frac{2\pi}{\beta_k})}{L_k(w)A_k^2(w) + L_k(w + \frac{2\pi}{\beta_k})A_k^2(w + \frac{2\pi}{\beta_k})}.$$

All terms in the right-hand side are positive, therefore we have got an orthogonal wavelet function $\psi_k$. If the Fourier transformation of the $L_k(w)$ possesses no zero points, the solution is unique up to a $(-1)$ factor. Here we should point out again that $\psi_k(x) \neq \psi_{k+1}(2x)$ in generally.

# 8 Appendix

We prove the determinant of the matrix $\det(e^{\lambda_j x_k}) \neq 0$ for $\{\lambda_j\}_{j=1}^n$ and $\{x_j\}_{j=1}^n$ pairwise distinct. Equivalently we prove inductively that the function $\sum_{j=1}^n a_j e^{\lambda_j x}$ does not have $n$ zeros except the zero function itself. It is trivial that the proposition holds for $n = 1$. If the proposition hold for $n = l - 1$, we assume that the function $\sum_{j=1}^l a_j e^{\lambda x}$ possesses $l$ zeros and $a_l \neq 0$. Then

$$\sum_{j=1}^{l-1} \frac{a_j}{a_l} e^{(\lambda_j - \lambda_l)x} + 1$$

possesses $l$ zeros and the derivative

$$\sum_{j=1}^{l-1} \frac{a_j}{a_l} (\lambda_j - \lambda_l) e^{(\lambda_j - \lambda_l)x}$$

possesses $l - 1$ zeros, these contradicted to the induction assumption. This proposition is already appeared in [2] as an exercise of chapter 1.

# References

[1] J.W. Chen, G.Z. Wang. A class of Bezier-like curves. *Computer Aided Geometric Design* **20**: 29-39, 2003.

[2] W. Cheney, W. Light. *A Course in Approximation Theory*, Thomson Learning, 2000.

[3] C.De Boor. *A practical guide to splines*, Springer-Verlag, New York, 1978.

[4] N. Dyn, D. Levin, A. Luzzatto. Exponentials reproducing subdivision schemes. *Foundations of Computational Mathematics* **3**: 187-206, 2003.

[5] G. Farin. *Curves and Surfaces in Computer Aided Geometric Design*, Academic Press, 1988.

[6] X.L. Han. Quadratic trigonometric polynomial curves with a shape parameter. *Computer Aided Geometric Design* **19**: 503-512, 2002.

[7] X.L. Han. Piecewise quadratic trigonometric polynomial curves. *Mathematics of Computation* **72**: 1369-1377, 2003.

[8] M.K. Jena, P.Shunmugaraj, P.C. Das. A subdivision algorithm for trigonometric spline curves. *Computer Aided Geometric Design* **19**: 71-88, 2002.

[9] M.K. Jena, P. Shunmugaraj, P.C. Das. A non-stationary subdivision scheme for generalizing trigonometric spline surfaces to arbitrary meshes. *Computer Aided Geometric Design* **20**: 61-77, 2003.

[10] I. Khalidov, M.Unser. From Differential Equations to the Construction of Wavelet-LIke Bases. *IEEE Transaction on Signal Processing*, to appear.

[11] Y.G. Lu, G.Z. Wang, X.N. Yang. Uniform trigonometric polynomial B-spline curves. *Science in China Series F* **45**: 335-343, 2002.

[12] Y.G. Lu, G.Z. Wang, X.N. Yang. Uniform hyperbolic polynomial B-spline curves. *Computer Aided Geometric Design* **19**: 379-393, 2002.

[13] E. Mainar, J.M. Pena. Basis of C-Bezier splines with optimal properties. *Computer Aided Geometric Design* **19**: 291-295, 2002.

[14] E. Mainar, J.M. Pena, J. Sanchez-Reyes. Shape preserving alternatives to the rational Bezier model. *Computer Aided Geometric Design* **18**: 37-60, 2001.

[15] G. Morin, J. Warren, H. Weimer. A subdivision scheme for surfaces of revolution. *Computer Aided Geometric Design* **18**: 483-502, 2001.

[16] J. Sanchez-Reyes. Harmonic rational Bezier curves, p-Bezier curves and trigonometric polynomials. *Computer Aided Geometric Design* **15**: 909-923, 1998.

[17] L. Schumaker, *Spline Functions*, John Wiley & Sons Newyork, 1981.

[18] G. Walz. Trigonometric Bezier and Stancu polynomials over intervals and triangles. *Computer Aided Geometric Design* **14**: 393-397, 1997.

[19] J.W. Zhang. Two different forms of C-B-splines. *Computer Aided Geometric Design* **14**: 31-41, 1997.

[20] J.W. Zhang. C-Bezier curves and surfaces. *Graphical Models and Image Processing* **61**: 2-15, 1999.

# Mathematical Problems in System-on-Chip Design and Manufacture

Xuan Zeng, Hengliang Zhu, Fan Yang, Jun Tao,
Yi Wang, Jintao Xue

*State Key Lab. of ASIC & System, Microelectronics Dept.*
*Fudan University, Zhangheng Road 825*
*Shanghai 201203, China*

**Abstract**

It is hard to image what life would be like if there were no electronic devices. Computers, cell phones, televisions, refrigerators, any electronic devices you can think of, would not exist without the development of Integrated Circuits (IC). Nowadays, the technique of System-on-Chip (SoC) enables billions of transistors to be integrated into a small tiny chip, and operated at the clock frequency of several GHz. The advanced process technology can produce the SoC with the minimum feature size of 45nm. In order to design and manufacture such highly complex SoC, the design, simulation and verification tools are indispensable to guarantee the functionality, performance and manufacturability of the IC design. These design, simulation and verification tools are fundamentally based on the mathematical methods for solving the large scale ODE/PDE (Ordinary Differential Equation/Paratial Differential Equation) problems, the stochastic ODE/PDE problems and the inverse problems. In this paper, we will give a survey of the mathematical problems and challenges in SoC design and manufacture, and focus on the applications of Static Timing Analysis (STA), parasitic extraction of interconnects, Model Order Reduction (MOR), lithography simulation, and Optical Proximity Correction (OPC). We welcome more mathematicians to join in this research field full of challenges.

## 1 Introduction

Integrated Circuits (IC) have been honored as the most significant invention in science and technology in the twentieth century. They are silicon chips composed of microscopic arrays of electronic components.

Transistors are the primary devices of IC chip. Metal interconnects are applied to link these primary devices to construct various kinds of digital and analog circuit blocks. Since their invention in 1960s, Integrated Circuits have pervaded into every aspects of human life. These miniature circuits have been built into the principal components of almost all the electronic devices, such as modern computers, communication and manufacturing systems, mostly due to low cost, high reliability, low power requirements, and other demonstrated advanced properties. Indeed, many scholars believe that it is the expeditious development of integrated circuit technology that arouses the third industrial revolution, which is regarded as one of the most significant occurrences in the history of mankind.

During the last forty years, IC technology has advanced at a remarkable pace. As predicted by Moore's Law, "the number of transistors that can be fit onto a chip doubles every 18 months". Today, the System-on-Chip (SoC) technique provides superior flexibility to integrate several functional blocks (e.g. memory, microprocessor, DSP and other IP components), consisting of billions of transistors and up to 12 layers of interconnects, into a single chip of the size no larger than 1 $cm^2$. In order to produce such highly complex SoC, the design, simulation and verification tools are indispensable for IC design and IC manufacture to guarantee the functional correctness and the manufacturability of the design.

For example, in order to guarantee that the circuit operating frequency meets the design specification, timing verification is an indispensable technique in IC design. The purpose of the timing verification is to gauge the maximum frequency that an IC can work at. The maximum operating frequency is determined by the maximum path delay of the combinational circuit blocks from input to output ports. The maximum path delay is calculated by the widely adopted Static Timing Analysis (STA) methods using the delay information of interconnects and gates, which are the basic logical function blocks in digital circuits [1–3]. It is realized that the interconnect delay accounts for 70% of the total delay time of a chip. Accurate model of interconnects is required to capture the electromagnetic effect of interconnects working at multi-GHz. Based on the numerical calculation of Maxwell's equations, parasitic extraction technique is generally adopted to build the equivalent RLC (resistance, inductance, capacitance) circuit model for interconnects [4–9]. Since the number of elements of the extracted RLC circuits goes beyond $10^6$, Model Order Reduction(MOR) technique is then applied to reduce the extracted RLC circuits, and reduce the computational cost of latter simulation and verification [10–15]. Consequently, STA, parasitic extraction and MOR are the three key techniques for timing verification of the whole SoC system.

Nowadays, as the IC process technology reaches 45 nanometer node and beyond, it becomes more and more difficult to manufacture the chips exactly matching the design specification. One of the major reasons for the difficulties comes from the so-called sub-wavelength lithography process for the IC manufacture, which is used to print the design patterns from the mask onto the silicon wafer. "Sub-wavelength" means that the illumination wavelength used in lithography process is much larger than the feature size on the circuit chips. Consequently, the patterns printed on the wafer will be distorted or even vanish due to optical diffraction and interference. As a result, the interconnects with the same structure but at different IC chips, or different locations of the same chip, will be produced with different line widths. These geometric variations of interconnects could deteriorate the circuit performance, and cause the yield loss. It is very important for simulation and verification tools in IC design and IC manufacture to consider the process variations in order to improve the circuit performance and enhance the yield.

In recent years, the Manufacturability-aware and yield-aware design methodology has been developed for the SoC design to predict the impact of the process variations on the circuit performance. Variation-aware parasitic exaction technique is proposed to "extract" a parameterized equivalent RLC circuit for interconnects in the present of process variations [16, 17]. The Parameterized MOR techniques seek for a small scale parameterized reduced order circuit that preserves input/output behavior of the original parameterized circuit [18–22]. STA is extended to Statistic STA (SSTA) to calculate the probability distribution of the maximum path delay [23–30].

It is also important to optimize the lithography system and enhance the lithography resolution in the IC manufacture. Lithography simulation is one of the most valuable tools for this purpose and exhibits great advantages on time and cost savings compared with the expensive experimental solutions. Modern lithography simulators consist of the modules for four major steps of a lithography process, i.e. the illumination of optical sources, the transmission of light through mask, the transmission of light through imaging lens and the light propagation through wafer. Until now, the simulation of light propagation through thick mask or non-planar wafer surface, formulated as a 3-D electromagnetic scattering analysis problem, still remains a very challenging problem in terms of both computational time and accuracy [31]. On the other hand, increasing geometric variations of interconnects and devices are induced by parameter variations in lithography process. Stochastic lithography simulation should be developed to predict the random distributions of geometric variations, which are critical for statistical process optimization [32] and statistical circuit analysis, such as variation-aware parasitic extraction, parametric MOR, and SSTA as mentioned above.

Optical Proximity Correction (OPC), as one of the most important resolution enhancement techniques, is to systematically correct the masks to compensate for the patterning distortions. In such a way, the patterns printed on the wafer from the corrected mask can have the exact shape as what they are designed. Mathematically, OPC can be regarded as the inverse problem of lithography simulation. Available OPC algorithms are challenged by billions of patterns to be corrected, which always results in very high computational complexity and excessive amount of produced data. It would take about two weeks, 100 computers running in parallel, and hundreds of Gigabits of memory storage to run one OPC on a 1 cm $\times$ 1 cm chip under the 65nm process. Therefore, novel OPC algorithms should be invented to deal with the increasing design complexity.

In this paper, the SoC design and manufacture problems and the related mathematical challenges will be presented. The research topics cover the Static Timing Analysis (STA) in Section 2, the parasitic extraction of interconnects in Section 3, the model order reduction in Section 4, and the lithography simulation and OPC in Section 5. We wish through this paper mathematicians would engage their attentions to the challenging problems in SoC design and manufacture. Fruitful and promising contributions are also expected for joint research efforts from both mathematics and microelectronics.

# 2 Static timing analysis

Static timing analysis (STA) is a matured technique with wide industry applications for more than 20 years. It can fast evaluate whether a digital chip meets the performance specification such as circuit speed. Nowadays high performance digital integrated circuits might contain billions of gates and nearly ten layers of interconnects. During signal processing in such a complex circuit, the time spent by the signal to transfer across those gates and interconnects determines the working frequency of the chip. As an input independent simulation method, STA algorithm can perform quick simulation based on gate level delay model and reduced-order interconnect delay model with linear computational complexity.

In this section, we will firstly give a brief introduction to the basic ideas of STA. The mathematical challenges in the newly arising SSTA due to process variations will be also presented.

## 2.1 Background of static timing analysis (STA)

STA provides a fast way to check the highest speed at which a synchronized digital circuit can operate. As shown in Figure 1 (a), a synchro-

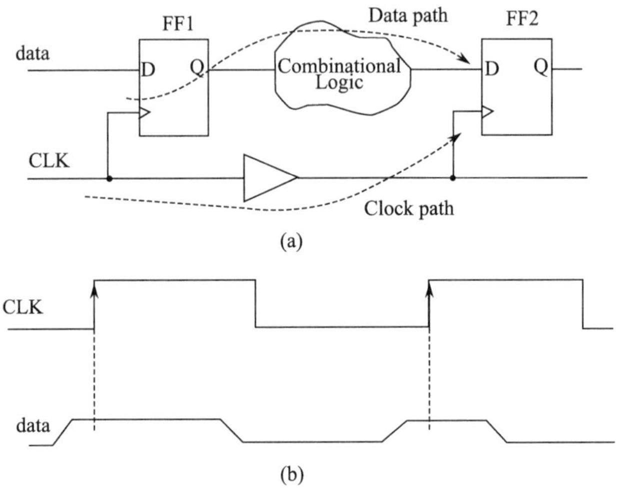


Figure 1 (a) Clock path and Data path; (b) Timing constraints

nized digital circuit consists of two parts, i.e. the combinational circuits and the sequential elements. The combinational logical circuits contain only logical gates (adder, inverter, etc.) connected by interconnects, which fulfil the specified logic operations (adding, inverting, etc.) to the input signals. The sequential elements, such as flip-flops FF1 and FF2 in Figure 1 (a), can sample the input data only at the rising edge of the clock signal (CLK) and hold the signal at the output during the rest time of the clock period. For the logic circuit in Figure 1, the input data is sampled by FF1 at the first rising edge of CLK, and then fed to the combinational logic circuit to perform the specific logic operations. The combinational logic circuit is required to finish its data processing and deliver its output data to the input port of FF2 within one clock period before FF2 samples the input signals at the second rising edge of CLK. So the longest delay among all combinational blocks in such a synchronized circuit determines the fastest clock frequency at which it can operate. Moreover, STA can also detect the critical path which causes the longest delay and is used for circuit optimization.

In order to calculate the longest delay by STA, the combinational circuit is transformed to a longest path problem of a graph. Taken a simple combinational circuit shown in Figure 2 (a) as an example, the signal delays across each logical gate and interconnects within this combinational circuit are modeled individually. Then the combinational circuit can be represented by a directed acyclic graph $G = (V, E)$ as shown in Figure 2 (b), where the vertex set of $V$ contains delay models for logical gates, interconnects and the primary inputs (PI), primary outputs (PO) of the circuit, and the edge set represents the connection among them. Note that in order to make illustration simple, delay models for interconnects have been omitted in this graph. A source/sink node is conceptually added before/after the PI/PO so that the timing graph can be analyzed as a *Single-Input Single-Output network.* STA algorithms try to find the longest path between source and sink node on this graph, which is a classical graph traversal problem and can be solved with breadth first (also named block-based) or depth first (also named path-based) travelling method. Since path-based methods need to travel all paths in the circuit whose number can grow exponentially with respect to the circuit size, block based algorithms have been used more widely due to the fact that its runtime is linear to the circuit size. In order to obtain the delay (named arrived time: AT) after each vertex in a block-based STA, two atomic operators SUM and MAX are needed. Apart from the calculation of arrived times at primary outputs, STA can easily identify critical path, which is the path with longest delay, by a re-traversal process for circuit optimization. STA is one of the most important fast simulation techniques in the last twenty years and lots of research results have been published [1–3].

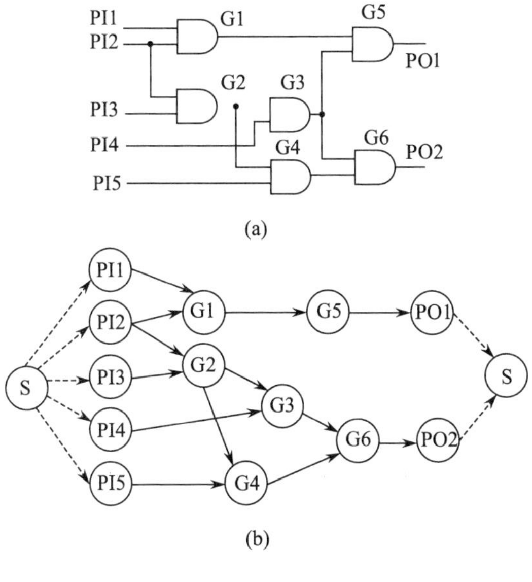


Figure 2 (a) Schematic of combinational circuit; (b) Timing graph

## 2.2 Statistical static timing analysis (SSTA) with parametric delay model

In nanometer technology, performance (e.g. delay) deviations of logical gates and interconnects due to the process variations become more and more significant, which should be carefully considered in timing analysis. Statistical STA (SSTA) is proposed in recent years to conduct more accurate timing analysis. Different from the traditional delay models with deterministic values, SSTA adopts polynomial delay models with respect to the process parameters [3]. In Section 3.6, interconnect statistical modeling will be presented. Here we assume that parametric delay models for gates and interconnects have been provided with a high-order① polynomial form [27], i.e.

$$Y\left(\vec{\zeta}\right) = \sum_j c_j \Gamma_j(\vec{\zeta}) = \vec{\zeta}^T A \vec{\zeta}^T + B\vec{\zeta}^T + \text{const}, \tag{2.1}$$

where $\Gamma_j(\vec{\zeta})$ is a set of polynomial bases and $\vec{\zeta}^T = \{\zeta_1, \zeta_2, ..., \zeta_n\}$ represents process variations as a set of random variables with independent identical distribution (i.i.d.)②. In order to adopt those delay models in STA framework, SUM and MAX operators on statistical inputs should be developed accordingly.

### 2.2.1 SUM and MAX in statistical timing analysis

As described previously,"SUM" and "MAX" are two atomic operators employed in block-based STA framework. Since SUM is a linear operator whose solution is straightforward, we only discuss MAX operator here. MAX operator is required for each multi-input vertex to determine the latest Arrived Time (AT) which should be propagated to the next stage. The problem can be mathematically defined in the following.

#### A. Statistical MAX problem

Provided Arrived Time (AT) $A_i$ at each input being modeled by (2.1) as $A_i = \sum_j a_{i_j} \Gamma_j\left(\vec{\zeta}\right)$, the output of MAX should also be modeled by the same set of polynomial bases,

$$C = MAX(A_1(\vec{\zeta}), \cdots, A_N(\vec{\zeta})) \approx \widehat{C}(\vec{\zeta}) = \sum_j c_j \Gamma_j\left(\vec{\zeta}\right), \tag{2.2}$$

①Linear delay models has been proved not accurate enough as in [23, 26–28].

②Correlation decomposition methods like Principal Component Analysis can been employed for iid generation with Gaussian distribution.

where $c_i$ are unknown coefficients of a polynomial. $\widehat{C}(\overrightarrow{\zeta})$ needs to hold the same polynomial form as $A_i$ in order to be accepted as input Arrived Time (AT) by the downstream connected nodes in the timing graph.

## B. Review of existing methods

For resolving the MAX problem, the Moment Matching technique [23, 26, 28] and the Stochastic Collocation Method [24] have been proposed and widely adopted in SSTA.

Moment Matching technique is first proposed by Clark [33] to deal with the statistical MAX problem with input delays modeled by linear polynomials, and then extended to input delays modeled by high-order polynomials in [26, 28]. We use a simple two-input MAX example, $C = MAX(A, B)$, to illustrate the idea of moment matching technique for quadratic polynomial delay model. The moment matching technique uses a quadratic polynomial expansion in (2.2) to approximate the $MAX(A, B)$ as

$$MAX(A, B) \approx \sum_j c_j \Gamma_j \left(\overrightarrow{\zeta}\right), \tag{2.3}$$

and calculates the unknown coefficients $\{c_j\}$ by matching the first and second order moments of the both sides of (2.3). Here the $n$-th moment of a random variable $X$ is defined as:

$$m_n = E\{X^n\} = \int_{-\infty}^{\infty} X^n f_X dx, \tag{2.4}$$

where $f_X$ is the probability density function (PDF) of $X$. The first moment $m_1$ and the second moment $m_2$ of $\sum_j c_j \Gamma_j \left(\overrightarrow{\zeta}\right)$ in (2.3) can be analytically calculated with (2.4). In order to calculate the moments of $MAX(A, B)$ using (2.4), the PDF of $MAX(A, B)$ needs to be calculated first by

$$f_{MAX(A,B)} = F_A f_B + F_B f_A, \tag{2.5}$$

provided $A$ and $B$ are two independent random variables [34], where $F$ and $f$ represents cumulative density function (CDF) and PDF respectively. The PDFs of polynomial $A$ and $B$ are calculated by the APEX algorithm in [35], which is based on the exponential function approximation of PDF. However, both the moment matching and the approximation of PDF in APEX algorithm may cause computation errors.

Stochastic Collocation Method (SCM) [24] is another useful approach for solving the statistical MAX problem in (2.2). The Polynomial Chaos $\{H_j(\overrightarrow{\zeta})\}$ have been used to approximate both $\widehat{C}(\overrightarrow{\zeta})$ and $A_i(\overrightarrow{\zeta})$ by replacing $\{\Gamma_j(\vec{\zeta})\}$ in (2.2). The unknown coefficients $c_i$ can then be easily calculated by Galerkin method,

$$\begin{aligned} c_j &= \langle MAX(A_1(\vec{\zeta}),\cdots,A_N(\vec{\zeta})), H_j\left(\vec{\zeta}\right)\rangle \\ &= \int_\Omega MAX(A_1(\vec{\zeta}),\cdots,A_N(\vec{\zeta}))\cdot H_j\left(\vec{\zeta}\right)\cdot PDF\left(\vec{\zeta}\right)d\vec{\zeta}. \end{aligned} \tag{2.6}$$

The computation of $\{c_j\}$ in (2.6) needs to solve a multi-dimensional integral problem. Stochastic Collocation Method employs numerical quadrature techniques in (2.7) to calculate the integration in (2.6).

$$\begin{aligned} &\int_\Omega MAX(A_1(\vec{\zeta}),\cdots,A_N(\vec{\zeta}))\cdot H_j\left(\vec{\zeta}\right)\cdot PDF\left(\vec{\zeta}\right)d\vec{\zeta} \\ &= \sum_k MAX(A_1(\vec{\zeta}^{(k)}),\cdots,A_N(\vec{\zeta}^{(k)}))\cdot H_j\left(\vec{\zeta}^{(k)}\right)\cdot w_k, \end{aligned} \tag{2.7}$$

where $\vec{\zeta}^{(k)}$ is the $k$-th collocation point and $w_k$ is the corresponding weight. $MAX(A_1(\vec{\zeta}^{(k)}),\cdots,A_N(\vec{\zeta}^{(k)}))$ becomes a deterministic MAX problem at each collocation point. The multi-dimensional collocation points in (2.7) are generally generated by the Tensor Product of one-dimensional Gaussian quadrature points. However, the number of collocation points in Tensor Product increases exponentially with respect to the dimensionality. Stochastic Collocation Method has very high accuracy, but it would lose computational efficiency when dimensions of the independent random variables $\vec{\zeta}$ increase too high.

### 2.2.2 Mathematical Challenges of (S)STA

During the last twenty years, the problem of STA has been well studied, problems and challenges are mostly arising from statistical algorithms in SSTA.

- For statistical MAX Problem, both moment matching method [23, 26, 28] and stochastic collocation method [24] are hard to achieve efficiency and accuracy simultaneously. Moments matching methods are fast since all operators are analytically computed. However their accuracy is lost due to the approximation of PDF in APEX algorithm. Moreover, they are only accurate for low frequency components of a distribution [35] even if the moments can be accurately computed. On the other hand, due to large number of collocation points generated by Tensor Product, application of stochastic collocation method is prohibited to SSTA problems with large dimension of independent random variables.
- Most of the existing SSTA algorithms are based on the Gaussian distribution assumption of process variations, for example, PCA

technique in statistical delay modeling [36], APEX method in moment matching method [35] and orthogonal polynomial construction in stochastic collocation method [24]. However the real process variations like via resistance are known as non-Gaussian in recent research [26]. Therefore, SSTA should be extended to accommodate process variations with arbitrary distribution.

- In traditional STA with deterministic delay, critical path can easily be found by re-traversal of the timing graph. However critical path in SSTA can hardly be defined, since it is difficult to determine the input port which causes the output Arrived Time (AT) on a multi-input vertex when the delays are random. It is still an open problem in SSTA to give definition of critical path in a statistical sense and provide methods to identify it.

# 3 Parasitic extraction of interconnects

Today's IC design is actually interconnect-centric. Analysis of interconnects is indispensable for designers to ensure the correctness of their designs. Instead of simulating directly by some electro-magnetic (EM) simulators, which can be very time-consuming, interconnects are generally modeled by equivalent resistance, inductance and capacitance (RLC) circuits, and incorporated with the simulation tools such as STA. These RLC circuits are termed as "parasitic parameters", and parasitic extraction, just as its name implies, "extracts" these "parasitic parameters" from interconnects based on the numerical calculation of Maxwell's Equations.

In this section, the parasitic extraction of interconnects and related mathematic challenges will be presented. We first introduce some fundamental concepts of parasitic extraction, including the Maxwell's equations, PEEC model and several quasi assumptions, and then discuss some new topics of parasitic extraction of interconnects, including the susceptance model and the variation-aware parasitic extraction.

## 3.1 Parasitic Extraction and Maxwell's Equations

Maxwell's equations (3.1-3.4) are the fundamental equations of parasitic extraction. These four equations describe how conductors and dielectrics interact with electric and magnetic fields.

$$\nabla \cdot \varepsilon \vec{E} = \rho, \tag{3.1}$$

$$\nabla \cdot \mu \vec{H} = 0, \tag{3.2}$$

$$\nabla \times \vec{E} + j\omega\mu\vec{H} = 0, \tag{3.3}$$

$$\nabla \times \vec{H} - j\omega\varepsilon\vec{E} = \vec{J}. \tag{3.4}$$

When the electric and magnetic fields themselves are simulated, the interconnects and all passive components are translated into conductors and dielectrics, with their associated geometries and material properties. Signal driven into the interconnects is converted into an incident electromagnetic wave, and Maxwell's Equations are used to predict how this wave interacts with the conductors and dielectrics. The material geometries and properties define the boundary conditions in which Maxwell's Equations are solved.

The research of parasitic extraction is therefore based on the mathematical background of solving the partial differential equations (PDE) problem in (3.1-3.4). Various numerical methods have been applied, including differential methods [37] and integral equation methods [5, 8]. Among these methods, integral equation methods are of the most popular, which mostly due to its much less number of unknowns compared with the differential methods.

## 3.2 Partial element equivalent circuit (PEEC) model

One of the most well-developed model based on integral equation method for parasitic extraction is Partial Element Equivalent Circuit (PEEC). It was first proposed by Ruehli in 1974 [4]. The idea of PEEC is exactly a numerical solution of the following mixed-potential integral equation (MPIE) derived from Maxwell's equations by Boundary Element Method (BEM).

$$\frac{\vec{J}(\vec{r},t)}{\sigma} + \frac{\partial}{\partial t}\left[\mu \int_{V'} G(\vec{r},\vec{r}')\vec{J}(\vec{r}',t-t_d)dv'\right] + \nabla\varphi(\vec{r},t) = 0, \quad (3.5)$$

$$\varphi(\vec{r},t) = \frac{1}{\varepsilon}\int_{S'} G(\vec{r},\vec{r}')q(\vec{r}',t-t_d)ds', \quad (3.6)$$

where $G(\vec{r},\vec{r}') = \frac{1}{4\pi|\vec{r}-\vec{r}'|}$ is the free space Green's function, $t_d = \frac{|\vec{r}-\vec{r}'|}{c}$ denotes the retarded time, and $c$ is the light speed. Using the piece-wise constant discretization scheme, the volume of conductors is discretized into filaments, and the surfaces of conductors are discretized into panels. The currents flowing in each filament as well as the charges on each panel are assumed to be uniform. After discretization, the mixed-potential integral equations in (3.5) and (3.6) become

$$\frac{l_i}{\sigma a_i}\vec{I}_i(t) + \sum_j \left(\frac{\mu \int_{V_i}\int_{V_j} G(\vec{r},\vec{r}')dV_i dV_j}{a_i a_j}\right)\frac{\partial \vec{I}_j(t-t_{ij})}{\partial t} + \varphi_{l(i)}(t) - \varphi_{r(i)}(t) = 0, \quad (3.7)$$

$$\varphi_m(t) = \sum_n \left(\frac{1}{\varepsilon a_m}\int_{S_m} G(\vec{r},\vec{r}')dS_m\right) q_n(t-t_{mn}), \quad (3.8)$$

where $a_i$ is the area of the $i$-th panel or the area of the cross section of the $i$-th filament, $t_{ij}$ represents the retarded time, $\varphi_{l(i)}(t)$ and $\varphi_{r(i)}(t)$ denote the potential at the left corner and the right corner of the $i$-th filament, respectively. Equations (3.7) and (3.8) can be represented by an equivalent RLC circuit, where the first term on the LHS of (3.7) corresponds to a resistance, the other terms correspond to inductance couplings and terms in (3.8) are the capacitance couplings.

The major problem of the PEEC model is the size of the equivalent circuit of (3.7) and (3.8) can be very large even for the simplest interconnect structure. This is why the PEEC model is not directly used by IC designers, nevertheless, PEEC formulation in (3.7) and (3.8) provides an alternative approach for simulating the behavior of interconnects instead of solving the full Maxwell's equations (3.1-3.4). Many parasitic extraction methods still adopt this PEEC formulation as the fundamental equations and further apply some quasi assumptions to simplify the computation [5,8].

## 3.3 Quasi assumptions for parasitic extraction

Several quasi assumptions can be made upon Maxwell's equations or PEEC formulation in order to simplify the calculation. These assumptions are Electro-Quasi-Static (EQS) assumption for the parasitic capacitance extraction [5–7], Magnetic-Quasi-Static (MQS) assumption for the parasitic inductance extraction [8] and Electro-Magnetic-Quasi-Static (EMQS) assumption for the impedance extraction [9].

The strictest assumption is EMQS assumption, in which only the retardation of EM wave is ignored. For small interconnect length comparable to the wavelength of EM wave, term $j\omega\varepsilon\vec{E}$ in (3.4) can be discarded, or equivalently, the retardation $t_d$ can be assumed zero in PEEC formulation [9,38]. Thereby, (3.5) and (3.6) can be transformed to the frequency domain,

$$\frac{\vec{J}(\vec{r})}{\sigma} + j\omega\left[\mu\int_{V'} G(\vec{r},\vec{r}')\vec{J}(\vec{r}')dv'\right] + \nabla\varphi(\vec{r}) = 0, \tag{3.9}$$

$$\varphi(\vec{r}) = \frac{1}{\varepsilon}\int_{S'} G(\vec{r},\vec{r}')q(\vec{r}')ds'. \tag{3.10}$$

The problem of parasitic extraction can be solved at the frequency point of interest. Using the similar discretization scheme as for (3.7) and (3.8),

(3.9) and (3.10) become

$$\frac{l_k}{\sigma a_k}\vec{I}_k + j\omega \sum_l \left( \frac{\mu \int_{V_k} \int_{V_l} G(\vec{r},\vec{r}')dV_k dV_l}{a_k a_l} \right) \vec{I}_l + \varphi_{l(k)} - \varphi_{r(k)} = 0, \tag{3.11}$$

$$\varphi_m = \sum_n \left( \frac{1}{\varepsilon a_m} \int_{S_m} G(\vec{r},\vec{r}')dS_m \right) q_n. \tag{3.12}$$

For even smaller interconnect length, which is at least one order smaller than the wavelength of EM wave, the coupling between the electronic field and magnetic field can be ignored, and one can further make an EQS assumption or a MQS assumption. Besides neglecting the retardation of EM wave in the Maxwell's equations, EQS assumption discards the current term $\vec{J}$ in (3.4), while MQS assumption discards the charge term $\rho$ in (3.1). PEEC formulation with EQS assumption is (3.12) without (3.11), which is generally used for the parasitic capacitance extraction [5–7]. PEEC formulation with MQS assumption is (3.11) without (3.12), which is for the parasitic inductance extraction [8].

## 3.4 Parasitic extraction methods

Parasitic extraction of interconnects involves numerical calculation of the PEEC formulation in (3.11) and (3.12), based on different quasi assumptions in Section 3.3 to simplify the calculation. Fast solvers are generally adopted to further accelerate the calculation. These fast solvers include Fast Multi-pole Method (FMM) [5,8], precorrected-FFT method [9,39], Hierarchical method [6,7], etc..

When the interconnect length is much smaller than the wave length, the capacitance and inductance parameters of interconnects are generally "extracted" separately. EQS assumption is generally adopted for capacitance extraction, while MQS assumption is for inductance extraction. MIT has proposed a capacitance extraction tool FastCap [5] and an inductance extraction tool FastHenry [8]. Both of these tools are based on the FMM for solving the mixed potential integral equation in (3.11) and (3.12). FMM is a kernel-dependent fast solver, meaning that the FMM in (3.11) and (3.12) is designed only for the free space Green's function. A kernel-independent fast solver namely Hierarchical method, is proposed in [6] for the capacitance extraction, which is called HiCap and claimed 60 times faster than FastCap. PhiCap, which is a combination of HiCap and the preconditioning technique, is proposed in [7] and is doubly faster than HiCap.

When the operation frequency increases beyond GHz, the wave length and the length of interconnects are in the same order, and the coupling between the electronic field and the magnetic field can no longer be ignored. Based on the EMQS assumption, FastImp is proposed in [9] using precorrected-FFT fast solver to "extract" the impedance parameters of interconnects, in which the capacitances and inductances are coupled and cannot be distinguished from each other. For even higher operation frequency, the retardation of EM wave should be taken into consideration, and full-wave analysis solving the full Maxwell's equation is required in order to get an accurate model for interconnects.

## 3.5 Resistance, capacitance and susceptance (RCS) model for interconnects

In order to simulate the delay of interconnects, the equivalent RLC network is generally formulated as an ordinary differential equation (ODE), where the resistances, inductances and capacitances (RLC) are represented in the matrix form, as will be illustrated in equation (4.4) in Section 4.1. The resistance matrix is diagonal. Since the capacitance interaction is local, meaning that the far-distance capacitance can be neglected, the capacitance matrix is sparse. However, unlike the capacitances, the interaction among inductances could be global. In consequence, the inductance matrix could be dense, which will burden the simulation of extracted RLC interconnect circuits. Furthermore, it is understood that making the matrix sparse by merely discarding the smallest terms can render the matrix positive indefinite and thereby introduce positive pole(s) in subsequent circuit simulations. In order to solve this problem, susceptance is proposed as an inversion of the inductance matrix to form a RCS model of the interconnects [40–43].

$$\mathbf{S} = \mathbf{L}^{-1}. \tag{3.13}$$

The advantage of using the susceptance model instead of the inductance model is the locality of susceptance parameters, meaning that the faraway mutual susceptances are much smaller compared with the faraway mutual inductances. By ignoring small mutual susceptance elements, the susceptance matrix can be much sparser than the inductance matrix. As shown in Figure 3, the inductance matrix is global and dense, while the susceptance matrix, as the inversion of inductance matrix, is diagonal dominant and sparse.

Susceptance is a very efficient model to capture the inductance effect of interconnects. The major problem of applying susceptance model is to prove the Symmetric Positive Definition (SPD) of the susceptance matrix for general interconnect structure. Several papers have been published discussing this issue [41–43]. Window-based methods are proposed in [42, 43] for the extraction of susceptance parameters, using iterative cutting

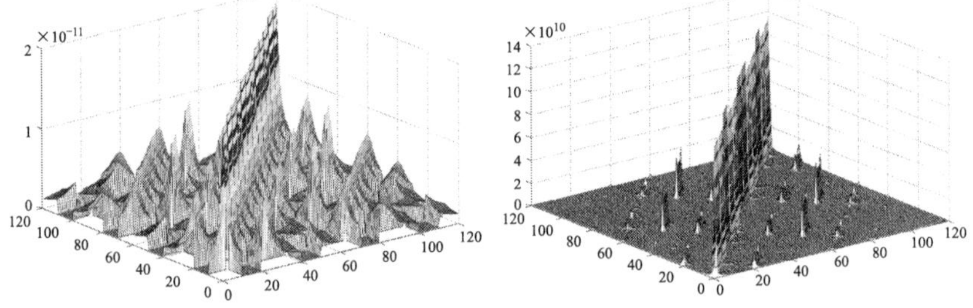


(a) Inductance matrix (b) Susceptance matrix

Figure 3 Locality of susceptance: an example.

to guarantee the SPD of susceptance matrix. However, the iterative cutting method results in larger scale RCS model of interconnects and costs higher computation time for susceptance extraction. More efficient extraction methods and strict proof of the SPD of susceptance matrix are still needed to apply the RCS model to interconnect analysis.

## 3.6 Variation-aware parasitic extraction

Geometric variation of interconnects caused by process variation is one of the major problems that IC designer encountered when process technology scales down to nanometer range. The shape of interconnects is no longer deterministic and is generally modeled by random fields. The resulting variation-aware parasitic extraction problem becomes a PDE problem with stochastic boundaries. Two kinds of methods have been proposed, so far, for solving such a stochastic PDE problem [16, 17].

Taking the capacitance extraction problem for example, we can still use the PEEC formulation in (3.10), while the difference is that the surfaces of conductors, denoted by $S'$, are now modeled by some random fields, which are typically Gaussian as defined by the probability density function in (3.14) and the correlation function in (3.15) [16, 17].

$$f\left(h(\vec{r})\right) = \frac{1}{\sqrt{2\pi\sigma^2}} \exp - \frac{h^2(\vec{r})}{2\sigma^2}, \tag{3.14}$$

$$\mathrm{Cov}\left(h(\vec{r}), h(\vec{r}\,')\right) = \sigma^2 \exp - \frac{|\vec{r} - \vec{r}\,'|^2}{\eta^2}. \tag{3.15}$$

Among the variation-aware capacitance extraction methods [16, 17], the first step is to simplify the random field model of stochastic boundaries. The random field, which means an infinite dimension random space, is reduced to a much smaller set of random variables by Principle Component Analysis (PCA) in [16] and by K-L expansion in [17]. Both PCA and K-L expansion are based on the eigen-decomposition of correlation function, while the latter one could be more efficient because the discretization is made according to the correlation length of the random field. As a result, the stochastic boundaries are represented by a small number of random variables, and parasitic capacitances can be modeled by high-order polynomial model with respect to these random variables in (2.1), which exactly meets the polynomial model requirement of Statistical Static Timing Analysis.

In order to calculate the coefficients of the polynomial model, a perturbation method is adopted in [16] and a stochastic collocation method (SCM) is proposed in [17]. Perturbation method assumes that the variation is small enough and computes the coefficients by a Taylor expansion of the potential coefficient matrix. Stochastic collocation method, on the other hand, is based on the stochastic spectral theory, which models the capacitances by Homogeneous Chaos expansion [44] and uses collocation method to compute the corresponding coefficients.

Although the stochastic spectral method can be very promising for stochastic PDE problems, challenges still remain for the variation-aware parasitic extraction. For example, SCM in [17] is based on the assumption of Gaussian random field (3.14), and has the optimal (exponential) convergence rate only when the geometric variations of interconnects are Gaussian. In fact the geometric variations from IC fabrication could have arbitrary distribution, which may strongly deviate from the Gaussian one. Therefore, it is desirable to extend the SCM in [17] to tackle with the real geometric variations considering arbitrary random probability.

## 3.7 Challenges of parasitic extraction

During the last twenty years, the problem of parasitic extraction has been well studied, and methods proposed so far are mostly based on the numerical methods for solving PDE problem. However, new challenges keep coming forth to the designers as the IC technology advances.

- "Frequency-dependent" is one of the requests for parasitic extraction. As the frequency increases, parasitic parameters become sensitive to the frequency parameter. Frequency-dependent parasitic extraction methods are needed to model the frequency dependency of the parasitic parameters of interconnects.
- "Full-wave extraction" is another important issue in parasitic extraction. As the frequency increases, the wave length is reduced to the same order of the length of interconnects. As a result, the EMQS assumption is no longer valid. Retardation of the EM wave

should be taken into account when modeling the interconnects with the equivalent RLC model [9].

- The third challenge is the multilayered Green's function when considering the multilayered structure of interconnects. The multilayered Green's function does not have a simple form as the free space Green's function, and has to be calculated according to the multilayered interconnect structure [45]. Furthermore, many fast solvers, like FMM [5, 8], are kernel-dependent, meaning that they are designed for the free space Green's function and could fail when coping with the multilayered Green's function.
- "Full-chip extraction", in which hundreds of thousands of interconnects are involved, is another hard nut in parasitic extraction. Solving Maxwell's equation for full-chip extraction is never feasible even with today's best hardware and software tools. An ad hoc approach is using pattern match technique [46]. As a result, the accuracy is lost. Effects like the far field coupling can no longer be taken into account in this approach.
- Finally, variation-aware parasitic extraction, stimulated by the challenge of process variations when IC design enters into Nano-era, is a stochastic PDE problem. Stochastic methods are looked forward to tackling with the real geometric variations considering arbitrary random probability.

# 4 Model order reduction

With the rapid increase of the signal frequency and decrease of the feature sizes of high-speed electronic circuits, interconnect has become a dominant factor in determining circuit performance in SoC design. In today's SoC, up to twelve layers of interconnects with several kilometers length are integrated in a single chip to complete the interconnect of the transistors. The equivalent circuits of the interconnects extracted by "Parasitic Extraction", either based on RLC or RCS model, tend to be in order of millions, which poses great challenges to interconnect analysis. Model order reduction is a necessity for efficient interconnect modeling, simulation, design and optimization. For most of the circuit blocks, only the signal behavior at the ports of the block is of interest. Model Order Reduction (MOR) techniques generate small scale reduced models of the large scale interconnect circuits that accurately approximate circuit behavior at the port terminals while sacrifice the modeling of behavior at internal nodes.

Process variation is an important problem encountered by IC design when process technology scaling down to nanometer regime. The indetermination in the manufacturing of IC chips may cause geometric variations of interconnects, which could result in the chip performance unpredictable and cause significant parametric yield loss. Moreover, during the circuit synthesis of large scale digital or analog applications, it is also crucial to evaluate the response of interconnect as functions of other design parameters, such as geometry and temperature. In these cases, parameterized model order reduction (PMOR) methods are considered as necessary techniques for analysis of parametric interconnect circuits. The PMOR techniques seek for a small scale parameterized reduced order circuit that preserves input/output behavior of the original parameterized circuit, and thus facilitate the simulation of large scale parameterized interconnect circuits.

In this section, we introduce the problem definition of MOR and PMOR and give a brief review of the MOR and PMOR techniques. More importantly, the mathematical challenges of MOR and PMOR will be presented.

## 4.1 Problem definition of model order reduction

After parasitic extraction, the interconnect can be modeled by either RLC or RCS model. The RLC model, consisting of linear elements such as resistors, capacitors and inductors etc., can be described by the following modified nodal analysis (MNA) equation

$$\begin{aligned} C_x\dot{x}(t) + G_x x(t) &= B_x u(t), \\ y(t) &= L_x^T x(t), \end{aligned} \tag{4.1}$$

where $t$ is the time variable, $x(t) \in R^N$ is the state vector, $y(t) \in R^q$ is output vector, $C_x, G_x \in R^{N\times N}$ are system matrices, $B_x \in R^{N\times p}$ and $L_x \in R^{N\times q}$ are input and output incidence matrices. By performing Laplace transform on (4.1), we have the MNA equation in frequency domain

$$\begin{aligned} (sC_x + G_x)x(s) &= B_x u(s), \\ y(s) &= L_x^T x(s), \end{aligned} \tag{4.2}$$

where $s$ is the Laplace variable, $x(s)$, $u(s)$ and $y(s)$ are the Laplace transforms of $x(t)$, $u(t)$ and $y(t)$, respectively. The MNA equation is a first order linear system. The transfer function, which can fully represent the input/output characteristics of the system, is defined as

$$H_f(s) = \frac{y(s)}{u(s)} = L_x^T(sC_x + G_x)^{-1}B_x. \tag{4.3}$$

Since (4.2) describes the physical interconnect circuits, the system matrices $C_x$, $G_x$ actually have special block structures. Therefore, equation (4.2) can be reformulated as

$$\begin{aligned}\left(s\begin{bmatrix} C & 0 \\ 0 & L \end{bmatrix} + \begin{bmatrix} G & E \\ -E^T & 0 \end{bmatrix}\right)\begin{bmatrix} v(s) \\ i(s) \end{bmatrix} &= \begin{bmatrix} Bu(s) \\ 0 \end{bmatrix}, \\ y(s) &= \begin{bmatrix} D^T & 0 \end{bmatrix}\begin{bmatrix} v(s) \\ i(s) \end{bmatrix},\end{aligned} \tag{4.4}$$

where $v(s)$, $i(s)$ represent the voltage and current variables in frequency domain, $C \in R^{M\times M}$, $G \in R^{M\times M}$ and $L \in R^{P\times P}$ represent the contributions of the capacitors, resistors and inductors, respectively. $E \in R^{M\times P}$ is the incidence matrix for inductors.

As an alternative approach, RCS model uses resistors, capacitors and susceptance elements to modeling the interconnect. Since the susceptance matrix $S$ can be regarded as the inverse of the inductance matrix $L$, the MNA equation of RCS circuit can be expressed as

$$\begin{aligned}\left(s\begin{bmatrix} C & 0 \\ 0 & I \end{bmatrix} + \begin{bmatrix} G & E \\ -SE^T & 0 \end{bmatrix}\right)\begin{bmatrix} v(s) \\ i(s) \end{bmatrix} &= \begin{bmatrix} Bu(s) \\ 0 \end{bmatrix}, \\ y(s) &= \begin{bmatrix} D^T & 0 \end{bmatrix}\begin{bmatrix} v(s) \\ i(s) \end{bmatrix},\end{aligned} \tag{4.5}$$

where $S = L^{-1}$. Compared with the inductance matrix $L$, the susceptance matrix $S$ exhibits superior properties in terms of the symmetry, diagonally dominance and sparsity.

In most applications, the auxiliary current variables $i(s)$ are generally intermediate variables. By eliminating $i(s)$ from (4.5), we obtain a second-order system

$$\begin{aligned}(sC + G + \tfrac{1}{s}\Gamma)v(s) &= Bu(s), \\ y(s) &= D^T v(s),\end{aligned} \tag{4.6}$$

where $\Gamma = ESE^T$. The transfer function of the second-order system can be expressed as

$$H_s(s) = D^T(sC + G + \frac{1}{s}\Gamma)^{-1}B. \tag{4.7}$$

The main goal of MOR techniques for interconnect circuits is to generate lower order reduced models of the large scale linear circuits, while guaranteeing the following properties, i.e., high accuracy, numerical stability, passivity preserving [12] and structure preserving [47]. Most of the MOR methods are based on the concept of projection. Specially, for the RLC circuit model described by (4.2), let $V \in R^{N\times n}$ and $W \in R^{N\times n}$ be two predefined projection subspaces with $n \ll N$, and let the state variable $x \approx V\tilde{x}$, where $\tilde{x}$ is a vector of dimension $n$. Substituting $x$ by

$V\tilde{x}$ into (4.2) and premultiplying the equation (4.2) by $W^T$, we have a model with reduced-order $n$:

$$\begin{aligned}(s\tilde{C}_x + \tilde{G}_x)\tilde{x}(s) &= \tilde{B}_x u(s),\\ \tilde{y}(s) &= \tilde{L}_x^T \tilde{x}(s),\end{aligned} \tag{4.8}$$

where $\tilde{C}_x = W^T C_x V$, $\tilde{G}_x = W^T G_x V$, $\tilde{B}_x = W^T B_x$ and $\tilde{L}_x = V^T L_x$. If $W \neq V$, the projection is termed "oblique projection". Otherwise, the projection is termed "orthogonal projection". With the similar projection strategy, the reduced-order model of the time-time domain equation (4.1) can also be obtained.

For the reduction of RCS circuit model (4.6), "orthogonal projection" are usually employed. With projection matrix $Q \in R^{M\times m}$, the reduced order system with order $m(m \ll M)$ of the original system (4.6) with order $M$ can be expressed in frequency domain as

$$\begin{aligned}(s\tilde{C} + \tilde{G} + \tfrac{1}{s}\tilde{\Gamma})\tilde{v}(s) &= \tilde{B}u(s),\\ \tilde{y}(s) &= \tilde{D}^T \tilde{v}(s),\end{aligned} \tag{4.9}$$

where $\tilde{C} = Q^T C Q$, $\tilde{G} = Q^T G Q$, $\tilde{B} = Q^T B$ and $\tilde{D} = Q^T D$.

In the following subsection, we will review the existing mode order reduction techniques.

## 4.2 Existing model order reduction techniques

Model order reduction can be carried out in either frequency domain or time domain. Correspondingly, the model order reduction methods can be classified as frequency-domain and time-domain methods. The frequency domain methods preserve the input/output behaviors by approximating the transfer function through moment matching, or by truncated balanced realization which provides provable Hankel norm error bounds [48,49]. The time domain methods directly approximate the impulse response of the system, which is the time domain counterpart of transfer function, by matching the expansion coefficients of appropriate basis functions.

### 4.2.1 Frequency-domain methods

According to the approximation method adopted, the frequency domain reduction methods can be classified as moment-matching based methods and Truncation Balanced Realization (TBR) based methods.

### A. Moment-matching based methods

The objective of the moment-matching based MOR is to guarantee the following four properties, i.e., moment matching, numerical stability, passivity preserving and structure preserving. The moments are defined as the Taylor expansion coefficients of the transfer function, such as (4.3) or (4.7), of linear system with respect to the Laplace variable around an expansion point. The accuracy of the moment-matching based methods is judged by the number of moments matched between the reduced order system and the original system. Numerical stability ensures the accuracy of the reduced order model can be steadily improved as the reduced order increases [11]. Passivity is an important property to satisfy because stable, but not passive reduced order model can produce unstable systems when connected to other stable, even passive, loads [12]. Structure preserving is also very important for MOR methods, since equivalent circuit can be synthesized from structure preserved reduced order model, which can facilitate the application of reduced order model for interconnect analysis.

For the first-order system, the Taylor expansion of the transfer function (4.3) can be expressed as

$$\begin{aligned} H_f(s) &= L_x^T(sC_x + G_x)^{-1}B_x \\ &= L_x^T R_x - sL_x^T A_x R_x + \cdots + (-1)^i s^i L_x^T A_x^i R_x + \cdots , \end{aligned} \tag{4.10}$$

where $R_x = G_x^{-1}B_x$ and $A_x = G_x^{-1}C$. Two Krylov subspaces can be derived from the Taylor expansion

$$\begin{aligned} V &= \mathrm{span}\{v_0, v_1, \cdots v_i \cdots\} \\ &= \mathbf{K}\{A_x; R_x\} = \mathrm{span}\{R_x, A_x R_x, \cdots A_x^i R_x \cdots\}, \end{aligned} \tag{4.11}$$

$$\begin{aligned} W &= \mathrm{span}\{w_0, w_1, \cdots w_i \cdots\} \\ &= \mathbf{K}\{A_x^T; L_x\} = \mathrm{span}\{L_x, A_x^T L_x, \cdots (A_x^T)^i L_x \cdots\}. \end{aligned} \tag{4.12}$$

The vectors in the Krylov subspace (4.11) and (4.12) have the following recursive relations

$$\begin{aligned} v_0 &= R_x, \\ v_i &= A_x v_{i-1}, \ \text{for } i \geqslant 1, \end{aligned} \tag{4.13}$$

$$\begin{aligned} w_0 &= L_x, \\ w_i &= A_x^T w_{i-1}, \ \text{for } i \geqslant 1. \end{aligned} \tag{4.14}$$

The pioneering work of MOR for the first-order system is AWE [10], which generates the reduced-order system by explicitly matching the moments of the transfer function. Since a vector iteration with matrix $A_x$, i.e., $A_x^i R_x$, is involved in the explicitly calculation of the moments of $H_f(s)$, AWE like methods [10, 50] suffer from numerical instability and cannot generate high-order models. To address this problem, MOR methods based on Krylov subspace techniques such as PVL [11], MPVL [51] and [52] were proposed to implicitly match moments. These PVL like methods employ Lanczos process to generate the projection matrices $V$ and $W$ derived from the Krylov subspaces in (4.11) and (4.12), and use "oblique projection" to achieve the reduced order models. Since the moments are implicitly matched, these methods can lead to numerically stable order reduction process, which is highly desired for practical applications. However, PVL and MPVL may result in poles on the right half plane, which causes the reduced order models unstable. SyPVL [53] and SyMPVL [54] were further proposed to eliminate the unstable poles. To maintain the passivity of the reduced order model, PVL$\pi$ [55] and PRIMA [12] were proposed. In PVL$\pi$ [55], post-processing procedure for PVL like methods was proposed to ensure the passivity of these methods. Hence, the properties such as accurate moment matching, numerical stability and passivity preserving can be guaranteed in PVL$\pi$ methods. Different from PVL like methods, PRIMA [12] employs Arnoldi process to generate the projection matrix $V$ in equation (4.11) derived from the Krylov subspaces, and uses "orthogonal projection" to obtain the reduced order models. Since the "orthogonal projection" is actually a congruence transform, PRIMA can preserve passivity. Furthermore, Arnoldi process further ensures the numerical stability and accurate moment matching of PRIMA. Recently, a structure-preserving reduction method SPRIM [47] has been proposed for the first order system, which preserves the crucial properties, such as numerical stability, passivity, second-order form and matches twice as many moments as PRIMA with the same computational work.

The MOR techniques for the second-order system in (4.6) have been developed in a similar strategy. For the second-order systems, ENOR [56] was first proposed to generate a passive reduced-order model by utilizing the symmetry positive definite (s.p.d.) property of the system matrices. However, like AWE [10], ENOR uses a recursive formula to explicitly calculate the moments of the original system, and is therefore numerically unstable. To address this issue, SMOR method was proposed in [57] trying to employ the Krylov subspace techniques. Based on a recursive relation similar to the one in ENOR, SMOR eliminates the auxiliary variables that are not orthonormalized in ENOR method, thus improves the numerical stability and accuracy. However, the projection subspace formed by SMOR method is only an approximation of the space spanned by the moments of the original system. As a result, the reduced-order system by SMOR cannot match the moments of the original system exactly. Recently, SAPOR [13] and Block SAPOR [58] have been proposed for the reduction of the second order system (4.6), which are based on a moment recurrence relation like

$$\begin{aligned} r_0 &= u, \\ r_1 &= F_1 r_0, \\ r_j &= F_1 r_{j-1} + F_2 r_{j-2} \quad \text{for } j \geqslant 2, \end{aligned} \tag{4.15}$$

where $\{r_j\}$ are moments of the state variables, $F_1$ and $F_2$ are matrices with order $M$. The second-order Krylov subspace [59] is firstly defined based on $F_1$, $F_2$, and $u$, i.e., $\mathbf{G}\{F_1, F_2; u\} = \text{span}\{r_0, r_1, \cdots, r_{m-1}\}$. Contrary to the Krylov subspaces $\mathbf{K}\{A_x; R_x\}$ and $\mathbf{K}\{A_x^T; L_x\}$ in (4.11) and (4.12) defined by one-term recursive relations in (4.13) and (4.14), the second-order Krylov subspace $\mathbf{G}\{F_1, F_2; u\}$ is defined by the two-term recursive relation in (4.15). A Second Order ARnoldi (SOAR) was proposed in [59] to construct the orthonormal basis of the second-order Krylov subspace, which is numerically stable. SAPOR [13] and Block SAPOR [58] were further proposed to use a generalized SOAR procedure to construct the projection matrix for "orthogonal projection" of (4.6), which can simultaneously guarantee all of the desired properties, i.e., accurate moment-matching, numerical stability, passivity preserving and block structure preserving.

The aforementioned methods are all based on the single-point expansion of the transfer functions. The single-point MOR methods only reserve high approximation accuracy near the specified expansion frequency point, but lose accuracy rapidly in the frequency region far away from the expansion point. In order to guarantee the reduction accuracy over a broad frequency range, multi-point moment matching techniques have been proposed to match moments around several expansion points. Complex frequency hopping (CFH) method [60] was proposed to select the expansion points by a heuristic binary search approach, whose accuracy lacks of a theoretic proof over a specified frequency range. A multi-point version Krylov subspace based MOR method [52] was proposed to select the expansion over the whole frequency range adaptively according to the estimation of the reduction error. Multi-point moment matching order reduction method with provable error bounds is still an open problem.

For interconnect circuits with large number of input/output terminals, such as power/ground network, clock distribution network and large data bus, etc., the moment matching based MOR methods can hardly achieve lower order reduced circuits, because when matching the same number of moments, the size of the reduced model is proportional to the number of inputs. In recent years, two classes of methodologies have been proposed to efficiently reduce interconnect circuits with large number of terminals. The first class of methods try to generate reduced models by approximating the output response under specified inputs [61, 62]. Therefore, the dimension of the moment subspace is no longer restricted by the number of inputs. As a result, the reduced model obtained by performing projection onto the moment subspace is much more compact compared with that by standard moment matching based MOR methods. However, this kind of methods cannot guarantee simulation accuracy when input signal is different from the specified one used by MOR. It is observed by the second class of methods [63,64] that in a large class of practical applications, there is a significant correlation between the entries of the matrix transfer function. This correlation is exploited to produce reduced order models that can be computed and stored with much lower complexity. The existing MOR methods for large number of terminals assume either the regularities of the interconnect structure or the specified signals of input. Efficient reduction of interconnect circuits with large number of terminals for general interconnect structures and arbitrary inputs is still an open problem.

**B. TBR based methods**

The classical Truncation Balanced Realization (TBR) methods [48, 49] employ the idea of balance and truncation, which arises from rich theories in control. The balanced truncation is based on the analysis of the controllability and observability Grammians $X$, $Y$ respectively. The Grammians are usually computed from the Lyapunov equations. Reduction is performed by projection onto the invariant subspaces associated with the dominant eigenvalues of the product of Grammians $XY$. One of the important features of TBR is an absolute error bound over the entire frequency range. Let $\sigma_i$ denote the square root of the $i$th largest eigenvalue of $XY$ ($XY$ always has real eigenvalues), the error in the transfer function of the order $q$ TBR approximation is bounded by $2\sum_{i=q+1}^{N}\sigma_i$ [49].

As the TBR methods are too expensive to directly apply to large scale interconnect circuit problems [48], various two-stage and iterative Krylov methods have been proposed [65–71] that combine Krylov subspace projection and TBR. While these hybrid techniques do a fairly good job of addressing the excessive order issue, the error-bound properties are weakened. Recently, an efficient TBR based method named "Poor man's TBR (PMTBR)" [15] has been developed to approximate the Grammians of the linear system by numerical integration in frequency domain. It can be computationally simple to implement, has near-optimal error properties, and possesses simple error estimation and order-control procedures. Recently, attentions have also been paid to preserve the passivity in TBR-like methods [67, 72], which is achieved by solving a pair of algebraic Riccati equations (AREs). However, for all the TBR-like methods, the block structures of the original system cannot be preserved by the reduced model.

### 4.2.2 Time-domain methods

When the inductive effect becomes more and more serious in todays technology, the waveform of the impulse response of interconnects may be very complicated. It is quite hard to predict the accuracy of the time-domain response of the reduced model, based on the accuracy of the frequency-domain response, and it is needed to do model-order reduction directly in the time domain. Time-domain reduction methods were proposed to preserve the impulse response of the system, which is the time domain counterpart of transfer function [14, 73]. In the following, we will review the time domain MOR methods [14, 73].

In time domain MOR methods, the impulse response of the state vector is approximated by proper basis functions. Taking the linear system (4.1) as example, $x(t)$ is expanded with some basis functions, i.e.,

$$x(t) = H\theta(t),$$

where $H \in R^{N\times K}$ is the coefficient matrix, $\theta(t) = [\theta_1(t), \theta_2(t), \cdots, \theta_K(t)]$ are basis functions, $K$ is the number of basis functions. By computing the coefficients of the basis functions i.e., $H$, a projection matrix is derived by orthonormalization of $H$ and a reduced model can be obtained by projecting the original model onto the projection matrix [14, 73].

In [73], Chebyshev functions and generalized orthonormal polynomials are taken as the basis functions to approximate the impulse response. The coefficient vectors are obtained by vector equation solver, which is time consuming and is prohibitive for large scale circuits. Due to the global support property of Chebyshev functions and generalized orthonormal polynomials, the impulse response in high speed interconnect circuits with strong singularities cannot be efficiently handled by this method. In [14], X. Zeng *et al.* propose to use local support wavelet functions to approximate the impulse response of state vector so that the impulse response with strong singularities can be approximated with very high accuracy. To compute the coefficient vectors more efficiently, a fast Sylvester equation solver is also proposed, which works more than one or two orders faster than the vector equation solver employed by the time domain MOR methods in [73]. The time domain wavelet order reduction method is very efficient and accurate for time domain model order reduction, especially when dealing with very large scale interconnect circuits with singularities.

Although the time domain MOR problems have been studied for the last decade, several research topics still remain open. For example, the second-order system reduction problem and the structure preserving problem have not been studied for the time domain MOR yet.

## 4.3 Problem definition of parameterized model order reduction

Parameterized linear system is introduced to model the interconnect circuits under process variations or design the interconnect circuits by optimizing the design parameters. Taking linear system (4.2) for example, the parameterized system can be expressed as

$$\begin{aligned}&(sC_x(\vec{\zeta}) + G_x(\vec{\zeta}))x(s,\vec{\zeta}) = B_x u(s),\\&y(s,\vec{\zeta}) = L_x^T x(s,\vec{\zeta}),\end{aligned} \tag{4.16}$$

where $\vec{\zeta}$ represents the parameters. The parameters $\vec{\zeta}$ could be random variables with specified distribution, which models the process variations, or general variables, which models the design parameters. The parameterized second-order system can be obtained similarly

$$\begin{aligned}&(sC(\vec{\zeta}) + G(\vec{\zeta}) + \tfrac{1}{s}\Gamma(\vec{\zeta}))v(s,\vec{\zeta}) = Bu(s),\\&y(s,\vec{\zeta}) = D^T v(s,\vec{\zeta}).\end{aligned} \tag{4.17}$$

The Parameterized MOR (PMOR) methods aim to seek for a parameterized reduced order model, which simultaneously guarantees the following properties, i.e., high accuracy, numerical stability, passivity and structure preserving. In the following, we will review the PMOR methods.

## 4.4 Existing parameterized model order reduction techniques

Most of the existing PMOR methods are the extensions of the traditional frequency domain MOR methods. The pioneering work of moment-matching based PMOR is the perturbation technique [18], which tries to match moments defined in the traditional MOR methods under small variation around the nominal circuit values. However, perturbation scheme becomes inefficient when modeling strong nonlinear effects caused by the intra-die variations [21]. An improved perturbation technique, called One-shot Projection Method (OPM), was proposed in [19] to handle PMOR of parameterized second-order system in (4.17). Compared with the perturbation method in [18], OPM can greatly reduce the computational cost of Monte Carlo analysis of the reduced order system by decoupling the projection matrix construction with the Monte Carlo analysis.

Recently, the variational MOR method PMTBR [74] has been derived from truncated balanced realization approach. However, the method has high computational complexity due to a large amount of sampling points typically required to calculate the system Grammians.

As an alternative approach to PMTBR, the multidimensional moment matching methods were proposed in [20, 75]. The multidimensional moments are defined as the coefficients of the Taylor expansion of the parameterized transfer function with respect to Laplace variable

$s$ and the parameters $\vec{\zeta}$. The system structure and the passivity are preserved by these methods. However, the projection matrix generation methods in [20, 75] are not numerical stable due to the fact that the multiparameter moments are explicitly calculated. Another difficulty with [20, 75] is that the tradeoff of error versus order is hard to control as moment matching in high dimensional space leads to projection space, whose size becomes exponentially large. In CORE method [21], the multidimensional moments are matched in a two-step explicit-and-implicit way. The CORE algorithm proposes a numerically stable way to construct the projection matrix by using the Arnoldi process. Unfortunately, CORE can not preserve the structure and therefore the passivity of the original system. A two-dimensional Krylov subspace (one dimension for frequency variable $s$, the other for parameter variables i.e. $\vec{\zeta}$) is first defined in [22] according to the two-dimensional moment recurrence relation. Meanwhile, a numerically stable Two-dimensional Arnoldi Process (TAP) is proposed in [22] to generate the projection matrix. The PIMTAP (Parameterized Interconnect Macromodeling via a Two-dimensional Arnoldi Process) method proposed in [22] is computationally stable and robust, and preserves the original structure and therefore the passivity of parameterized interconnect systems. Besides, an adaptive scheme is also proposed in [22] to match a desired number of the multiparameter moments. Most of the aforementioned PMOR methods only consider design parameters. The PMOR of parameterized system with random process parameters is still an open problem.

## 4.5 Mathematical challenges for model order reduction of interconnect

Since last decade, MOR techniques have been developed for the reduction of large scale interconnect circuits. Parameterized MOR techniques have also been derived to evaluate the response of interconnects as functions of design parameters or process parameters. Although there were a great deal of methods proposed, mathematical challenges still remain for the MOR and parameterized MOR of interconnects.

- For frequency domain MOR, the state-of-the-art moment-matching methods can guarantee the following four properties, i.e., moment matching, numerical stability, passivity preserving and structure preserving. However the accuracy of the moment-matching methods is not guaranteed for the entire frequency range. On the other hand, TBR methods are much less computational efficient than moment-matching methods, but have an error bound over entire frequency range. How to combine these advantages of moment-matching methods and TBR methods, i.e. guaranteeing error bound over entire frequency range, numerical stability, passivity preserving, structure preserving and computational efficiency, is still an open problem.

- The existing methods for reduction of system with large number of terminals either assume the matrix transfer functions to be strongly correlated based on the regularities of the interconnect structure [64] or assume specified input signals [62]. The efficient reduction of system with large number of terminals for general interconnect structures and arbitrary inputs is still an open problem.
- Although the time domain MOR problems have been studied for the last decade, several research topics still remain open. For example, the second-order system reduction problem and the structure preserving problem haven't been studied for the time domain MOR yet.
- PMOR for the reduction of interconnect circuits considering design parameters and process parameters has become a research intensive area in recent years. For the design parameters, PIMTAP (Parameterized Interconnect Macromodeling via Two-dimensional Arnoldi Process) is the most efficient PMOR method guaranteeing moment-matching, structure-preserving, passivity-preserving and numerical stability. However, the PMOR of parameterized system with random process parameters is still an open problem.

# 5 Lithography simulation and optical proximity correction (OPC)

For IC manufacture from 90nm down to 32nm technology node, 193nm wavelength stepper is employed by lithography process to transfer the design patterns from mask to wafer. The sub-wavelength lithography has caused severe geometric distortions during pattern transfer. As illustrated in Figure 4 (a), the 180nm design patterns can be printed on wafer accurately for a 193nm stepper, while 130nm and 90nm design patterns may be distorted or even disappear on wafer. Various kinds of techniques, such as OPC, PSM etc., have been invented to improve the printability. Optical Proximity Correction (OPC) aims to systematically correct the masks to compensate for the patterning distortions due to optical diffraction effects, so that the patterns printed on the wafer from the corrected mask can have the exact shape as what they are designed, as shown in Figure 4 (b). Phase Shift Mask (PSM) introduces a $180°$ phase shift between adjacent features on the mask, as illustrated in Figure 5, to destroy the interferes in the light diffracted into nominally dark spaces from adjoining clear openings, and thus improves feature resolution, as shown in Figure 4 (b).

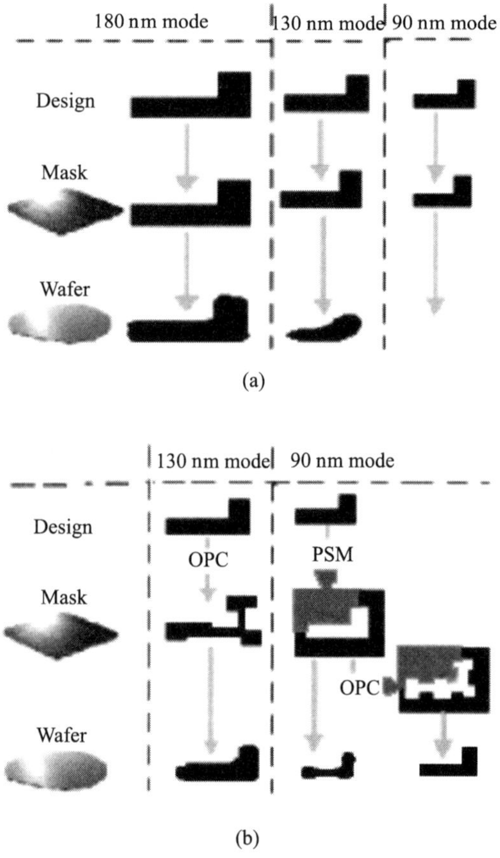


Figure 4 Resolution enhancement techniques for a 193nm stepper

OPC and PSM turn to be tremendously challenging due to the ever increasing design complexity. Lithography simulation, being an indispensable tool for developing OPC, PSM techniques, exhibits great advantages on time and cost savings compared with process experiments. In this section, we focus on 3-D lithography simulation, statistical lithography simulation and OPC because of their important roles in nano-scale chip design and manufacture. The physical problems, mathematical models and available solutions for these three issues are presented. Mathematical challenges are summarized in the end.

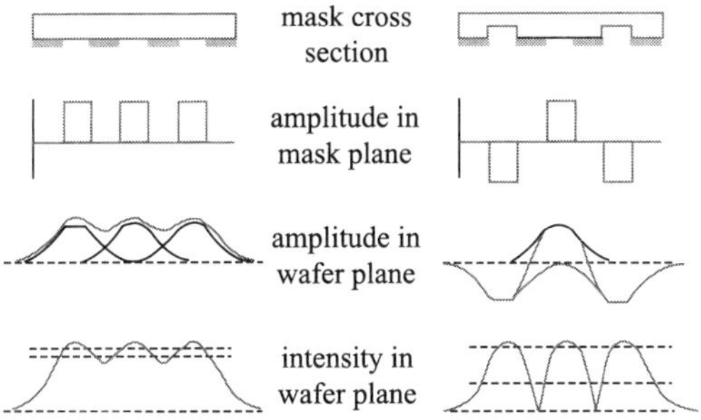


Figure 5 Schematic diagram comparing conventional binary mask lithography (left) with phase-shifted-mask lithography (right)

## 5.1 3-D lithography simulation

The 193nm optical lithography system is illustrated in Figure 6. To simulate this lithography system, the modern lithography simulators include four modules for modelling the illumination, the transmission of light through the mask, the transmission of light through aberrating optical system and the optical field propagation inside wafer surface. Traditionally, the mask is treated as an infinite thin planar grating and Kirchhoff boundary condition can be used to model the field just behind the mask. A stratified medium as a pile of thin homogeneous films is utilized to model the wafer surface, and closed form formulations can be used to describe the field in the medium. However, on account of the increasing verticality of mask structure and non-planarity of wafer surface, light propagation through mask and wafer surface turns to be 3-D electromagnetic scattering analysis problem.

The 3-D electromagnetic scattering analysis is also needed for extreme ultraviolet (EUV) lithography simulation, as shown in Figure 7. Different from the state-of-the-art transmissive optical lithography in Figure 6, EUV lithography employs illumination of 13.5nm wavelength and transfers the patterns from mask to wafer by reflective imaging principle. Hence, the reflected fields should be rigorously calculated for the EUV mask scattering problem.

### A. Physical problem and mathematic model

Mathematically, rigorous electromagnetic scattering analysis for both mask and wafer surface turns to be the same problem. In this section, we take mask scattering as an example to introduce 3-D electromagnetic analysis problem in lithography simulation. The cross section of a slanted PSM is illustrated in Figure 8. Chrome lines and phase shift material are deposited upon a glass substrate. The chrome lines absorb incident

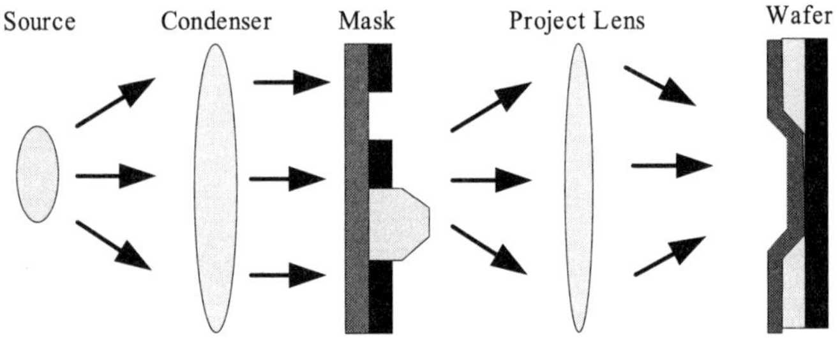


Figure 6 A 193nm lithography system

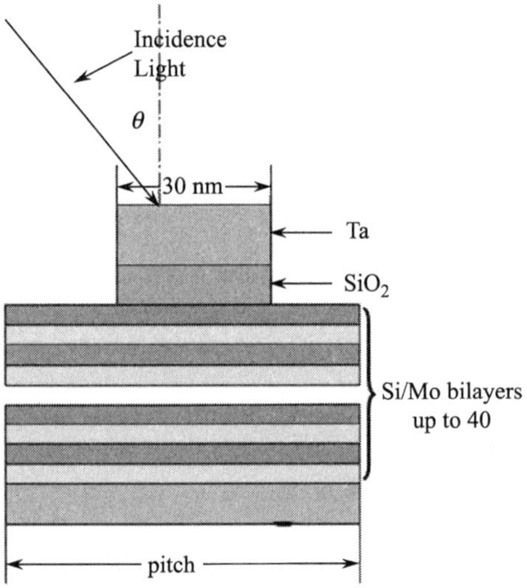


Figure 7 A 13.5nm EUV mask where reflected fields should be rigorously calculated

light, and phase shift material changes the phase of light by 180° (Figure 8).

Maxwell equations are the governing equations for the field analysis within the mask structure (for example, Figure 8) and can be written in time domain as,

$$\bigtriangledown \times \vec{E}(\vec{r},t) = -\frac{\partial \vec{B}(\vec{r},t)}{\partial t}, \tag{5.1}$$

$$\bigtriangledown \times \vec{H}(\vec{r},t) = \frac{\partial \vec{D}(\vec{r},t)}{\partial t} + \vec{J}(\vec{r},t), \tag{5.2}$$

$$\bigtriangledown \cdot (\varepsilon \vec{E}(\vec{r},t)) = 0, \tag{5.3}$$

$$\bigtriangledown \cdot \vec{B}(\vec{r},t) = 0 \tag{5.4}$$

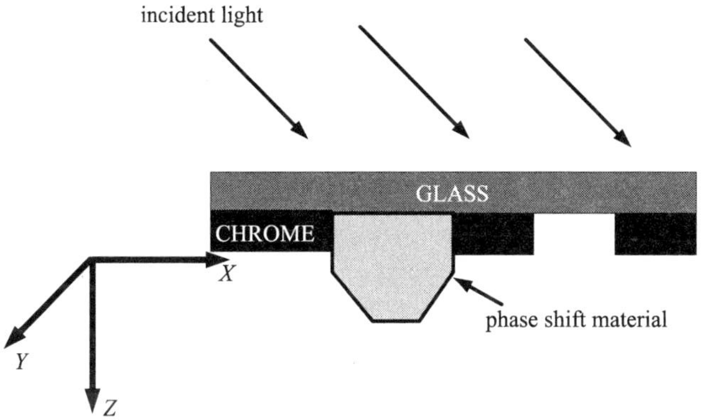


Figure 8 2-D cross section of a slanted phase shift mask

supplemented with the constitutive relations,

$$\vec{D} = \varepsilon \vec{E}, \quad \vec{B} = \mu \vec{H}, \quad \vec{J} = \sigma \vec{E},$$

where $\varepsilon$, $\mu$ and $\sigma$ are real numbers corresponding to electric permittivity, magnetic permeability and conductivity respectively. Incident monochromatic light of frequency $\omega$ is assumed on the top of the mask structure. The Maxwell equations combined with the incident light and boundary conditions around the mask are solved to obtain the electromagnetic field just up or below the mask. Then the field calculated can be used to form aerial images on the wafer surface by Hopkins' imaging theory [76].

## B. Available solutions for 3-D lithography simulation

The rigorous electromagnetic modeling techniques for mask and wafer surface scattering include finite-difference-time-domain method (FDTD) [77, 78], finite element method (FEM) [79], modal expansion method such as waveguide method [80] or closely related rigorous coupled wave analysis (RCWA) [81]. These methods differ in the approaches by which Maxwell equations are numerically solved and the boundary conditions above and below the mask structure are established. A simple review of these methods is given in the following.

### *B.1 Finite-difference time-domain method (FDTD)* [77]

FDTD method solves Maxwell's equations in time domain [77]. The first step for FDTD approach is to discretize Maxwell's equations (5.1, 5.2) in both time and space domain. Then the radiation conditions above and below the mask are implemented by second order absorbing boundary condition. Enhanced periodical conditions are applied on transverse boundaries where both the field amplitude and phase are identical. The equations obtained are solved by simulating the field evolution through time until time-harmonic steady-state is reached.

Accuracy and speed of the FDTD method depend on the space and time discretization, as well as the total periods to reach time-harmonic steady-state. Usually, to achieve acceptable accuracy, about 15 simulation nodes per wavelength are required for one dimension [77]. To keep the FDTD algorithm stable, the temporal discretization $\triangle t$ and the spatial discretization $\triangle x$ should satisfy the following relation for the 3-D simulation region,

$$\triangle t \leqslant \frac{\triangle x}{v\sqrt{3}},$$

where $v$ is the light propagation velocity in mask. The constraint on time step and space lattice makes FDTD method very cost intensive. As shown in [82], simulation of an EUV mask of a contact hole with length and width of 200 nm × 200 nm and height of 320 nm needs 30 hours when the FDTD algorithm runs on a standard 2.8 GHz personal computer.

*B.2 Finite-element method* [79]

Finite-element method solves the following wave equation derived from Maxwell's equations in frequency domain,

$$\nabla \times \frac{1}{\hat{\varepsilon}} \nabla \times \vec{H}(\vec{r}) - \omega^2 \mu \vec{H}(\vec{r}) = 0, \tag{5.5}$$

where $\hat{\varepsilon} = \varepsilon + i\frac{\sigma}{\omega}$ is complex permittivity. The radiation conditions on the vertical boundary are realized with PML method, and periodical conditions on transverse boundaries are assumed. Equation (5.5) is solved in its weak form together with boundary conditions. The computational domain is discretized with tetrahedral patches. The functional spaces are discretized using Nedelec's edge elements. The discretization of computational domain and functional spaces leads to a large sparse matrix equation which is solved by LU-factorization or iterative methods.

The flexibility of triangulations shared with FEM approach allows for the simulation of masks with sloped etch profiles such as the PSM in Figure 8, which is not easy for FDTD method because of its orthogonal grids. In virtue of adaptive mesh and multi-grid techniques, FEM based mask simulation tool JCMHarmony [79] exhibits efficient computational time and memory requirements which grow linearly with the number of unknowns.

*B.3 Waveguide method (WG)* [80]

Waveguide method solves the light scattering problem at discrete spatial frequencies instead of discrete spatial points in finite difference or finite element methods [80]. In the following, a simply introduction of this method is given by a 2-D mask scattering problem①. The 3-D version [80, 82] is referred for interested readers.

As shown in Figure 9, the mask is divided into layers along $z$ direction, and *Manhattan geometry* format can be used to describe the materials in each layer. For TE mode, Maxwell's equation governing the electric field $E^j(x,z)$ for layer $j$ is

$$\nabla^2 E^j - \mu_0\varepsilon_0\hat{\epsilon}^j(x)\frac{\partial^2 E^j}{\partial t^2} = 0, \tag{5.6}$$

where $\hat{\epsilon}^j(x) = \frac{\varepsilon^j(x)}{\varepsilon_0}$ is relative complex dielectric permittivity in layer $j$. Then, the wave equation (5.6) is seperated into two ordinary differential equations using separation of variables $E^j(x,z) = X^j(x)Z^j(z)$.

$$\frac{d^2X^j}{dx^2} + [k_0^2\hat{\epsilon}^j(x) + (\alpha^j)^2]X^j = 0, \tag{5.7}$$

$$\frac{d^2Z^j}{dz^2} - (\alpha^j)^2 Z^j = 0. \tag{5.8}$$

Assuming $\hat{\epsilon}^j(x), X^j(x)$ have the general form

$$\hat{\epsilon}^j(x) = \sum_{q=-M}^{M} \hat{\epsilon}_q^j \exp(i2\pi qbx), \tag{5.9}$$

$$X^j(x) = \sum_{l=-M}^{M} B_l^j \exp(i2\pi lbx), \tag{5.10}$$

where $b$ is the inverse of the mask period along $x$ direction, we obtain an eigenvalue problem for layer $j$ by substituting $\hat{\epsilon}^j(x), X^j(x)$ in (5.7) with (5.9) and (5.10). The field $E^j(x,z)$ can be written as,

$$E^j = \sum_{m=-M}^{M} [(A_m^{+j}\exp(\alpha_m^j z) + A_m^{-j}\exp(-\alpha_m^j z))\sum_{l=-M}^{M} B_{l,m}^j \exp(i2\pi lbx)], \tag{5.11}$$

where $\alpha_m^j, \mathbf{B}_m^j = [B_{-l,m}^j \cdots B_{l,m}^j]^T$ are the $m$-th eigenvalue and $m$-th eigenvector of the eigenvalue problem of layer $j$ derived above. Coefficients $A_m^{+j}, A_m^{-j}$ in (5.11) are determined by continuous conditions between different layers.

Compared with FDTD and FEM methods which need to solve a large sparse matrix equation, waveguide method only needs to solve small or medium sized full matrix equation, which makes waveguide method accessible on computing workstations, even on personal computers.

① All components are assumed to be constant in $y$ direction for Figure 8.

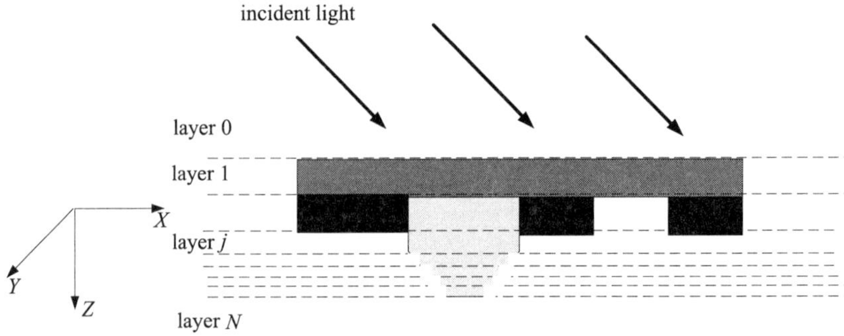


Figure 9 Mask discretization for waveguide method

## C. Mathematical challenges in 3-D lithography simulation

Rigorous analysis for electromagnetic scattering of transmissive mask (Figure 8) and wafer surface has been studied for more than 10 years. FDTD method, FEM method, WG method and closely related RCWA method have been developed for solving the problem. However, there still exists two challenges according to International Technology Roadmap of Semiconductors (ITRS) 2005 [31].

1. The performance of FEM method, FDTD method, WG method and RCWA method has to be critically evaluated in terms of accuracy, memory requirement and computing speed for industrial application. In [79], FEM method is demonstrated to converge faster than FDTD and WG methods for a 2-D mask example. However, more careful comparison between FEM method and other methods should be conducted for complex 3-D structures.
2. More efficient 3-D scattering analysis methods are still needed for large scale and complex structures, such as optimization of mask related optical resolution enhancements, description of light scattering from mask defects etc..

Recently, the reflective EUV mask scattering problem attracts more attentions. As shown in Figure 7, absorbing materials (Ta and $SiO_2$ in the figure) are deposited upon a substrate composed of up to 40 Si/Mo bilayers. The structure gets more complex when a defect is deposited in the substrate. Figure 10 shows the cross section of a deformed EUV mask with a Gaussian shaped defect. Currently, modal expansion methods, such as WG method [82] and RCWA method [83], have been applied for the problem. However, for modal methods, hundred of layers should be divided for the deformed multilayer substrate, and high order Fourier series should be expanded for the permittivity function (5.9) in each

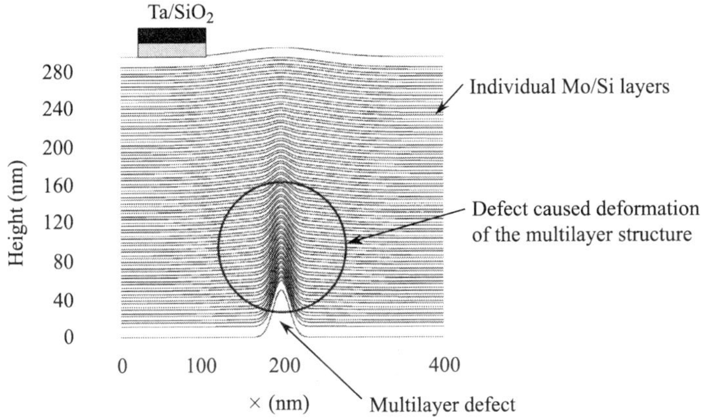


Figure 10 2-D cross section of an EUV mask with a defect on the bottom

layer to account for abruptly changed dielectrics. These two problems in modal methods increase the computation complexity dramatically. More efficient methods for defective EUV mask simulation are in urgent need.

## 5.2 Statistical lithography simulation

As the IC technology reaches 65nm and beyond, random variations of process parameters, such as lens abberation levels, defocus, dose, resist thickness etc., affect the outputs of a lithography system significantly. The line width on wafer, which is usually called Critical Dimension (CD), is no longer a deterministic number matched the design specification, but has a random variation. The CD variation will affect the electrical performance of transistors and interconnects, and causes the manufacturing yield loss. The prediction and control of CD variation have become a vital task for IC design and manufacturing. Statistical lithography simulation, which aims to accurately predicting CD variation on wafer, plays an important role in the lithography simulation.

A lithography system shown in Figure 6 includes a mask module and an imaging system which consists of illumination, projection system and wafer modules. The pattern transfer process of the lithography system can be mathematically described as,

$$G(x, y) = T(F(x', y'))(x, y). \tag{5.12}$$

Here, $F(x', y') \in \mathbf{L}^2(\mathbb{R}^2)$ is the mask transmission function. For a binary mask, $F(x', y')$ takes 0 for opaque regions and 1 for transparent ones. For a phase shift mask shown in Figure 8, $F(x', y')$ is a complex transmission function. The nonlinear operator $T$ denotes the model of the imaging system, which provides a mapping from the mask transmission function $F(x', y')$ to light intensity (also called as image) $G(x, y) \in \mathbf{L}^2(\mathbb{R}^2)$ on wafer surface. Based on the above statements, statistical lithography simulation is defined as,

$$G(x, y, \lambda_1, \lambda_2, ...) = T(F(x', y'), \lambda_1, \lambda_2, ...)(x, y), \tag{5.13}$$

where $\lambda_1, \lambda_2, ...$ correspond to random process parameters in a lithograph system. The CD distribution can be extracted from obtained image distribution $G(x, y, \lambda_1, \lambda_2, ...)$.

Earlier work on statistical lithography simulation was based on the Response Surface Method (RSM) [84]. The response surface is built by simulating CD dependence on lithography process variables. The process parameters are then sampled from Gaussian distribution to generate the distribution of the resulting CDs. Since a large number of process parameters which affect the CD variation significantly should be considered in the construction of the response surface, the computational runtime is rapidly increased for Response Surface Method. For example, for six process parameters with ten points for each parameter, a total amount of $10^6$ deterministic lithography simulations are required to build the response surface.

Recently, an improved Monte Carlo method was proposed in [85] to simulate the CD variation. The computational complexity of Monte Carlo method can be independent of the number of process parameters. However, the convergence rate of Monte Carlo method is pretty low. As a result, hundreds of thousands of deterministic lithography simulations are still needed in Monte Carlo method for CD variation simulation. In our experiment, the total time of Monte Carlo simulation with $10^4$ sample points amounts to 60 days on a 2.0GHz workstation, which prohibits its extensive use for IC yield analysis and statistical design. Highly efficient statistical algorithms are urgently required for statistical lithography simulation.

## 5.3 Optical proximity correction (OPC)

### A. Mathematical description for OPC

The OPC problem is considered as an inverse problem of the lithography simulation. The lithography simulation has been modeled mathematically by (5.12). The OPC problem is defined by the inverse process, i.e. how to find the optimized mask $F^d(x', y')$ provided a desired image $G^d(x, y)$ on wafer and a specific imaging system $T$.

### B. Existing OPC algorithms

OPC algorithms have been developed for more than two decades and can be divided into two different categories, i.e. the rule based and model based methods. Rule based OPC algorithms build a large database containing various kinds of mask modification rules which are constructed by data from experiments or lithography simulations. The mask is meshed into pieces which are corrected by looking up the rule database. On the other hand, model based methods use a mathematical system model to decide modifications to the mask. The system model always includes a lithography model and a control module which decides the mask modification strategy from the outputs of lithography simulation. The lithography model can be a physical model [86], a lumped-behavioral model constructed from experimental data [87] or a parametric physical model supplemented by experimental data [88]. Rule based OPC usually is more efficient than model based OPC. However, OPC rules can not cover all the complex design patterns which grow explosively in sub-90nm designs. Model based OPC methods therefore become dominant for their accuracy and flexibility in dealing with complex designs. In this section, we take model based OPC flow as an example to give readers a rough idea about OPC algorithms.

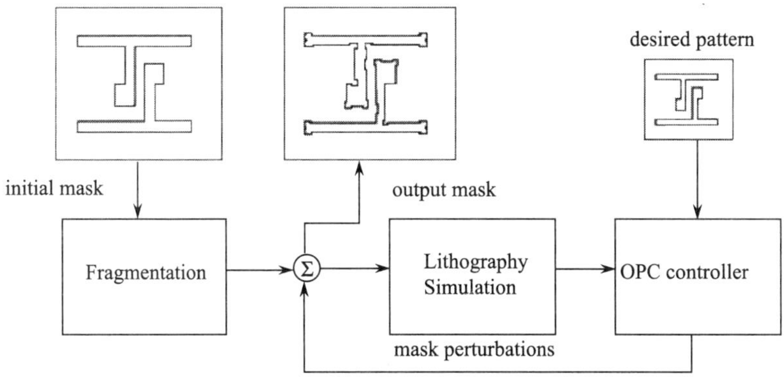


Figure 11 Model based OPC flow

The model based OPC algorithm is illustrated in Figure 11 and consists of the following steps.

1. *Segment the initial mask into edge objects and corner objects.*
2. *For all mask objects, compute cost function for an object each time, then move the object backward or forward until the minimum cost function is obtained.*
3. *Iterates step 2 several times.*

In this flow, the cost function for an object is defined as the difference between computed intensity and desired intensity at chosen points around the object within a certain fixed window. The lower the cost function, the closer the actual graph on wafer approaches the desired graph.

As design complexity continues to increase, current model based OPC algorithms are challenged by two issues. One is *computational complexity*. It will take several days for commercial OPC tools to complete a full chip OPC flow even using modern parallel computing system. The other one is the complexity of the corrected mask. Typically, the data of a single layer mask corrected by OPC tools easily exceeds several gigabytes which increase the mask manufacturing cost dramatically. How to get simplified mask structure while keeping imaging accuracy becomes a vital problem for next generation OPC algorithms. Recently, cell based OPC and design aware OPC have been proposed to address this problem.

To reduce the complexity of OPC algorithms, standard cell based OPC algorithms are presented. The layouts of standard cells are precorrected and stored, then full-chip correction algorithms are applied to the boundaries between different cells. Cao et al. [89] suggest using dummy poly insertion to shield inter-cell optical interference. However, the dummy polys may potentially induce parasitic capacitance and reduce the performance of the transistors. Pawlowski et al. [90] use different corrected cell versions depending on adjacent cells in a row, and so stresses the influence of adjacent cells. The experimental results show that, compared with a commercial tool, this method has up to 100X speedup and 35X reduction in mask data size.

To reduce the data of corrected masks, design-aware OPC algorithms emerges as a good solution. The mask objects are labeled different accuracy levels for correction, for example, the objects lie at a critical path may be labeled high levels. The mask objects with higher level are segmented and corrected more finely, which generates more mask data but produces closer desired graphs on wafer. This entails passing performance analysis and functional intentions from logic-layout synthesis to physical verification. The required flow integrations must span library creation, detailed routing, and physical verification. Such passing of designers' intent to OPC tools can lead to a functionally better OPC result in addition to huge mask cost savings.

### C. Challenges for next generation OPC

In spite of the efforts on the acceleration of OPC algorithms and simplification of OPC outputs, more efficient OPC algorithms are still requested because billions of mask shapes should be corrected in reasonable time. The next generation OPC algorithms, such as rule and model based hybrid OPC algorithms, cell based OPC algorithms and design-aware OPC algorithms etc., need to be further improved in time, memory, and accuracy requirements.

As the technology node continues scaling down, random process variations lay great challenges on the state-of-the-art OPC algorithms. The masks corrected by next generation OPC algorithms should not only produce functional chips at nominal process conditions, but also be insensitive to more and more severe process variations. Process variation aware OPC emerging as a leading-edge research direction still remains an open problem.

# 6 Summary

As the IC technology advances, the design, simulation and verification techniques and the related mathematical problems still remain a very challenging and interesting research area. In this paper, we have presented the mathematical problems encountered in the research areas for the SoC design and manufacture, including Static Timing Analysis (STA), parasitic extraction, Model Order Reduction (MOR), lithography simulation and OPC.

Static Timing Analysis (STA), as a key technique to evaluate the Integrated Circuit (IC) performance, is now facing with the statistical delay calculation difficulties in Statistical STA (SSTA) considering process variations. MAX operator, which calculates the maximum delay among a set of statistical delay distributions as given in equation (2.1), is a fundamental mathematical problem in SSTA. Fast and accurate numerical methods are needed to deal with the MAX operator for large dimension of independent random variable problems. Furthermore, in order to consider the real process variations with arbitrary random probability, SSTA should be extended in both statistical modeling and statistical MAX algorithms. Finally, identification of the critical paths with longest delays in SSTA is still an open problem when all delays are provided with random distributions.

Mathematic challenges for the parasitic extraction are mostly driven by the need of the accuracy of interconnect modeling, and by the need of handling the more and more complex interconnect structure. These challenges include the "frequency-dependent extraction" and the "full-wave extraction" in order to further consider the effect of the circuit frequency on the interconnects, and the multilayered Green's function and the "full-chip extraction" in order to handle the complexity of interconnect structure. Furthermore, when IC technology reaches nanometer range, it is desirable to address the stochastic problem of variation-aware parasitic extraction of interconnects considering the geometric variations.

MOR and PMOR techniques reduce large scale extracted RLC/RCS interconnect circuits to a lower order model, thus enable fast simulation and verification of the interconnect circuits. The mathematical challenges for MOR and PMOR include: 1. It still remains a very challenging problem for frequency domain MOR methods to simultaneously guarantee the following properties, i.e. error bound over the entire frequency range, computational efficiency, structure preserving, passivity preserving and numerical stability. 2. MOR for interconnect circuits with large number of terminals is still an open problem. 3. Time-domain reduction methods need to be developed to address the structure preserving problem and the second-order system problem. 4. Numerically stable and computationally efficient PMOR method that simultaneously guarantees accuracy, structure-preserving and passivity needs to be developed for multiple design parameters. More importantly, the PMOR of parameterized system with random parameters is also very desired.

Lithography simulation, as an indispensable tool for the lithography system optimization and the lithography resolution enhancement, is challenged by two issues. One is the rigorous electromagnetic modeling for field propagation through 3-D mask and non-planar wafer surface, which is governed by Maxwell equations. The other one is the statistical lithography simulation, which demands for efficient statistical method to calculate the geometric variations needed by statistical circuit analysis and statistical process optimization. Optical Proximity Correction (OPC), as the inverse problem of lithography simulation, are challenged by how to deal with billions of mask shapes to be corrected in tolerable time, how to reduce excessive amount of data generated, and how to become insensitive to process variations.

# Acknowledgements

This work is supported partly by the National Basic Research Program of China under the grant 2005CB321701, partly by NSFC research project 90307017 and 60676018, partly by the doctoral program foundation of Ministry of Education of China 20050246082.

This paper is based on the invited talk given by Prof. Xuan Zeng at the 9th Annual Meeting of China Society for Industrial and Applied Mathematics, which was invited by Academician Prof. Ta-Tsien Li. We would first like to give our particular thanks to Academician Prof. Ta-Tsien Li for his invitation and his valuable advices to the talk. A large number of valuable researches that contribute to this paper are benefited from the joint research of mathematics and microelectronics. We would like to express our great gratitude to Prof. Yangfeng Su from Fudan University and Prof. Zhaojun Bai from University of California Davis for their great helps in our joint research of MOR, Prof. Wei Cai from University of North Carolina at Charlotte for his great efforts at developing the original ideas of wavelet and stochastic spectral method in the application of circuit analysis, variation-aware parasitic extraction, and lithography simulation, Prof. Zhiming Chen from Institute of Computational Mathematics, Chinese Academy of Sciences and Prof. Wenbin Chen from Fudan University for their valuable advices on the parasitic extraction and lithography simulation research topics.

# References

[1] N.P. Jouppi. Timing analysis and performance improvement of mos VLSI design. *IEEE Transactions on Computer-Aided Design of Integrated Circuits and Systems* 9(4), 650 665, July 1987.

[2] S. Zhou. Static timing analysis in VLSI design, Ph.D. dissertation, University of California, San Diego, 2006.

[3] S. Sapatnekar. *Timing.* New York: Kluwer Academic Publishers, 2004.

[4] A.E. Ruehli. Equivalent circuit models for three-dimensional multiconductor systems. *IEEE Transactions on Microwave Theory and Techniques* 22(3), 216-221, March 1974.

[5] K. Nabors and J. White. Fastcap: A multipole accelerated 3-D capacitance extraction program. *IEEE Transactions on Computer-Aided Design of Integrated Circuits and Systems* 10(11), 1447-1459, Nov. 1991.

[6] W. Shi, J. Liu, and N. Kakani. A fast hierarchical algorithm for 3-D capacitance extraction. *IEEE Transactions on Computer-Aided Design of Integrated Circuits and Systems* 21(3), 330-336, March 2002.

[7] S. Yan, V. Sarin, and W. Shi. Sparse transformations and preconditioners for 3D capacitance extraction. *IEEE Transactions on Computer-Aided Design of Integrated Circuits and Systems* 24(9), 1420-1426, Sept. 2005.

[8] J.K. White, M. Kamon, and M.J. Tsuk. FASTHENRY: a multipole-accelerated 3-D inductance extraction program. *IEEE Transactions on Microwave Theory and Techniques* 42(9), 1750-1758, Sept. 1994.

[9] Z. Zhu, B. Song, and J. White. Algorithms in FastImp: a fast and wide-band impedance extraction program for complicated 3-D geometries. *IEEE Transactions on Computer-Aided Design of Integrated Circuits and Systems* 24(7), 981-998, July 2005.

[10] L.T. Pillage and R.A. Rohrer. Asymptotic waveform evaluation for timing analysis. *IEEE Transactions on Computer-Aided Design of Integrated Circuits and Systems* 9(4), 352-366, April 1990.

[11] P. Feldman and R.W. Freund. Efficient linear circuit analysis by padé approximation via the Lanczos process. *IEEE Transactions on Computer-Aided Design of Integrated Circuits and Systems* 14(5), 639-649, May 1995.

[12] A. Odabasioglu, M. Celik, and L. Pileggi. PRIMA: Passive reduced-order interconnect macromodeling algorithm. *IEEE Transactions on Computer-Aided Design of Integrated Circuits and Systems* 17(8), 645-654, Aug. 1998.

[13] Y.F. Su, J. Wang, X. Zeng, Z. Bai, C. Chiang, and D. Zhou. SAPOR: Second-order arnoldi method for passive order reduction of RCS circuits. in *Proceedings of IEEE/ACM International Conference on Computer-Aided Design*, Nov. 2004.

[14] X. Zeng, L. Feng, Y. Su, W. Cai, D. Zhou, and C. Chiang. Time domain model order reduction by wavelet collocation method. in *Proceedings of IEEE/ACM Design, Automation and Test in Europe*, 21-26, 2006.

[15] J.P. et al. Poor man's TBR: A simple model reduction scheme. in *Proceedings of IEEE/ACM Design, Automation and Test in Europe*, 2004.

[16] R. Jiang, W. Fu, J.M. Wang, V. Lin, and C.C.-P. Chen. Efficient statistical capacitance variability modeling with orthogonal principle factor analysis. in *Proceedings of IEEE/ACM International Conference on Computer-Aided Design*, 2005.

[17] H. Zhu, X. Zeng, W. Cai, J. Xue, and D. Zhou. A sparse grid based spectral stochastic collocation method for variations-aware capacitance extraction of interconnects under nanometer process technology. in *Proceedings of IEEE/ACM Design, Automation and Test in Europe*, 2007.

[18] Y. Liu, L. Pileggi, and A. Strojwas. Model order-reduction of rc(l) interconnect including variational analysis. in *Proceedings of IEEE/ACM Design Automation Conference*, 201-206, 1999.

[19] J. Tao, X. Zeng, F. Yang, Y. Su, L. Feng, W. Cai, D. Zhou, and C. Chiang. A one-shot projection method for interconnects with process variations. in *Proceedings of IEEE International Symposium on Circuits and Systems*, 2006.

[20] L. Daniel, O.C. Siong, L.S. Chay, K.H. Lee, and J. White. A multi-parameter moment-matching model-reduction approach for generating geometrically parameterized interconnect performance mod-

els. *IEEE Transactions on Computer-Aided Design of Integrated Circuits and Systems* 23(5), 678-693, May 2004.

[21] X. Li, P. Li, and L. Pileggi. Parameterized interconnect order reduction with explicit-and-implicit multi-parameter moment matching for inter/intra-die variations. in *Proceedings of IEEE/ACM International Conference on Computer-Aided Design*, 2005.

[22] Y.-T. Li, Z. Bai, Y. Su, and X. Zeng. Parameterized model order reduction via a two-dimensional arnoldi process. in *Proceedings of IEEE/ACM International Conference on Computer-Aided Design*, 2007.

[23] J. Singh and S. Sapatnekar. Statistical timing analysis with correlated non-gaussian parameters using independent component analysis. in *Proceedings of IEEE/ACM Design Automation Conference*, 155-160, 2006.

[24] S. Bhardaj, P. Ghanta, and S. Vrudhula. Framework for statistical timing analysis using non-linear delay and slew models. in *Proceedings of IEEE/ACM International Conference on Computer-Aided Design*, 3C(2), 2006.

[25] C.S. Amin, N. Menezes, K. Killpack, and F. Dartu. Statistical static timing analysis: How simple can we get. in *Proceedings of IEEE/ACM Design Automation Conference*, 652-657, 2005.

[26] H. Chang, V. Zolotov, S. Narayan, and C. Visweswariah. Parameterized block-based statistical timing analysis with non-gaussian parameters, nonlinear delay functions. in *Proceedings of IEEE/ACM Design Automation Conference*, 71-76, 2005.

[27] Y. Zhan, A.J. Strojwas, X. Li, and L.T. Pileggi. Correlation-aware statistical timing analysis with non-gaussian delay distribution. in *Proceedings of IEEE/ACM Design Automation Conference*, 77-82, 2005.

[28] L. Zhang, W. Chen, Y. Hu, J.A. Gubner, and C.C.-P. Chen. Correlation-preserved non-gaussian statistical timing analysis with quadratic timing model. in *Proceedings of IEEE/ACM Design Automation Conference*, 83-88, 2005.

[29] V. Khandelwal and A. Srivastava. A general framework for accurate statistical timing analysis considering correlations. in *Proceedings of IEEE/ACM Design Automation Conference*, 89-94, 2005.

[30] C. Visweswariah, K. Ravindran, K. Kalafala, S.G. Walker, and S. Narayan. First order incremental blockbased statistical timing analysis. in *Proceedings of IEEE/ACM Design Automation Conference*, 331-336, 2004.

[31] *International technology road of semiconductors, http://www.itrs.net/Common/2005ITRS/Home2005.htm*, 2005.

[32] A. Zeidler, et. al.. Advanced statistical process control: controlling sub-0.18um lithography and other process. in *Proceedings of SPIE* 4344, 312-322, 2001.

[33] C.E. Clark. The greatest of a finite set of random variables. *Operations Research.*

[34] S. Sapatnekar. *Probability, Random Variables and Stochastic Processes.* New York: Kluwer Academic Publishers, 2004.

[35] X. Li, J. Le, P. Gopalakrishnan, and L.T. Pileggi. Asymptotic probability extraction for non-normal distributions of circuit performance. in *Proceedings of IEEE/ACM International Conference on Computer-Aided Design*, 2-9, 2004.

[36] D. Morrison. Multivariate statistical methods. 1976.

[37] J. Zhao and Z. Li. FDTD analysis of the electrical performance for interconnection lines in multichip module (MCM) with perforated reference planes components, packaging, and manufacturing technology. *IEEE Transactions on Advanced Packaging* 20(1), 34-41, Jan. 1997.

[38] N.A. Marques, M. Kamon, L.M. Silveira, and J.K. White.. Generating compact, guaranteed passive reduced-order models of 3-D RLC interconnects. *IEEE Transactions on Advanced Packaging* 27(4), 569-580, April 2004.

[39] J.R. Phillips and J.K. White. A precorrected-FFT method for electrostatic analysis of complicated 3-D structures. *IEEE transactions on Computer-Aided Design of Integrated Circuits and Systems* 16(10), 1059-1072, Oct. 1997.

[40] A. Devgan, H. Ji, and W. Dai. How to efficiently capture on-chip inductance effect: Introducing a new circuit element K. in *Proceedings of IEEE/ACM Intenational Conference on Computer Aided Design*, 2000.

[41] H. Ji, A. Devgan, and W. Dai. Ksim: A stable and efficient RKC simulator for capturing on-chip inductance effect. in *Proceedings of IEEE/ACM Asia and South Pacific Design Automation Conference*, 2001.

[42] T.-H. Chen, C. Luk, H. Kim, and C.C.-P. Chen. INDUCTWISE: Inductance-wise interconnect simulator and extractor. in *Proceedings of IEEE/ACM International Conference on Computer-Aided Design*, 2002.

[43] Y. Du and W. Dai. Partial reluctance based circuit simulation is efficient and stable. in *Proceedings of IEEE/ACM Asia and South Pacific Design Automation Conference*, 2005.

[44] N. Wiener. The homogeneous chaos. *American Journal of Mathematics* 60(4), 897-936, April 1938.

[45] Z. Ye, W. Yu, and Z. Yu. Efficient 3-D capacitance extraction considering lossy substrate with multilayered green's function. *IEEE Transactions on Microwave Theory and Techniques* 54(5), 2128-2137, May 2006.

[46] W. Kao, C.-Y. Lo, M. Basel, and R. Singh. Parasitic extraction: current state of the art and future trends. *Proceedings of the IEEE* 89(5), 729-739, 2001.

[47] R.W. Freund. SPRIM: Structure-preserving reduced-order interconnect macromodeling. in *Proceedings of IEEE/ACM International Conference on Computer-Aided Design*, 80-87, Nov. 2004.

[48] B.C. Moore. Principal component analysis in linear systems: Controllability, observability and model reduction. *IEEE Transactions on Automatic Control* 35(1), 17-32, Feb. 1981.

[49] K. Glover. All optimal hankel-norm approximations of linear multivariable systems and their l-error bounds. *International Journal of Control*, 39(6), 1115-1193, June 1984.

[50] C.L. Ratzlaff and L.T. Pillage. RICE: Rapid interconnect circuit evaluation using AWE. *IEEE Transactions on Computer-Aided Design of Integrated Circuits and Systems* 13(6), June 1994.

[51] R.W. Freund and P. Feldmann. Reduced-order modeling of large linear subcircuits via a block lanczos algorithm. in *Proceedings of IEEE/ACM Design Automation Conference*, 1995.

[52] E.J. Grimme. Krylov projection methods for model reduction. Ph.D. dissertation, Univ. Illinois, Urbana-Champaign, 1997.

[53] R. Freund and P. Feldmann. Reduced-order modeling of large passive linear circuits by means of the SyPVL algorithm. in *Proceedings of IEEE/ACM International Conference on Computer-Aided Design*, 1996.

[54] R.W. Freund and P. Feldmann. The SyMPVL algorithm and its applications to interconnect simulation. in *Proceedings of IEEE/ACM International Conference on Computer-Aided Design*, 1997.

[55] Z. Bai, P. Feldmann, and R. Freund. Stable and passive reduced-order models based on partial padé approximation via the Lanczos process. Bell Laboratories, Lucent Technologies, Numerical Analysis Manuscript 97/3-10, November 1997.

[56] B.N. Sheehan. ENOR: Model order reduction of RLC circuits using nodal equations for efficient factorization. in *Proceedings of IEEE/ACM Design Automation Conference*, 17-21.

[57] H. Zheng and L. Pileggi. Robust and passive model order reduction for circuits containing susceptance elements. in *Proceedings of IEEE/ACM International Conference on Computer-Aided Design*, Nov. 2002.

[58] B. Liu, X. Zeng, and Y.F. Su. Block SAPOR: Block second-order arnoldi method for passive order reduction of multi-input multi-output RCS interconnect circuits. in *Proceedings of IEEE/ACM Asia and South Pacific Design Automation Conference*, 244-249, 2005.

[59] Z. Bai and Y. Su. SOAR: A second-order arnoldi method for the solution of the quadratic eigenvalue problem. *SIAM Journal on Matrix Analysis and Applications* 26(3), 640-659, Mar. 2005.

[60] E. Chiprout and M.S. Nakhla. Analysis of interconnect networks using complex frequency hopping (CFH). *IEEE Transactions on Computer-Aided Design of Integrated Circuits and Systems* 14(2), Feb. 1995.

[61] J. Wang and T.V. Nguyen. Extended Krylov subspace method for reduced order analysis of linear circuits with multiple sources. in *Proceedings of IEEE/ACM Design Automation Conference.* Los Angeles, 247-252, June 2000.

[62] Y.-M. Lee, Y. Cao, T.-H. Chen, W.J.M., and C.-P. Chen. HiPRIME: Hierarchical and passivity reserved interconnect macromodeling engine for RLKC power delivery. *IEEE Transactions on Computer-Aided Design of Integrated Circuits and Systems* 24(6), 797-806, June 2005.

[63] P. Feldmann. Model order reduction techniques for linear systems with large numbers of terminals. in *Proceedings of IEEE/ACM Design, Automation and Test in Europe*, 2004.

[64] P. Feldmann and F. Liu. Sparse and efficient reduced order modeling of linear subcircuits with large number of terminals. in *Proceedings of IEEE/ACM International Conference on Computer-Aided Design*, Nov. 2004, 88-92.

[65] P. Rabiei and M. Pedram. Model order reduction of large circuits using balanced truncation. in *Proceedings of IEEE/ACM Asia and South Pacific Design Automation Conference*, 1998.

[66] M. Kamon, F. Wang, and J. White. Generating nearly optimally compact models from Krylov-subspace based reduced-order models.

*IEEE Transactions on Circuits and Systems II* 47(4), 239-248, Feb. 2000.

[67] J.R. Phillips, L. Daniel, and L.M. Silveira. Guaranteed passive balancing transformations for model order reduction. *IEEE Transactions on Computer-Aided Design of Integrated Circuits and Systems* 22(8), 1027-1041, Aug. 2003.

[68] J.-R. Li, F. Wang, and J. White. An efficient Lyapunov equation-based approach for generating reduced-order models of interconnect. in *Proceedings of IEEE/ACM Design Automation Conference*, 1999.

[69] P. Heydari and M. Pedram. Model reduction of variable-geometry interconnects using variational spectrally-weighted balanced truncation. in *Proceedings of IEEE/ACM International Conference on Computer-Aided Design*, 2001.

[70] X. Zeng, D. Zhou, and W. Cai. An efficient dc-gain matched balanced truncation realization for VLSI interconnect circuit order reduction. *Microelectronic Engineering* 60(1-2), 2-15, Jan. 2002.

[71] I. Jaimoukha and E.M. Kasenally. Oblique projection methods for large scale model reduction. *SIAM Journal on Matrix Analysis and Applications* 16, 602-627, 1995.

[72] N. Wong and V. Balakrishnan. Multi-shift quadratic alternating direction implicit iteration for high-speed positive-real balanced truncation. in *Proceedings of IEEE/ACM Design Automation Conference*, 2006.

[73] J.M. Wang, C.-C. Chu, Q. Yu, and E.S. Kuh. On projection-based algorithms for model-order reduction of interconnects. *IEEE Transactions on Circuits and Systems-I* 49(11), 1563-1585, Nov. 2002.

[74] J.R. Phillips. Variational interconnect analysis via PMTBR. in *Proceedings of IEEE/ACM International Conference on Computer-Aided Design*, 872-879, 2004.

[75] R. Khazaka and M. Nakhla. Analysis of transmission line circuits using multidimensional model reduction techniques. *IEEE Transactions on Advanced Package* 25(2), 174-180, May 2002.

[76] H.H. Hopkins. On the diffraction theory of optical images. *Proceedings of Royal Society A* 217(1130), 408-432, 1953.

[77] A.K. Wong and A.R. Neureuther. Rigorous three-dimensional time-domain finite-difference electromagnetic simulation for photolithographic applications. *IEEE Transactions on Semiconductor Manufacturing*, 8(4), 419-431, Apr. 1995.

[78] A. Taflove and S. Hagness. *Computational Electromagnetics: The Finite-Difference Time Domain Method*, 2nd ed. Artech House, 2000.

[79] S. Burger, R. Köhle, L. Zschiedrich, and W. Gao. Benchmark of FEM, waveguide and FDTD algorithms for rigorous mask simulation. in *Proceedings of SPIE* 5992, 378-389, 2005.

[80] K.D. Lucas, H. Tanabe, and A.J. Strojwas. Efficient and rigorous three-dimensional model for optical lithography simulation. *Journal of the Optical Society of America* 13(11), 2187-2199, Oct. 1996.

[81] M.G. Moharam. Coupled-wave analysis of two-dimensional dielectric gratings. in *Proceedings of SPIE* 883, 8-11, 1988.

[82] P. Evanschitzky and A. Erdmann. Three dimensional EUV simulations - a new mask near field and imaging simulation system. in *Proceedings of SPIE* 5992, 59925B, 1-9, 2005.

[83] R. Smaali, M. Besacier, and P. Schiavone. Three-dimensional rigorous simulation of EUV defective masks using modal method by fourier expansion. in *Proceedings of SPIE* 6151, 615124, 1-10, 2006.

[84] E. Charrier and C. Mack. Yield modeling and enhancement for optical lithography. in *Proceedings of SPIE* 2440, 435-447, 1995.

[85] S. Postnikov, K. Lucas, K. Wimmer, V. Ivin, and A. Rogov. Monte Carlo method for highly efficient and accurate statistical lithography simulations. in *Proceedings of SPIE* 4691, 1118-1126, 2002.

[86] N. Cobb and A. Zakhor. Fast, low-complexity mask design. in *Proceedings of SPIE* 2440, 313-327, 1995.

[87] J. Stirniman and M. Rieger. Fast proximity correction with zone sampling. in *Proceedings of SPIE* 2197, 294-301, 1994.

[88] Y. Granik. Calibration of compact OPC models using SEM contours. in *Proceedings of SPIE* 5992, 59921V, 2005.

[89] K. Cao, S. Dobre, and J. Hu. Standard cell characterization considering lithography induced varations. in *Proceedings of IEEE/ACM Design Automation Conference*, 801-804, 2006.

[90] D.M. Pawlowski, L. Deng, and M.D.F. Wong. Fast and accurate OPC for standard cell layouts. in *Proceedings of IEEE/ACM Asia and South Pacific Design Automation Conference*, 2007.

# A New Reconstruction Algorithm for Cone-beam CT with Unilateral Off-centered RT Multi-scan*

Weiwei Qi, Ming Chen, Huitao Zhang, Peng Zhang†
*Computer Tomography Laboratory*
*Capital Normal University, Beijing 100037, China*

**Abstract**

In order to enlarge the horizontal field of view of cone-beam CT in industry, several scanning modes have been proposed, of which RT (rotation-translation) multi-scan mode is more practical in engineering. In this paper, we develop a BPF (backprojection-filtration)-based reconstruction algorithm for cone beam CT with unilateral off-centered RT scan mode. One feature of the algorithm is that the algorithm is able to reconstruct image directly without explicit projection data rebinning process. Therefore, it not only speeds up the image reconstruction, but also improves the spatial resolution of the reconstructed image. Another feature of the algorithm is that the projection data required for image reconstruction is approximately reduced by half. In addition, the algorithm is fit to be accelerated by graphic processing unit. The numerical experiment results verify the validity of the proposed algorithm.

# 1 Introduction

In the recent years, cone-beam computer tomography (CT) has been a topic of great interest in the areas like non-destructive testing (NDT) and medical imaging, because it can make use of the X-ray effectively

*This work was supported by the National Natural Science Foundation of China (Grant No.60472071 and No.60532080) and Beijing Natural Science Foundation (Grant No.4051002)

†Corresponding author. E-mail:pzhang@mail.cnu.edu.cn

and shorten the scanning time. However, one obstacle to the practical applications of the cone-beam CT is the small size of the panel detector. Therefore, to enlarge the field of view for cone-beam CT is a very important issue.

In order to enlarge the longitudinal field of view, the helical cone-beam scan was first proposed in 1991 and has been extensively studied since then. A number of approximate and exact algorithms for image reconstruction from standard and non-standard helical cone-beam projections have been developed [1]. The first significant breakthrough for image exact reconstruction from helical cone-beam projections was achieved by Katsevich [2], who showed an exact reconstruction algorithm with filtered backprojection (FBP) format which is highly efficient in computation. Based on Katsevich's work, Zou and Pan found a backprojection filtration (BPF) formula [3], which is another significant breakthrough. The main idea of BPF is to obtain the so-called differentiated backprojection (DBP) image by backprojecting a partial derivative in the projection space. Since the DBP image is related to the Hilbert image along some direction [4], the CT image can be recovered from the Hilbert image using the finite inverse formula of Hilbert transform.

On the other hand, several scanning configurations were proposed to enlarge the horizontal field of view. One obvious option is simply to translate the detector horizontally within its plane. However, the flux output from a conventional X-ray source is not isotropic, which will usually result in the difference of X-ray intensities detected by the units of panel detector, especially under X-ray accelerator. To reduce the difference, it is often required to translate both the detector and the source simultaneously, or to rotate both the detector and the source around the vertical axis of the source. But those are difficult in practice. One way instead in industrial cone-beam CT is to rotate and translate the scanned object under the fixed source and detector. The so-called rotation-translation-translation (RTT) scan and rotation-translation (RT) scan for cone-beam CT were introduced in [5–7]. The difference of them is that the CT turntable translation is required in two orthogonal directions for RTT while only in one direction for RT. So, RT scan is easier to be carried out in engineering. The existing methods to reconstruct CT image for cone-beam RTT and RT scan are all based on filtered-backprojection (FBP) algorithm [5–7]. As a common feature of those methods, a data rebinning process is inevitably introduced, which not only increases computation cost, but also degrades image spatial resolution. The reconstruction algorithm for cone-beam circular scan with a panel detector displaced on one side of middle X-ray was discussed in [8], which only can increase the horizontal field at most one time.

A reconstruction algorithm for fan-beam CT with unilateral off centered RT multi-scans was proposed recently in [9]. In the present paper, we develop the BPF-based reconstruction algorithm for cone-beam CT with unilateral off-centered RT scan. Our algorithm is able to reconstruct image directly without explicit projection data rebinning process. Therefore, it not only speeds up the image reconstruction, but also improves the spatial resolution of the reconstructed image. Another feature of our algorithm is that the projection data required for image reconstruction is approximately reduced by half. In addition, the algorithm is fit to be accelerated by GPU (graphic processing unit).

Our algorithm is based on BPF reconstruction algorithm. The key point of the present paper is how to establish the entire DBP formula for each slice of the reconstructed image vertical to rotation axis from multiple sets of multi-scan projections with CT turntable translated unilaterally. Without losing generality, we limit our derivation for the DBP formula to the cone-beam RT two-scan with CT turntable translating unilaterally, which can be extend to the general RT multi-scan without any difficulty. In fact, the cone-beam CT with RT two-scan mode often meets the requirement in practical NDT, since it is able to increases the horizontal field of view up to almost 4 times. Although our algorithm is approximated, it usually reaches a satisfactory CT image when the vertical angle of cone-beam is small (less than 6 degree).

## 2 RT two-scan mode with unilateral off-centered turntable and its virtual equivalence

We first introduce the RT two-scan mode for cone-beam CT. Such scan mode requires an X-ray source and a panel detector, as well as a turntable to be translated along the direction which is perpendicular to its own rotation axis and parallel to the panel detector. As shown in Figure 1, we establish a 3D coordinate system $(\xi, \eta, z)$ with origin $O$. We call $z = 0$ the mid-plane and $z \neq 0$ off-mid-plane. The X-ray source $S_0$ lies on the $\eta$ axis; the panel detector $ABCD$ with center $O_D$ is parallel to plane $\xi Oz$. $R_1 = |S_0O_D|$ represents the distance from X-ray source to panel detector, and $R_0 = |S_0O|$ represents the distance from X-ray source to $\xi$ axis. $l$ represents the width of the panel detector. The rotation axis of the turntable is parallel to $z$ axis. Two points $O_1$ and $O_2$ lie on $\xi$ axis.

RT two-scan mode with unilateral off-centered turntable is as follows. The scanned object is put on the turntable. As shown in Figure 1(a), the turntable is translated along $\xi$ axis to the position, so that the rotation axis of the turntable passes through point $O_1$. Then, when the turntable is rotating in full turn around its own rotation axis, the source at $S_0$ produces X-rays and the panel detector collects X-rays which penetrate

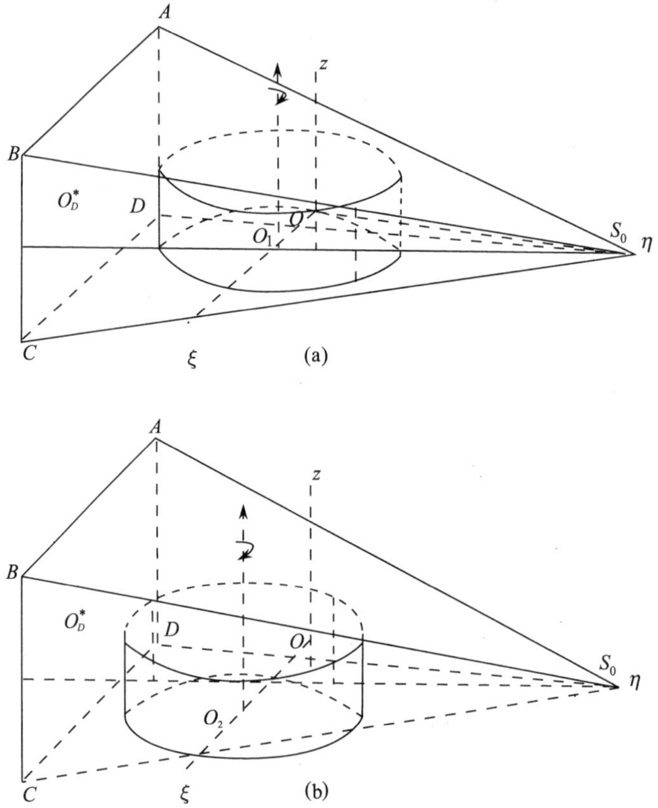


Figure 1 RT two-scan mode with turntable placed unilaterally on $O_1(a)$ and $O_2(b)$

through the scanned object. After finishing the first scan, as shown in Figure 1(b), the turntable is translated along $\xi$ axis to the position, so that the rotation axis of the turntable passes through point $O_2$. Then the second scan is carried out similar to the first scan. After finishing two scans, we obtain two sets of cone-beam projections corresponding to $O_1$ and $O_2$, respectively. In order to cover the scanned object by cone-beam RT two-scan for all views, we choose $O_1$ and $O_2$ so that $h_1 = |OO_1|$ and $h_2 = |OO_2|$ satisfy $0 < h_1 < R_0 l/(2R_1)$, $h_2 - h_1 < R_0 l/R_1$, and $h_1 < h_2$.

For convenience of the derivation in section 3, we introduce a virtual scan mode, which is equivalent to the RT two-scan mode for cone-beam with unilateral off-centered turntable. It is obvious that a turntable translation along the $\xi$ axis corresponding to a cone-beam is equivalent to the cone-beam translation along the opposite direction corresponding to the turntable. So, as shown in Figure 2, one can imagine that there are two cone-beams: one is formed by virtual source $S_1$ and virtual

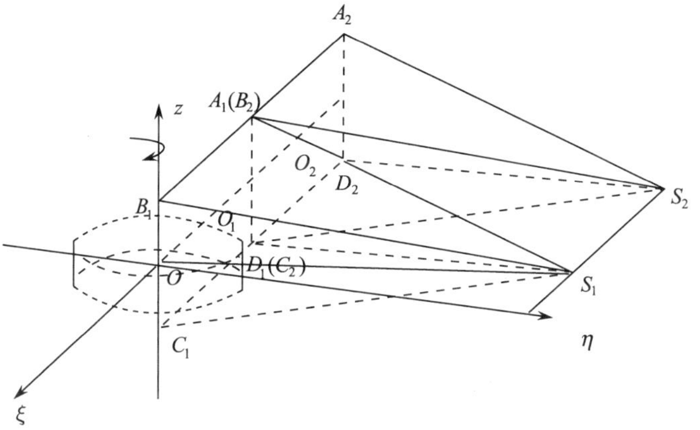


Figure 2 A virtual scan mode equivalent to the cone-beam RT two-scan mode with unilateral off-centered turntable

panel detector $A_1B_1C_1D_1$, and the other is formed by virtual source $S_2$ and virtual panel detector $A_2B_2C_2D_2$. $S_1$ and $S_2$ are located at $(-h_1, R_0, 0)$ and $(-h_2, R_0, 0)$, respectively. Two virtual panel detectors are in plane $\xi Oz$, where their centers $O_1$ and $O_2$ are located at $(-h_1, 0, 0)$ and $(-h_2, 0, 0)$, respectively. The rotation axis of the turntable is the $z$ axis. When the turntable is rotating in full turn around $z$ axis, two virtual detectors collect X-rays that are emitted respectively from their corresponding sources and penetrate the scanned object. It is obvious that RT two-scan mode for cone-beam with unilateral off-centered turntable in Figure 1 is equivalent to the virtual scan mode in Figure 2.

# 3 BPF algorithm for parallel-beam scan mode

As a foundation of next section we introduce the 2D BPF reconstruction algorithm for the parallel-beam scan mode in this section.

The Hilbert transform on 1D function (say $g(s)$) is defined by a convolution with the kernel function $1/(\pi s)$,

$$Hg(s) = \int_{-\infty}^{\infty} \frac{g(s')}{\pi(s - s')} ds'.$$

If there are $L$ and $U$ (satisfying $U > L$) so that $Hg(s)$ is known inside the interval $[L, U]$, and if there is some small positive $\varepsilon$ so that $g(s) = 0$ for $s \notin [L+\varepsilon, U-\varepsilon]$, then $g(s)$ for all $s \in [L+\varepsilon, U-\varepsilon]$ can be recovered by the following finite Hilbert inverse formula given by Mikhlin [10],

$$g(s) = \frac{-1}{\sqrt{(s-L)(U-s)}} \left( \int_L^U \sqrt{(s'-L)(U-s')} \frac{Hg(s')}{\pi(s-s')} ds' + C \right),$$

where $C$ is a constant which can be determined from knowledge of $g(s)$ at some $s \in [L, L+\varepsilon] \cup [U-\varepsilon, U]$.

Let $f(\mathbf{x})$ denote a 2D function, where $\mathbf{x} = (x, y)$. The Hilbert transform of $f(\mathbf{x})$ along lines at angle $\theta$ measured from the $y$ axis anti-clockwise is defined as

$$H_\theta f(\mathbf{x}) = \int_{-\infty}^{\infty} \frac{f((\mathbf{x}\cdot\boldsymbol{\theta})\boldsymbol{\theta} + s\boldsymbol{\theta}^\perp)}{\pi(\mathbf{x}\cdot\boldsymbol{\theta}^\perp - s)} ds, \tag{3.1}$$

where $\boldsymbol{\theta} = (\cos\theta, \sin\theta)$ and $\boldsymbol{\theta}^\perp = (-\sin\theta, \cos\theta)$.

We now describe differentiate backprojection (DBP) of parallel-beam. Let $p(\varphi, r)$ denote the parallel-beam projections of $f(\mathbf{x})$ defined by

$$p(\varphi, r) = \int_{-\infty}^{\infty} f(r\boldsymbol{\Phi} + s\boldsymbol{\Phi}^\perp) ds,$$

where $\boldsymbol{\Phi} = (\cos\varphi, \sin\varphi)$, $\boldsymbol{\Phi}^\perp = (-\sin\varphi, \cos\varphi)$, $\varphi$ is the angle of the normal of an X-ray measured from the $x$ axis anti-clockwise, and $r$ is the distance from the origin $O$ to the X-ray. The DBP formula $b_\theta(\mathbf{x}_0)$ of the parallel-beam is expressed in [4] as

$$b_\theta(\mathbf{x}_0) = \int_0^\pi \int_{-\infty}^{\infty} p(\varphi, r) sgn(\sin(\varphi - \theta)) \delta'(\mathbf{x}_0 \cdot \boldsymbol{\Phi} - r) dr d\varphi, \tag{3.2}$$

where $\delta'(r)$ is the derivative of the Dirac function.

It was proved in [4] that

$$b_\theta(\mathbf{x}_0) = -2\pi H_\theta f(\mathbf{x}_0). \tag{3.3}$$

For fixed $\theta$ and $t$, if there are $L_t$ and $U_t$ (satisfying $U_t > L_t$) so that $H_\theta f(\mathbf{x})$ is known for all $\mathbf{x}\cdot\boldsymbol{\theta}^\perp \in [L_t, U_t]$, and if there is some small positive $\varepsilon_t$ so that $f(\mathbf{x}) = 0$ for all $\mathbf{x}\cdot\boldsymbol{\theta}^\perp \notin [L_t + \varepsilon_t, U_t - \varepsilon_t]$, then $f(\mathbf{x}_0)$ for any $\mathbf{x}_0$ on line $\mathbf{x}\cdot\boldsymbol{\theta} = t$ satisfying $\mathbf{x}\cdot\boldsymbol{\theta}^\perp \in [L_t + \varepsilon_t, U_t - \varepsilon_t]$ can be recovered by the following finite Hilbert inverse formula given by Mikhlin [10],

$$f(\mathbf{x}_0) = \frac{-1}{\sqrt{(\mathbf{x}_0\cdot\boldsymbol{\theta}^\perp - L_t)(U_t - \mathbf{x}_0\cdot\boldsymbol{\theta}^\perp)}} \times \left( \int_{L_t}^{U_t} \sqrt{(s-L_t)(U_t-s)} \frac{H_\theta f((\mathbf{x}_0\cdot\boldsymbol{\theta})\boldsymbol{\theta} + s\boldsymbol{\theta}^\perp)}{\pi(\mathbf{x}_0\cdot\boldsymbol{\theta}^\perp - s)} ds + C_t \right), \tag{3.4}$$

where the constant $C_t$ can be calculated by use of the line integral (the projection data) $p(\theta, t)$, referring to the papers [4] and [11] in detail.

# 4 BPF-based algorithm for cone-beam RT two-scan mode with unilateral off-centered turntable

As mentioned in the section above, the key point to the BPF algorithm is to obtain the Hilbert image along some direction from cone-beam projection data. By the expression (3.3), the DBP image provides a link between cone-beam projections and the Hilbert transform of $f(\mathbf{x})$. In this section, for cone-beam RT two-scan mode with unilateral off-centered turntable, we derive 3D approximate DBP formula, similar to the idea of FDK algorithm in [12].

## 4.1 The derivation of the 3D DBP formula

First we describe some denotations for cone-beam RT two-scan mode with unilateral off-centered turntable. For simplicity, we will derive its DBP formula from its virtual equivalent scan mode in Figure 2. Apart from introducing the same symbols with Figure 2, we also introduce the following symbols in Figure 3. $(x, y, z)$ is a rotation coordinate system fixed on the turntable with the $z$ axis as its rotation axis, where $\beta$ is the rotation angle formed from the $\xi$ axis to the $x$ axis clockwise. Let $u_iO_iv_i \quad (i = 1, 2)$ denote the 2D coordinate system on the virtual panel detector, where the direction of $u_i$ in the rotation system $(x, y, z)$ is $\boldsymbol{\beta} = (\cos\beta, \sin\beta, 0)$, and the direction of $v_i$ is the same as $z$ axis. Let $p_i(\beta, u_i, v_i)$ denote the cone-beam projection data corresponding to $S_i$. Let $b_\theta(\mathbf{x}_0)$ denote the 3D DBP image of the projections of density function $f(\mathbf{x}_0)$, where $\mathbf{x}_0 = (x_0, y_0, z_0)$ in the rotation system $(x, y, z)$. Let $(u_{0,i}, v_{0,i})$ denote the projection position of $\mathbf{x}_0$ on system $u_iO_iv_i$ corresponding to $S_i (i = 1, 2)$.

For the pixel $\mathbf{x}_0$ lying on the mid-plane, the DBP formula $b_\theta(\mathbf{x}_0)$ is given from two sets of projection data $p_1(\beta, u_1, 0)$ and $p_2(\beta, u_2, 0)$ in [9].

$$b_\theta(\mathbf{x}_0) = b_{\theta,1}(\mathbf{x}_0) + b_{\theta,2}(\mathbf{x}_0), \tag{4.1}$$

where

$$\begin{aligned} b_{\theta,1}(\mathbf{x}_0) = &\int_0^{2\pi} \frac{R_0}{(R_0 - \mathbf{x}_0 \cdot \boldsymbol{\beta}^{\perp})^2} \\ &\times \frac{\mathrm{d}}{\mathrm{d}u_1}\left\{\left(k_\varepsilon\Big(\frac{R_0(u_1+h_1)}{\sqrt{R_0^2+u_1^2}}\Big) - k_\varepsilon\Big(\frac{R_0(u_1+h_1)}{\sqrt{R_0^2+u_1^2}} - T_1\Big)\right)\frac{R_0^2 - h_1u_1}{\sqrt{R_0^2+u_1^2}}\right. \\ &\left.\times\, p_1(\beta, u_1, 0)\operatorname{sgn}\left(\sin(\beta + \tan^{-1}\frac{u_1}{R_0} - \theta)\right)\right\}\Big|_{u_1=u_{0,1}}\, \mathrm{d}\beta, \end{aligned} \tag{4.2}$$

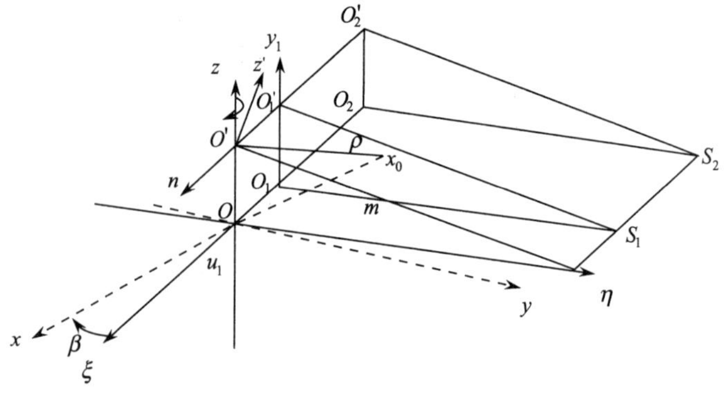


Figure 3 The geometric parameters of the derivation of the 3D DBP formula in the tilted plane

and

$$\begin{aligned} b_{\theta,2}(\mathbf{x}_0) = & \int_0^{2\pi} \frac{R_0}{(R_0 - \mathbf{x}_0 \cdot \boldsymbol{\beta}^{\perp})^2} \\ & \times \frac{\mathrm{d}}{\mathrm{d}u_2} \left\{ \left( k_\varepsilon(\frac{R_0(u_2+h_2)}{\sqrt{R_0^2+u_2^2}}) - T_1) \right) \frac{R_0^2 - h_2 u_2}{\sqrt{R_0^2+u_2^2}} p_2(\beta, u_2,0) \right. \\ & \left. \times \operatorname{sgn}\left( \sin(\beta + \tan^{-1}\frac{u_2}{R_0} - \theta) \right) \right\} \Big|_{u_2 = u_{0,2}} \mathrm{d}\beta, \end{aligned} \tag{4.3}$$

where

$$\boldsymbol{\beta}^{\perp} = (-\sin\beta, \cos\beta, 0), \quad u_{0,i} = \frac{R_0(\mathbf{x}_0 \cdot \boldsymbol{\beta} - h_i)}{R_0 - \mathbf{x}_0 \cdot \boldsymbol{\beta}^{\perp}},$$

$$T_1 = \frac{R_0(h_2 + h_1)}{\sqrt{4R_0^2 + (h_2 - h_1)^2}} \quad (i = 1, 2),$$

and $k_\varepsilon(r)$ is an infinitely differentiable function satisfying conditions: (i) $k_\varepsilon(r) = 1$, for $r \geqslant \varepsilon$; (ii) $k_\varepsilon(r)$ is monotone increasing for $-\varepsilon < r < \varepsilon$; (iii) $k_\varepsilon(r) = 0$, for $r < -\varepsilon$; $\varepsilon$ is a small positive number.

To reconstruct $f(\mathbf{x}_0)$ at $\mathbf{x}_0$ in off-mid-plane ($z_0 \neq 0$), we derive an approximate 3D DBP formula $b_\theta(\mathbf{x}_0)$ from the exact DBP formula (4.1), similar to the idea of the derivation of the FDK algorithm. For a fixed $\beta$, there exists a tilted plane passing through $S_1$, $S_2$ and $\mathbf{x}_0$. We need determine the incremental contribution $\delta b_\theta(\mathbf{x}_0) = \delta b_{\theta,1}(\mathbf{x}_0) + \delta b_{\theta,2}(\mathbf{x}_0)$ to $\mathbf{x}_0$ from the projection data for a small increment $\delta\beta$ (an actual rotation around $z$ axis). The steps of the derivation of $b_\theta(\mathbf{x}_0)$ are as follows: (i) rewrite the DBP formula in the tilted plane based on the formulas (4.2) and (4.3); (ii) find the relation between $\delta\beta$ and $\delta\beta'$ (an equivalent rotation about the normal to the tilted plane); (iii) find the coordinate denotation (under the system $(x, y, z)$) of $\mathbf{x}_0$ in the tilted plane, and calculate the source-to-detector distance $R_0'$ in the tilted plane; (iv) obtain the total $b_\theta(\mathbf{x}_0)$ by calculating the sum of the incremental contribution $\delta b_\theta(\mathbf{x}_0)$ from each rotation angle that passes through $\mathbf{x}_0$. Without loss of generality, we only derive the formula for $b_{\theta,1}(\mathbf{x}_0)$. The formula for $b_{\theta,2}(\mathbf{x}_0)$ can be similarly derived without any difficulty.

First, we establish a coordinate system $mO'n$ of the tilted plane passing through $S_1$, $S_2$ and $\mathbf{x}_0$, where the coordinate of the origin $O'$ is $(0, 0, v_1)$, the direction of $m$ axis and $n$ axis are the same as the vector $\mathbf{O_1{}'S_1}$ and $u_1$ axis, respectively. The normal to the tilted plane $mO'n$ is given by $\mathbf{z}' = \mathbf{m} \times \mathbf{n}$.

Now, we rewrite the DBP formula (4.2) for the pixel $\mathbf{x}_0$ as follows:

$$\delta b_{\theta,1}(\mathbf{x}_0) = \delta\beta' \frac{R_0'}{(R_0' - \boldsymbol{\rho}\cdot\mathbf{m})^2}$$
$$\times \frac{\mathrm{d}}{\mathrm{d}u_1}\left\{\left(k_\varepsilon\Big(\frac{R_0(u_1+h_1)}{\sqrt{R_0^2+u_1^2}}\Big) - k_\varepsilon\Big(\frac{R_0(u_1+h_1)}{\sqrt{R_0^2+u_1^2}} - T_1\Big)\right) p_1(\beta, u_1, v_1)\right.$$
$$\left.\times \frac{R_0'^2 - h_1 u_1}{\sqrt{R_0'^2 + u_1^2}} \operatorname{sgn}\left(\sin\Big(\beta + \tan^{-1}\frac{u_1}{R_0} - \theta\Big)\right)\right\}\Big|_{u_1=u_{0,1}, v_1=v_{0,1}}, \tag{4.4}$$

where $\boldsymbol{\rho}$ is a vector from $O'$ to $\mathbf{x}_0$ in the tilted plane $mO'n$. If $z_0 = 0$, $\boldsymbol{\rho}$ is identical with $\mathbf{x}_0$ in formula (4.2).

From the geometric relationship in Figure 3, we can obtain the relationship between a small range of rotation $\delta\beta$ about $z$ axis and a rotation of $\delta\beta'$ about $z'$ axis

$$\delta\beta' = \delta\beta \frac{R_0}{\sqrt{R_0^2 + v_1^2}}. \tag{4.5}$$

Any pixel $\mathbf{x}_0$ to be reconstructed in the tilted plane $mO'n$ can be written as

$$\mathbf{x}_0 = \boldsymbol{\rho} + v_1 \mathbf{z}. \tag{4.6}$$

Since $\boldsymbol{\rho}$ is a vector from $O'$ to $\mathbf{x}_0$, $\boldsymbol{\rho}$ lies in the tilted plane $mO'n$. So

$$\boldsymbol{\rho}\cdot\mathbf{z}' = 0. \tag{4.7}$$

We easily obtain the source-to-detector distance $R_0'$ in the tilted plane $mO'n$,

$$R_0' = |\mathbf{S}_1\mathrm{O}_1| = \sqrt{R_0^2 + v_1^2}. \tag{4.8}$$

Combining (4.4)–(4.8), by calculating the sum of the incremental contribution from all planes (one for a rotation angle $\beta$) that pass through

$S_1$, $S_2$ and $\mathbf{x}_0$, we can obtain $b_{\theta,1}(\mathbf{x}_0)$,

$$\begin{aligned} b_{\theta,1}(\mathbf{x}_0) = & \int_0^{2\pi} \frac{R_0^2}{(R_0 - \mathbf{x}_0 \cdot \boldsymbol{\beta}^{\perp})^2} \\ & \times \frac{\mathrm{d}}{du_1} \left\{ \left( k_\varepsilon(\frac{R_0(u_1+h_1)}{\sqrt{R_0^2+u_1^2}}) - k_\varepsilon(\frac{R_0(u_1+h_1)}{\sqrt{R_0^2+u_1^2}} - T_1) \right) \right. \\ & \times \frac{(R_0^2+v_1^2)R_0 - R_0 h_1 u_1}{(R_0^2+v_1^2)\sqrt{R_0^2+v_1^2+u_1^2}} p_1(\beta, u_1, v_1) \\ & \left. \times \operatorname{sgn}\left( \sin(\beta + \tan^{-1}\frac{u_1}{R_0} - \theta) \right) \right\} |_{u_1=u_{0,1}, v_1=v_{0,1}} \, d\beta. \end{aligned} \tag{4.9}$$

Similarly,

$$\begin{aligned} b_{\theta,2}(\mathbf{x}_0) = & \int_0^{2\pi} \frac{R_0^2}{(R_0 - \mathbf{x}_0 \cdot \boldsymbol{\beta}^{\perp})^2} \\ & \times \frac{\mathrm{d}}{du_2} \left\{ \left( k_\varepsilon(\frac{R_0(u_2+h_2)}{\sqrt{R_0^2+u_2^2}} - T_1) \right) \frac{(R_0^2+v_2^2)R_0 - R_0 h_2 u_2}{(R_0^2+v_2^2)\sqrt{R_0^2+v_2^2+u_2^2}} \right. \\ & \left. \times p_2(\beta, u_2, v_2) \operatorname{sgn}\left( \sin(\beta + \tan^{-1}\frac{u_2}{R_0} - \theta) \right) \right\} |_{u_2=u_{0,2}, v_2=v_{0,2}} \, d\beta, \end{aligned} \tag{4.10}$$

where $u_{0,i}$ and $T_1$ are identical with the above, and $v_{0,i} = \dfrac{R_0 \cdot z_0}{R_0 - \mathbf{x}_0 \cdot \boldsymbol{\beta}^{\perp}}$, $i = 1, 2$. When $z_0 = 0$, the formulas (4.9) and (4.10) are identical with the formulas (4.2) and (4.3) in the mid-plane, respectively.

## 4.2 Hilbert inverse formula

Now, we explain how to obtain a 3D CT image from an entire DBP image calculated by the formulas (4.1), (4.9) and (4.10).

First, we can obtain an approximated Hilbert image for the slice to be reconstructed from a DBP image by the following modified expression of the formula (3.3),

$$H_\theta f(\mathbf{x})\,|_{z=z_0} = -\frac{b_\theta(\mathbf{x})\,|_{z=z_0}}{2\pi}, \tag{4.11}$$

where $\mathbf{x} = (x, y, z)$.

Then, by use of the virtual trajectory and virtual PI-line segments [11] corresponding to the parameter $z_0$, we can modify the finite Hilbert inverse formula(3.4), and obtain the following formula to recover$f(\mathbf{x})$ to the slice $z = z_0$,

$$f(\mathbf{x})\,|_{z=z_0} = \frac{-1}{\sqrt{(\mathbf{x}\cdot\boldsymbol{\theta}'^{\perp} - L_{t,z})(U_{t,z} - \mathbf{x}\cdot\boldsymbol{\theta}'^{\perp})}} \left( \int_{L_{t,z}}^{U_{t,z}} \sqrt{(s - L_{t,z})(U_{t,z} - s)} \times \frac{H_\theta f((\mathbf{x}\cdot\boldsymbol{\theta}')\boldsymbol{\theta}' + s\boldsymbol{\theta}'^{\perp} + z(0,0,1))}{\pi(\mathbf{x}\cdot\boldsymbol{\theta}'^{\perp} - s)} ds + C_{t,z} \right) |_{z=z_0}, \tag{4.12}$$

where $\boldsymbol{\theta}' = (\cos\theta, \sin\theta, 0)$, $\boldsymbol{\theta}'^{\perp} = (-\sin\theta, \cos\theta, 0)$. The constants $L_{t,z}$, $U_{t,z}$ and $C_{t,z}$ are not only related to $\mathbf{x}\cdot\boldsymbol{\theta}' = t$, but also related to $z$. We know that $C_{t,z}$ can be calculated by use of the line integral (the projection data) of the object function $f(\mathbf{x})$ along the PI-line [11]. For a PI-line in the mid-plane($z = 0$), we can obtain $C_{t,z}$ from the line integral along the PI-line. However, for a virtual PI-line in an off-mid-plane ($z \neq 0$), the line integral along the virtual PI-line does not exist. So we cannot obtain an exact $C_{t,z}$ for an off-mid-plane. But we can use the method in [11] to approximate this line integral, and obtain an approximate $C_{t,z}$.

### 4.3 Implementation of the proposed algorithm

In the numerical implementation, we adopt the projection-driven method to calculate the DBP image $b_{\theta,i}(\mathbf{x}_0)$ from the projection data $p_i(\beta, u_i, v_i)$. This method especially fits to the implementation of Graphics Processing Unit, which can speed up the reconstruction process dramatically.

The numerical implementation of the proposed algorithm consists of the following four steps:

**Step 1:** To weight each projection data $p_i(\beta, u_i, v_i)$, and then differentiate the weighted projection in $u_i (i = 1, 2)$;

**Step 2:** To backproject the derivative of the weighted projection data to obtain $b_{\theta,i}(\mathbf{x}_0)$ of each slice of the scanned object;

**Step 3:** To add two partial DBP images $b_{\theta,1}(\mathbf{x}_0)$ and $b_{\theta,2}(\mathbf{x}_0)$ together, and obtain the entire DBP image $b_\theta(\mathbf{x}_0)$ of each slice;

**Step 4:** To obtain the Hilbert image from the DBP image for each slice by the expression (4.11), and then reconstruct the CT image for each slice by the Hilbert inverse transform (4.12).

## 5 Numerical simulations and conclusion

In this section, we perform numerical simulations to verify our algorithm for cone-beam RT two-scan mode.

Table 1 3D Shepp-Logan Phantom Description

| $N$ | $x_c$ | $y_c$ | $z_c$ | $a$ | $b$ | $c$ | $\alpha$ | Density value |
|---|---|---|---|---|---|---|---|---|
| 1 | 0 | 0 | 0 | 97.9 | 73.4 | 26.8 | $0^o$ | 1000.0 |
| 2 | 0 | 0 | 0 | 93.0 | 70.5 | 26.2 | $0^o$ | 300.0 |
| 3 | −23.4 | 0 | 0 | 17.0 | 43.6 | 20.8 | $108^o$ | 200.0 |
| 4 | 23.4 | 0 | 0 | 11.7 | 33.0 | 20.8 | $72^o$ | 200.0 |
| 5 | 0 | 37.2 | 0 | 26.6 | 22.4 | 14.9 | $0^o$ | 200.0 |
| 6 | 0 | 10.6 | 0 | 4.9 | 4.9 | 1.4 | $0^o$ | 200.0 |
| 7 | −6.4 | −69.2 | 0 | 2.5 | 4.9 | 0.6 | $0^o$ | 200.0 |
| 8 | 6.4 | −69.2 | 0 | 2.5 | 4.9 | 0.6 | $90^o$ | 200.0 |
| 9 | 8.5 | 0 | 18.6 | 5.9 | 4.3 | 6.0 | $0^o$ | 200.0 |
| 10 | 0 | 0 | 18.6 | 5.9 | 5.9 | 6.0 | $0^o$ | 200.0 |

We use the 3D Shepp-Logan phantom which consists of ten ellipsoids, as shown in Table 1, where $(x_c, y_c, z_c)$ represents the coordinate of the centers of the ellipsoids, $a$, $b$ and $c$ represent the length of three half axis of the ellipsoids corresponding to $x$, $y$ and $z$ axis, and $\alpha$ represents the rotation angle of the ellipsoids about the $z$ axis anti-clockwise. In numerical experiments, scanning geometric parameters are as follows: the distance between the source and the origin is $R_0$ =1426.0mm, the distance from X-ray source to panel detector is $R_1$ =1800.0mm, the panel detector consists of 257×257 detector cells, and the size of each cell is 0.3mm×0.3mm.

From the scanning parameters, we calculate that the radius of the field of view of the cone-beam formed by the X-ray source and the panel detector is 60.82mm along $x$ axis. However, from Table 1, the long half axis (along $x$ axis) of the largest ellipsoid in the phantom is 97.9mm. So the phantom cannot be completely covered within the field of view. Using the scan mode in Figure 1, the translation distance $h_1 = 30.42$ mm and $h_2 = 91.26$ mm. Two sets of projections of the phantom are acquired, of which each has 720 projections over 360 degree. The DR data under 100th projection angle are shown in Figure 4. In CT image reconstruction, we choose $\theta = 0$ as Hilbert filtering direction and $\varepsilon = 0.24$. We reconstruct the 200 slices of CT images of the scanned object with pixels 701×701 along $z$ axis, as shown in Figure 5 and Figure 6. From results of reconstruction, the validity of our algorithm is verified.

According to the Tuy data sufficiency condition, the cone-beam projection data under the circular cone-beam RT scan is not sufficient to exactly reconstruct 3D CT image. So our algorithm is approximated in off-mid-plane. But it usually reaches a satisfactory CT image when the vertical angle of cone-beam is small (usually less than 6 degree). For exactly reconstructing image of an object with large field of view in both longitudinal and horizontal directions, reconstruction algorithms for cone-beam CT with helical RT multi-scan mode need to be studied further.

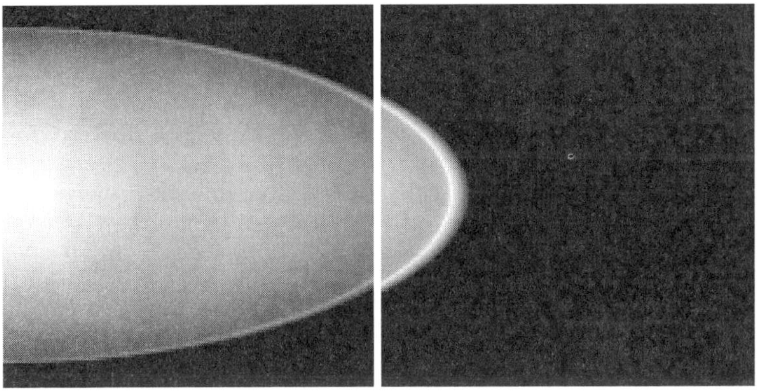

Figure 4 The DR image under 100th projection angle generated from cone-beam RT two-scan mode

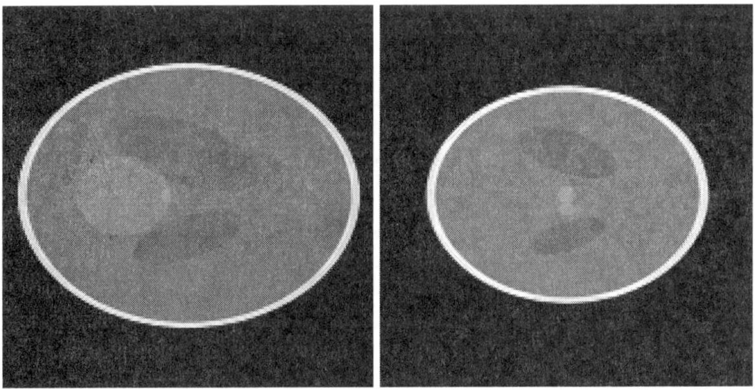

Figure 5 2D slices in the planes of $z = 0, -15.0$ mm

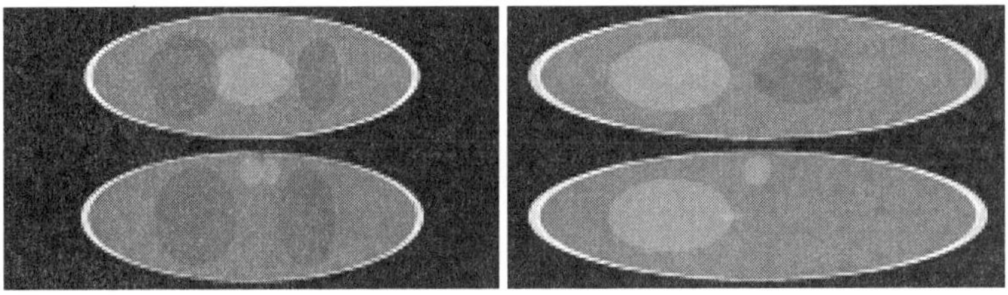

Figure 6 2D slices in 3D images reconstructed by use of the proposed algorithm. Two columns represent the 2D-slices in the planes of $x = -20.4$ mm, 2.1 mm (left), $y = -6.0$ mm, 0.9 mm (right) respectively

# References

[1] G. Wang, Y. Ye, H.Yu. Approximate and exact cone-beam reconstruction with standard and non-standard spiral scanning. *Phys. Med. Biol.* **52**: R1-R13, 2007.

[2] A.I. Katsevichy. Theoretically exact filtered back-projection-type inversion algorithm for spiral CT. *SIAM J. Appl. Math.* **62**: 2012-2026, 2002.

[3] Y. Zou, X. Pan. Exact image reconstruction on PI-lines from minimum data in helical cone-beam CT. *Phys. Med. Biol.* **49**: 941-959, 2004.

[4] F. Noo, R. Clackdoyle, J.D. Pack. A two-step Hilbert transform method for 2D image reconstruction. *Phys. Med. Biol.* **49**: 3903-3923, 2004.

[5] E A. Use of multiple CT scans to accommodate large objects and stretch dynamic range of detectability. *Nuclear Instruments and Methods in Physics Research* B(99): 761-764, 1995.

[6] F. Zhao, H.N. Lu, C.L. Sun. New scan mode for 2D-CT and its reconstruction algorithm. *Optical Technique* 32(2): 812-817, 2006.

[7] F. Jian, H.N. Lu, L. Gong. Research on cone-beam ray three dimensional large field of view industrial CT imaging method. *Optical Technique* 32(2): 209-212, 2006.

[8] Li Liang. Z.Q. Chen, et al. A new cone-beam X-ray CT system with a reduced size panel detector. *High Energy Physics and Nuclear Physics* 30(8): 812-817, 2006.

[9] Chen Ming, H.T. Zhang et al. *Reconstruction algorithm for unilateral off-centered RT multi-scans*, to submit.

[10] S.G. Mikhlin. Integral equations and their applications to certain problems in mechanics. *Mathematical Physics and Technology*, Pergamon, New York, 1957.

[11] Yu Lifeng, Pan Xiaochuan, et al. Region of interest reconstruction from truncated data in circular cone-beam CT. *IEEE Transactions on Medical Imaging* 25(7): 869-881, 2006.

[12] L.A. Feldkamp, L.C. Dabis, J.W. Kress. Practical cone-beam algorithm. *J. Opt. Soc. Amer.*, Vol. Al: 612-619, 1984.

# Bioluminescence Tomography Reconstruction by Radial Basis Function Collocation Method

Tie Zhou, Jiantao Cheng, Ming Jiang
*LMAM, School of Mathematical Sciences, Peking University, China*

**Abstract**

As a new molecular imaging technique, Bioluminescence tomography (BLT) has been developed rapidly and attracted more and more attention in recent years. Traditionally, Bioluminescence image (BLI) is a highly sensitive tool for monitoring molecular events in intact living animals, however it only works in two-dimension model and is incapable of obtaining depth information of light source distribution, generated by luciferase induced by reporter genes. In contrast to this active imaging mode, BLT reconstructs an internal bioluminescent source distribution and produces localized and quantitative analysis from data measured on the external surface of small animals. Mathematically, BLT can be formed as a highly ill-posed inverse problem which is to recover an internal bioluminescent source distribution subject to Cauchy data for the diffusion equation. In the paper, we apply the RBF-based collocation method to solve the inverse BLT problem and propose two reconstruction approaches. The first approach is that we transform the BLT problem into a matrix equation problem with nonnegative constrain, and then solve it with the well known algebraic EM method. The second one is to transform the BLT problem to an operator equation and solve it with the variational EM method. Initial numerical experiments are made to verify the utility of our approaches in tissue-like media.

## 1 Introduction

Bioluminescence tomography (BLT) is a technique that can reconstruct the distributions of the bioluminescent sources when the optical properties of tissue and the data gathered from the animal surface are known. Since its first introduction in year 2003 [18], BLT has attracted much attention from many researchers and is a rapidly developing area for molecular imaging [3, 8–10, 16, 17, 21]. Traditionally, Bioluminescence image (BLI) is a highly sensitive tool for monitoring molecular events in intact living animals, however it only works in two-dimension model and is incapable of obtaining depth information of light source distribution, generated by luciferase induced by reporter genes. The introduction of BLT is a substantial event in molecular imaging studies. With BLT, quantitative and localized analysis on a bioluminescent source distribution becomes feasible inside small animals, which reveals deep molecular and cellular signatures. BLT could be not only applied to study almost all diseases in every small animal model but also has great potentials in various other biomedical applications as well, its application including gene therapy, regenerative medicine, developmental therapeutics, treatment of residual minimal disease, and the concept of the cancer stem cell [4, 14].

Mathematically, BLT is based on a Boltzmann-type nonlinear transport equation for the photon intensity [1, 8]. This equation is often impossible to solve analytically, and is usually simplified by a diffusion equation. Therefore, BLT can be formed as an inverse problem to recover an internal bioluminescent source distribution subject to Cauchy data for the diffusion equation. Unfortunately, the inverse problem from BLT is a highly ill-posed problem and has not unique solution. To obtain a physically favorable unique solution, adequate prior knowledge must be utilized. In some special restrictions, it is satisfied that the uniqueness has been established [19]. The constrained iterative approach provides a mechanism for incorporating prior knowledge as constraints and has been widely used in practice. In the following algorithm, we use the non-negative and source support as a prior knowledge. According to the formal study [8–10], the BLT problem based on the diffusion model can be stated as follows.

For the complete measurement on the whole boundary $\Gamma$, Given the incoming light $f$ and outgoing radiance $g$ on $\Gamma$, find a source $q$ with the corresponding diffusion approximation $u$ such that

$$\textbf{BLT}(P)\begin{cases} -\nabla\cdot(D\nabla u)+\mu_a u = q, & \text{in } \Omega,\\ u+2D\dfrac{\partial u}{\partial \nu} = f, & \text{on } \Gamma,\\ D\dfrac{\partial u}{\partial \nu} = -g, & \text{on } \Gamma_P. \end{cases} \tag{1.1}$$

In a typical BLT configuration $f = 0$, for there is no incoming light. The third equation in BLT problem is called the measurement equation. We can get the optical parameters $D$ and $\mu_a$ from a database of the optical properties, or use the DOT technique [18].

In the paper, we applied the radial basis function (RBF) based collocation method to the solve BLT problem. Compared with finite element methods, this is a new method which has been developed in these years [2, 11–13]. The method works with points scattered throughout the domain of interest, it forms the RBF centered at each point and the RBF interpolation is a linear combination of these RBFs. Applying it to a partial differential equation (PDE) by Galerkin method or collocation method, we obtain an algebraic equation system and solve it to get the solution of the PDE. Since the solution uniqueness for BLT problem holds for the case of linear combination of RBFs, we will develop the reconstruction algorithm based on the RBF collocation approximation to the diffusion equation in this paper. This makes the implementation quite simple, even for the complex problem domain. The paper is organized as follows. 1. We solve the forward BLT problem by RBF collocation method. 2. We propose two kinds of methods which are based on matrix equation and operator equation respectively and give some relevant implement issues. 3. We report the numerical experiments: 3D simulation results. 4. We conclude the paper and discuss some future works.

# 2 RBF collocation method

A radial basis function (RBF) depends on the point-wise distance $r = \|x - x_I\|$ to a center point $x_I$ and is of the form $\phi_I(r, \epsilon)$, where $\epsilon$ is a shape parameter. We will use the following Multiquadrics (MQ) RBFs

$$\phi_I = \sqrt{1 + \epsilon^2 r^2}. \tag{2.1}$$

For a given set of nodes $\{x_j\}$, the RBF interpolation is a linear combination of RBFs centered at the scattered nodes $x_j$,

$$u^h(x, \epsilon) = \sum_{j=1}^{N} \lambda_j \phi_j(r, \epsilon). \tag{2.2}$$

In RBF based collocation PDE method, we get the coefficients $\lambda_j$ by letting the PDE and the boundary condition satisfied by collocation. Here, we will solve the diffusion equation

$$-\nabla \cdot (D\nabla u) + \mu_a u = q, \quad x \in \Omega, \tag{2.3}$$

$$u + 2D\frac{\partial u}{\partial \nu} = f, \quad x \in \Gamma. \tag{2.4}$$

For simplicity, we define two differential operators in $\Omega$ and on $\partial\Omega$ respectively:

$$L[u] = -\nabla \cdot (D\nabla u) + \mu_a u, \quad K[u] = u + 2D\frac{\partial u}{\partial \nu}. \tag{2.5}$$

We need to solve the following elliptical boundary value problem

$$L[u] = q, \quad x \in \Omega, \tag{2.6}$$
$$K[u] = f, \quad x \in \Gamma. \tag{2.7}$$

For collocation, we put some nodes on the boundary: $x_j, j = 1, 2, \ldots, N_B$ and in the interior: $x_j, j = N_B + 1, N_B + 2, \ldots, N_B + N_I$. After collocation at these nodes, we have

$$L[u^h](x_j) = q(x_j), \quad j = N_B + 1, N_B + 2, \ldots, N = N_B + N_I, \tag{2.8}$$
$$K[u](x_j) = f(x_j), \quad j = 1, 2, \ldots, N_B. \tag{2.9}$$

Introduce vectors

$$\boldsymbol{\lambda} = \begin{bmatrix} \lambda_1 \\ \lambda_2 \\ \vdots \\ \lambda_N \end{bmatrix}, \quad \boldsymbol{q} = \begin{bmatrix} g(x_1) \\ g(x_2) \\ \vdots \\ g(x_{N_B}) \end{bmatrix}, \quad \boldsymbol{f} = \begin{bmatrix} f(x_{N_B+1}) \\ f(x_{N_B+2}) \\ \vdots \\ f(x_{N_B+N_I}) \end{bmatrix} \tag{2.10}$$

(2.9) and (2.8) can be written as a linear system

$$\begin{bmatrix} K[\phi_j] \\ L[\phi_j] \end{bmatrix} \boldsymbol{\lambda} = \begin{bmatrix} \boldsymbol{q} \\ \boldsymbol{f} \end{bmatrix}. \tag{2.11}$$

Solving this linear algebraic system, we obtain coefficients $\lambda_j$, and get the approximation function $u^h(x)$ as

$$u^h(x) = C \begin{bmatrix} \boldsymbol{q} \\ \boldsymbol{f} \end{bmatrix} = AB^{-1} \begin{bmatrix} \boldsymbol{q} \\ \boldsymbol{f} \end{bmatrix}, \quad \text{with } A = [\phi_i(x_j)], B = \begin{bmatrix} K[\phi_j] \\ L[\phi_j] \end{bmatrix}. \tag{2.12}$$

We note that in the BLT regime the incoming light flux $\boldsymbol{f} = \mathbf{0}$, and the body of the small animal is segmented into severe anatomy parts, such as lung, muscle and heart. In other words, the coefficient $D(x)$ and $\mu_a(x)$ are piecewise constant functions. Therefore we use the sub-domain method in the collocation process. In each sub-domain, we use boundary value problem (2.3) and (2.4) to represent the values of $u$ in each sub-domain. On the inner boundary, the following conditions are set

$$u_2^h(x) = u_1^h(x), \quad D_2 \frac{\partial u_2^h}{\partial n}(x) = D_1 \frac{\partial u_1^h}{\partial n}(x). \tag{2.13}$$

# 3 BLT reconstruction method

## 3.1 The algebraic EM method

To solve the inverse problem, we discretize the boundary value problem into a linear system with nonnegative constrain, then solve it with the well known algebraic EM method. From the previous results, we know that the BLT inverse problem has no unique solution and is highly ill-posed. To do the image reconstruction, we need some prior knowledge on the light sources, for example, the source permissible region is known beforehand. Therefore, the source vector $\boldsymbol{q}$ could be divided into two parts: $\boldsymbol{q}_p$ in the permissible region and $\boldsymbol{q}_f$ in the forbidden region which means $\boldsymbol{q}_f = \boldsymbol{0}$. Changing the row order of matrix $C$, we can get the following formula

$$u^h(x) = [C_1, C_2] \begin{bmatrix} \boldsymbol{q}_p \\ \boldsymbol{0} \end{bmatrix}. \tag{3.1}$$

That is

$$u^h(x) = C_1 \boldsymbol{q}_p. \tag{3.2}$$

To solve the BLT problem, we need to reconstruct the source $q$ from the measurement data $g$. Here we know $g = D\frac{\partial u}{\partial \nu}$ for $f = 0$, i.e., $u = 2g$ on the boundary and need to reconstruct $q$. This means the BLT problem can be transformed into the following ill-conditional linear system

$$M\boldsymbol{q}_p = \boldsymbol{b}, \tag{3.3}$$

where $\boldsymbol{b} = 2[g(x_j)]^T$, and $M = C_{11}$ with $C_1 = \begin{bmatrix} C_{11} \\ C_{12} \end{bmatrix}$ is corresponding to (3.1). Suppose $M$ is an $m \times n$ matrix, $m$ and $n$ are the numbers of collocation nodes on the boundary and in the permission region respectively. By the maximum principle of elliptical partial differential equations [5], it follows that $q > 0$ if $q > 0$ . Hence the matrix $M$ is positive pointwise and we can employ a nonlinear iterative method which is known as algebraic EM method [15] to solve it, the formula is as follows

$$\boldsymbol{q}_p^{(k+1)} = \frac{1}{ME_{1n}} \boldsymbol{q}_p^{(k)} M^T \left[ \frac{\boldsymbol{b}}{M\boldsymbol{q}_p^{(k)}} \right], \tag{3.4}$$

where $E_{1n} = [1, 1, \ldots, 1]^T$ and all arithmetic operation between vectors are component-wise.

## 3.2 The variational EM method

Based on the previous works [9, 10], we can also transform the BLT problem into a linear operator equation and solve it with the variational

EM method. Let $\gamma_0$ and $\gamma_1$ be the boundary value maps

$$\gamma_0[u] = u|_\Gamma, \quad \gamma_1[u] = D\frac{\partial u}{\nu}\bigg|_\Gamma, \tag{3.5}$$

and $L$ be the differential operator in (2.5).

Given $f \in H^{\frac{1}{2}}(\Gamma_P)$, let $w_1 \in H^1(\Omega)$ be the solution of the following boundary value problem [19]

$$L[w_1] = 0, \quad \text{in } \Omega, \tag{3.6}$$
$$\gamma_0[w_1] = f, \quad \text{on } \Gamma. \tag{3.7}$$

We define a linear operator $N_\Gamma$ from $H^{\frac{1}{2}}(\Gamma)$ to $H^{-\frac{1}{2}}(\Gamma)$ by

$$N_\Gamma[f] = \gamma_1[w_1]|_\Gamma, \tag{3.8}$$

where $N_\Gamma$ is an extension of the well-known Dirichlet-to-Neumann map [7]. On the other hand, for $q$, we consider the following boundary value problem

$$L[w_1] = q, \quad \text{in } \Omega, \tag{3.9}$$
$$\gamma_0[w_1] = 0, \quad \text{on } \Gamma, \tag{3.10}$$

and define another linear operator $\Lambda$ from $L^2(\Omega)$ to $H^{\frac{1}{2}}(\Gamma)$ by

$$\Lambda[q] = -\gamma_1[w_2]|_\Gamma. \tag{3.11}$$

It is proved in [1] that $q$ is a solution to the BLT problem if and only if it is a solution to the following equation

$$\Lambda[q] = b, \tag{3.12}$$

where $b = N_\Gamma[f + 2g] + g$ on the boundary $\Gamma$.

Let us define

$$F[q] = \int_\Gamma \{b \log \Lambda[q] - \Lambda[q]\}\, d\Gamma, \tag{3.13}$$

which is the log likelihood function when the measured data $b$ is subject to Poissonian distribution. By the maximum principle of elliptical partial differential equations [5], it follows that $\Lambda[q] > 0$ and $q > 0$. Therefore we will find a solution for the BLT problem by performing the following optimization

$$\mathop{\mathbf{arg\,max}}_{q \geqslant 0} F[q]. \tag{3.14}$$

To solve this optimization problem, we need to find the Frechet derivative of $F$. Let

$$f(t) = F[q + tv], \quad \text{for } t \text{ around } 0, \tag{3.15}$$

where $v$ is an arbitrary bounded function in $L^2(\Gamma)$, and compute

$$\frac{d}{dt}f(t)\bigg|_{t=0} = \int_\Gamma \left\{ b\frac{1}{\Lambda[q]} - 1 \right\} \Lambda[v]\, d\Gamma = \int_\Omega \Lambda^* \left[ \frac{b}{\Lambda[q]} - 1 \right] v\, dx.$$

Hence, the Frechet derivative of $F$ is

$$F'[q] = \Lambda^* \left[ \frac{b}{\Lambda[q]} - 1 \right]. \tag{3.16}$$

If $q > 0$ is a solution of (3.14), it follows that $F'[q_0] = 0$. The general case of $q \geqslant 0$ is given by the Kuhn-Tucker condition [15]

$$q \cdot \Lambda^* \left[ \frac{b}{\Lambda[q]} - 1 \right] = 0. \tag{3.17}$$

Rewrite formula (3.17), we get the variational EM method

$$q^{(k+1)} = \frac{q^{(k)}}{\Lambda^*[1]} \cdot \Lambda^* \left[ \frac{b}{\Lambda[q^{(k)}]} - 1 \right], \tag{3.18}$$

where the adjoint operator $\Lambda^* : \Lambda^*[\phi] = \psi$ of $\Lambda$ is an operator from $H^{\frac{1}{2}}(\Gamma)$ to $L^2(\Omega)$, which is defined as

$$L[\psi] = 0, \quad \text{in } \Omega, \tag{3.19}$$

$$\gamma_0[\psi] = \phi, \quad \text{on } \Gamma. \tag{3.20}$$

Obviously, the implementation of this method needs to solve the boundary value problem twicely for each iteration.

## 3.3 Some relevant issues

1) To avoid inverse crime [22]. Numerical test of reconstruction methods usually make use of modeling data from the numerical solution of forward problem. One typical issue is coined as the notorious inverse crime. To avoid it, we employed different point distribution or PDE solving techniques in modeling and inversion. There are two approaches to get the modeling data: one method is to use different point distribution and solve the boundary value problem (2.3) and (2.4) by the RBF collocation method; the other one is to solve (2.3) and (2.4) by the finite element method.

2) Choice of the initial value $q^{(0)}$ for iteration. From the iterative EM iterative formula (3.4) and (3.18), we must have that $q^{(0)} \neq 0$, since $q^{(0)} = 0$ will leads to $q^{(k)} = 0$ for any $x$ and $k$. Hence we set $q^{(0)} = C \neq 0$

when $x \in \omega_p$ for both algorithms. By the Green's formula, the constant $C$ can be determined by the following formula [10]

$$C \geqslant \frac{\int_\Gamma g\, d\Gamma + \int_\Omega \mu_a w_1\, dx}{|\Omega_p|}. \tag{3.21}$$

3) Choice of the parameters of radial basis functions. Past experiments showed that the accuracy of RBF-based collocation method is very sensitive to the choice of those free parameters in many RBFs. Here we use the MQ RBF and choose $\epsilon = 0.15$ in our numerical experiments.

4) Convergence criteria. The convergence criteria for both algorithms may include (1) when the iteration number $k$ reaches an assumed maximum number; (2) when the successive incremental $|q^{(k+1)} - q^{(k)}|$ is smaller than an assumed error level.

# 4 Numerical experience

To demonstrate the feasibility of our algorithms, we carried out an experiment by using a heterogeneous tissue-like cylinder phantom, which is the same as the previous work [6]. The phantom of 20mm height and 10mm radius was set up and consisted of three kinds of materials: muscle (M), lung (L) and heart (H), as shown in Figure 1.

$$\begin{aligned}
\mathrm{M} &= \{x \in \Omega | 6 \leqslant \sqrt{x_1^2 + x_2^2} \leqslant 10\}, \\
\mathrm{L} &= \{x \in \Omega | 3 \leqslant \sqrt{x_1^2 + x_2^2} \leqslant 6\}, \\
\mathrm{H} &= \{x \in \Omega | \sqrt{x_1^2 + x_2^2} \leqslant 3\}.
\end{aligned}$$

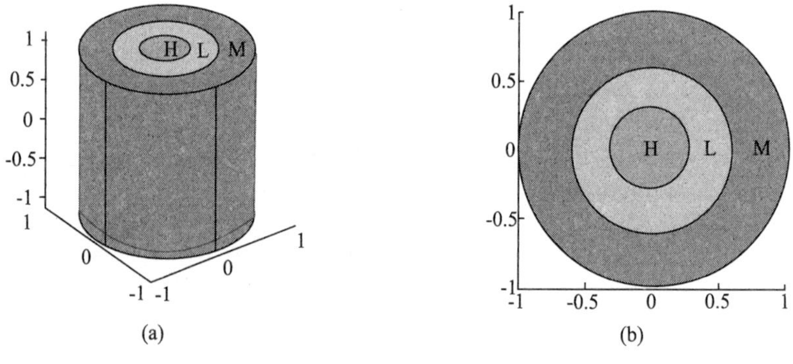


Figure 1 The phantom used to test the algorithm, (a) is the 3D figure, (b) is the cross section

Table 1 Optical parameters of the tissue-like phantom

| Material | Muscle (M) | Lung (L) | Heart (H) | Bone (B) |
|---|---|---|---|---|
| $\mu_a[\mathrm{cm}^{-1}]$ | 0.07 | 0.23 | 0.11 | 0.01 |
| $\mu_s[\mathrm{cm}^{-1}]$ | 10.31 | 20.00 | 10.96 | 0.60 |

The optical parameters documented the literature [20] are assigned to each of the three components, as listed in Table 1.

Two spherical sources of the form

$$q_i = A_i \chi_{\Omega_i}(x), \quad \Omega_i = \{x \| \|x - x_i\| < r\}, \quad i = 1, 2 \tag{4.1}$$

are embedded in the phantom, where $\Omega_1, \Omega_2$ are balls centered at $x_1 =$ (0.5 cm, 0.0 cm, 0.0 cm), $x_2 =$ (−0.5 cm, 0.0 cm, 0.0 cm) and the radius $r$ is 0.1 cm. The two sources are of the same intensity $A_i = 500$. We use MQ radial basis functions with $\epsilon = 0.15$ to solve the boundary value problem. We code the two algorithms in Matlab 7.1, and test with above numerically simulated phantom. The reconstruction time by algebraic EM is about 10 seconds, by variational EM is about 10 minutes. The reconstruction results are showed in Figure 2. We can see that the reconstructed locations are correct, but the intensities are not.

Comparing the reconstruction figures by both methods, we find that the algebraic EM method can reconstruct better than variational EM method, since the reconstructed source intensities are closer to the real case. Moreover, in the algebraic method we only need to solve the boundary value problem one time while in the variational EM method we have to do that many times. This means the algebraic EM method is more efficient than the variational EM method.

# 5 Conclusion

The image reconstruction in bioluminescence tomography can be modeled by an inverse source problem of diffusion equation. The theoretical investigation has proved that the bioluminescent sources in the form of RBF can be uniquely determined by the boundary measurement data. This is the main reason why we use the RBF-based collocation method to find the numerical solutions. In this paper, we have developed two RBF based EM reconstruction algorithms for BLT. Numerical experiments show the feasibility of these algorithms. By the numerical experiments, we find that the algebraic EM method is more efficient since it does not need to solve the diffusion equation in each EM iteration step. Compared with the finite element method, the RBF method does not need to generate a mesh in the domain, this gives facilities for the complicated

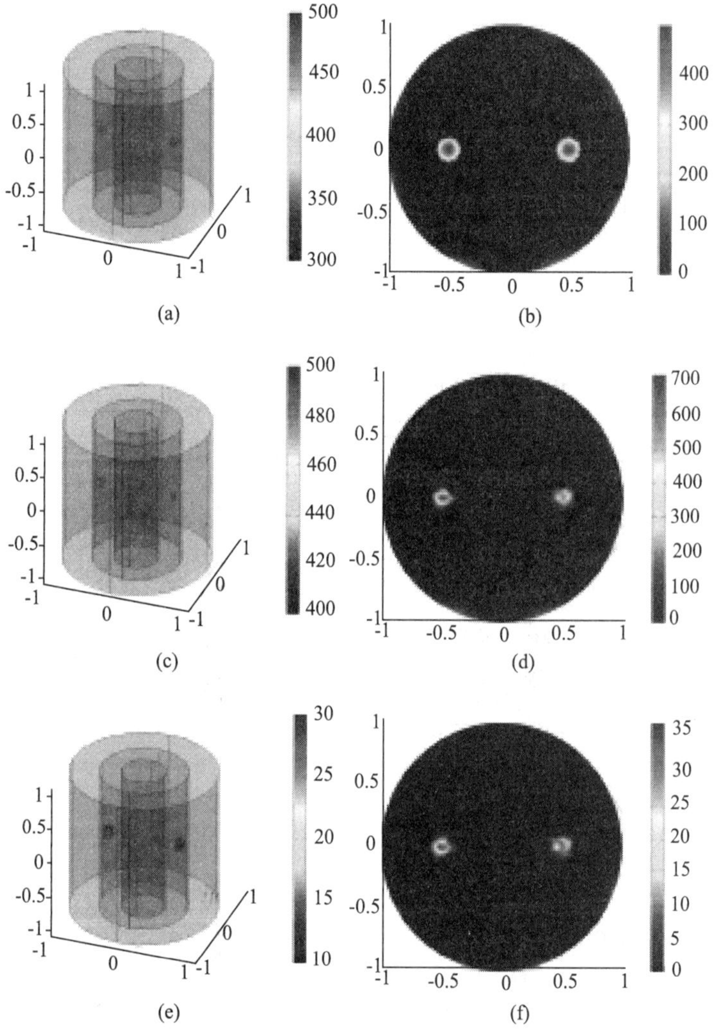


Figure 2 The reconstructed figures of two sources, which are centered at (0.5; 0; 0) and (−0.5; 0; 0), respectively. (a) is the three dimension source figures. (b) is the cross-section source figure at $z = 0$. (c) and (d) are the reconstruction figures by algebraic EM method. (e) and (f) are the reconstruction figures by variational EM method. All the spatial units are in cm.

domain. But the linear systems resulting from RBF collocation methods are dense and ill-conditioned. We are working on the pre-conditioning and other numerical issues to improve its performance.

# Acknowledgements

The support of NKBRSF (2003CB716101) and NSFC (60325101, 60272018), Chinese Ministry of Education (306017), Engineering Research Institute of Peking University, and Microsoft Research of Asia are gratefully acknowledged.


# References


[1] S.R. Arridge. Optical tomography in medical imaging. *Inverse Problems* **15**: 41-93, 1999.

[2] A.H.D. Cheng, M.A. Golberg, E.J. Kansa, G. Zammito. Exponential convergence and H-c multiquadric collocation method for partial differential equations. *Numerical Methods for Partial Differential Equations* **19**: 571-594, 2003.

[3] W.X. Cong, G.Wang, D. Kumar, Y. Liu, M. Jiang, L.V. Wang, E.A. Hoffman, G. McLennan, P.B. McCray, J. Zabner, A. Cong. Practical reconstruction method for bioluminescence tomography. *Optics Express* **13**: 6756-6771, 2005.

[4] M. Edinger, Y.A. Cao, Y.S. Hornig, D.E. Jenkins, M.R. Verneris, M.H. Bachmann, R.S. Negrin, C.H. Contag. (2002) Advancing animal models of neoplasia through in vivo bioluminescence imaging. *European Journal Of Cancer* **38**: 2128-2136, 2002.

[5] D. Gilbarg, N.S. Trudinger, *Elliptic Partial Differential Equations of Second Order*, Springer-Verlag, Berlin-Heideberg-New York, 1983.

[6] W.M. Han, W.X. Cong, G. Wang. Mathematical theory and numerical analysis of bioluminescence tomography. *Inverse Problems* **22**: 1659-1675, 2006.

[7] V. Isakov. *Inverse Problems for Partial Differential Equations*, Springer, New York-Berlin-Heideberg, 1998.

[8] M. Jiang, G. Wang. (2004) Image Reconstruction for Bioluminescence Tomography. *Proceedings of SPIE* **5535**: 335-351, 2004.

[9] M. Jiang, T. Zhou, J.T. Cheng, W.X. Cong, G. Wang. Development of bioluminescence tomography. *Proceeding of SPIE's Optics & Photonics*, 2006.

[10] M. Jiang, T. Zhou, J.T. Cheng, W.X. Cong, G. Wang. Image reconstruction for bioluminescence tomography from partial measurement. *Optics Express* **15**: 11095-11116, 2007.

[11] E. Larsson, B. Fornberg. A numerical study of some radial basis function based solution methods for elliptic PDEs. *Computers & Mathematics with Applications* **46**: 891-902, 2003.

[12] G. Liu. *Mesh Free Methods: Moving Beyond the Finite Element Method*, CRC Press, 2002.

[13] N. Mai-Duy, T. Tran-Cong. Approximation of function and its derivatives using radial basis function networks. *Applied Mathematical Modelling* **27**: 197-220, 2003.

[14] A. McCaffrey, M.A. Kay, C.H. Contag. Advancing molecular therapies through in vivo bioluminescent imaging. *Molecular Imaging* **2**: 75-86, 2003.

[15] F. Natterer, F. Wbbeling. *Mathematical methods in image reconstruction*, Society for Industrial and Applied Mathematics, Philadelphia, 2001.

[16] V. Ntziachristos, J. Ripoll, L.H.V. Wang, R. Weissleder. Looking and listening to light: the evolution of whole-body photonic imaging. *Nature Biotechnology* **23**: 313-320, 2005.

[17] A. Soling, N.G. Rainov. Bioluminescence imaging in vivo-application to cancer research. *Expert Opinion On Biological Therapy* **3**: 1163-1172, 2003.

[18] G. Wang, E.A. Hoffman, G. McLennan, F. Bohnenkamp, F. Colliso, W.X. Cong, M. Jiang, D. Kumar, H. Li, Y. Li, P. McCray, J.F. Meinel, E. Ritman, M. Suter, P. Taft, J. Tian, L.H. Wang, J. Zabner, F.P. Zhu. Development of the first bioluminescent CT scanner. *Radiology* 229(P): 566, 2003.

[19] G. Wang, Y. Li, M. Jiang. Uniqueness theorems in bioluminescence tomography. *Medical Physics* **31**: 2289-2299, 2004.

[20] A.J. Welch, M.J.C. van Gemert. *Optical-thermal response of laser-irradiated tissue (Lasers, Photonics, and Electro-Optics)*, Plenum Press, New York, 1995.

[21] J.C. Wu, I.Y. Chen, G. Sundaresan, J.J. Min, A. De, J.H.Qiao, M.C. Fishbein, S.S.Gambhir. Molecular imaging of cardiac cell transplantation in living animals using optical bioluminescence and positron emission tomography. *Circulation* **108**: 1302-1305, 2003.

[22] J. Kaipio, E. Somersalo. *Computational and Statistical Inverse Problems*, Springer-Verlag, 2004.